Communications in Computer and Information Science   **2362**

AF173480

Rationale

The CCIS series is devoted to the publication of proceedings of computer science conferences. Its aim is to efficiently disseminate original research results in informatics in printed and electronic form. While the focus is on publication of peer-reviewed full papers presenting mature work, inclusion of reviewed short papers reporting on work in progress is welcome, too. Besides globally relevant meetings with internationally representative program committees guaranteeing a strict peer-reviewing and paper selection process, conferences run by societies or of high regional or national relevance are also considered for publication.

Topics

The topical scope of CCIS spans the entire spectrum of informatics ranging from foundational topics in the theory of computing to information and communications science and technology and a broad variety of interdisciplinary application fields.

Information for Volume Editors and Authors

Publication in CCIS is free of charge. No royalties are paid, however, we offer registered conference participants temporary free access to the online version of the conference proceedings on SpringerLink (http://link.springer.com) by means of an http referrer from the conference website and/or a number of complimentary printed copies, as specified in the official acceptance email of the event.

CCIS proceedings can be published in time for distribution at conferences or as post-proceedings, and delivered in the form of printed books and/or electronically as USBs and/or e-content licenses for accessing proceedings at SpringerLink. Furthermore, CCIS proceedings are included in the CCIS electronic book series hosted in the SpringerLink digital library at http://link.springer.com/bookseries/7899. Conferences publishing in CCIS are allowed to use Online Conference Service (OCS) for managing the whole proceedings lifecycle (from submission and reviewing to preparing for publication) free of charge.

Publication process

The language of publication is exclusively English. Authors publishing in CCIS have to sign the Springer CCIS copyright transfer form, however, they are free to use their material published in CCIS for substantially changed, more elaborate subsequent publications elsewhere. For the preparation of the camera-ready papers/files, authors have to strictly adhere to the Springer CCIS Authors' Instructions and are strongly encouraged to use the CCIS LaTeX style files or templates.

Abstracting/Indexing

CCIS is abstracted/indexed in DBLP, Google Scholar, EI-Compendex, Mathematical Reviews, SCImago, Scopus. CCIS volumes are also submitted for the inclusion in ISI Proceedings.

How to start

To start the evaluation of your proposal for inclusion in the CCIS series, please send an e-mail to ccis@springer.com.

Prasanna Devi Sivakumar · Raj Ramachandran ·
Chitra Pasupathi · Prabha Balakrishnan
Editors

Computing Technologies for Sustainable Development

First International Research Conference, IRCCTSD 2024
Chennai, India, May 9–10, 2024
Proceedings, Part III

 Springer

Editors
Prasanna Devi Sivakumar ⓘ
SRM Institute of Science and Technology
Chennai, Tamil Nadu, India

Raj Ramachandran
Cardiff Metropolitan University
Cardiff, UK

Chitra Pasupathi ⓘ
SRM Institute of Science and Technology
Chennai, Tamil Nadu, India

Prabha Balakrishnan ⓘ
SRM Institute of Science and Technology
Chennai, Tamil Nadu, India

ISSN 1865-0929   ISSN 1865-0937 (electronic)
Communications in Computer and Information Science
ISBN 978-3-031-82385-5   ISBN 978-3-031-82386-2 (eBook)
https://doi.org/10.1007/978-3-031-82386-2

Preface

SRM Institute of Science and Technology (Vadapalani Campus), India, a prominent unit of the prestigious SRM Group of Institutions located in Chennai, India, is known for its commitment to fostering academic excellence and innovation. SRMIST (Vadapalani Campus) is distinguished by the emphasis placed on experiential learning, industry partnerships, and research initiatives. With specialized research centers and active industry collaborations, the campus provides an ideal environment for innovative research and learning.

SRMIST (Vadapalani Campus) takes immense pride in presenting the proceedings of the International Research Conference on Computing Technologies for Sustainable Development (IRCCTSD 2024). This distinguished event was jointly organized by SRMIST (Vadapalani Campus) and Cardiff Metropolitan University, UK, on May 9th and 10th, 2024. This Research Conference brought together a wide range of global scholars, industry professionals, and innovative researchers. The event served as an energetic platform for impactful discussions, groundbreaking ideas, and forward-thinking innovations in computing technologies focused on sustainable development. We are confident that the ideas shared here will make a lasting contribution to both academic research and industrial practices, shaping the future of sustainable technology solutions.

The conference received an outstanding 264 research articles. After a rigorous peer-review process, 82 articles were selected for presentation and publication. The conference featured a diverse range of tracks, including AI Solutions, Machine Learning Strategies, Deep Learning Techniques, Industrial Innovation, Image Processing Applications and IoT Innovations related to Computing Technologies domains, namely Healthcare, Security, Agriculture and Climate Mitigation. Each track provided a vibrant and inclusive space for academicians and professionals to exchange ideas, address challenges and contribute to the conversation on leveraging computing technologies to solve today's most pressing sustainability challenges.

Eminent industry expert P Ilango, HCL Technologies, India delivered the keynote address in the inauguration ceremony emphasizing the role of technological innovation in sustainable development; Raj Ramachandran, Cardiff Metropolitan University, UK; Deepan Raj, HCL Technologies, India; and Mohanraj Vengadachalam, Standard Chartered Bank, India delivered the session keynote addresses.

The research outcomes shared during this conference foster continuous innovation and collaboration in the field of computing technology for sustainable development.

November 2024

Prasanna Devi Sivakumar
Raj Ramachandran
Chitra Pasupathi
Prabha Balakrishnan

Acknowledgement

The SRM Institute of Science and Technology (Vadapalani Campus), India, being a premier Institution, focuses on diverse and interdisciplinary research activities emphasizing both applied and fundamental research, aligning with National and Global priorities for technological advancement and self-reliance.

The apex management of SRMIST hones the academicians and students to actively engage in pioneering research projects to address real-world challenges, particularly in the fields of Computing Technologies, Engineering, and Sustainable Development. The commitment to high-quality research output is reflected in the numerous research articles published by the faculty and students, contributing to both national and international academic communities.

SRMIST (Vadapalani campus) recognizes the exceptional contributions of all individuals and institutions who helped in making the International Research Conference on Computing Technologies for Sustainable Development (IRCCTSD 2024) such a remarkable success. This conference showcased the productive partnership between SRMIST, India and Cardiff Metropolitan University, UK. The collaboration between these renowned institutions fostered an environment of innovation and academic excellence, advancing the frontiers of sustainable technology development.

We express our sincere gratitude to the management of SRMIST for their continuous support and encouragement in advancing academic research and knowledge dissemination. The resources, facilities, and collaborative environment of the institution were instrumental in the successful completion of this work. We extend our heartfelt thanks to the administration of SRMIST and the faculty, research scholars, and staff of the CSE department for their guidance and support throughout the execution of this conference and publication of the research proceedings.

Our deepest thanks also go to Cardiff Metropolitan University, UK, for their incredible support and partnership throughout the conduct of the conference. The guidance and insights provided by the distinguished faculty from Cardiff Metropolitan University played a crucial role in shaping the conference's trajectory. Collaborating with such a prestigious institution has been a truly rewarding experience, and we are thankful for the dedication of the team to promoting academic growth and innovation.

We sincerely thank Springer for their outstanding support in publishing this proceedings. It is a privilege to collaborate with such a prestigious publishing platform, renowned for its dedication to high-quality publications.

We sincerely appreciate the invaluable contributions of all the authors for their research efforts and the reviewers whose meticulous evaluations were pivotal in finalizing the selection of articles. We also wish to acknowledge the tireless work and commitment of the organizing and technical program committees, whose efforts were crucial to the conference's success.

In conclusion, we extend our deepest gratitude to all individuals and institutions whose contributions made this conference a success. The collaboration and support provided by the authors, reviewers, organizing committee, and sponsors were pivotal in advancing key subtopics such as AI solutions for sustainability, IoT applications for environmental monitoring, green computing practices, and smart city innovations. These proceedings reflect the ongoing exploration of cutting-edge Computing Technologies for Sustainable Development, and we are confident that the insights shared will inspire further innovation and drive meaningful progress in this essential field.

Organization

Chief Patrons

Ramasamy Paarivendhar	SRM Institute of Science and Technology, India
Ravi Pachamuthu	SRM Institute of Science and Technology, India
Pachamuthu Sathyanarayanan	SRM Institute of Science and Technology, India
Harini Ravi	SRM Group, India

Patrons

Chellamuthu Muthamizhchelvan	SRM Institute of Science and Technology, India
Suruttaiyaudiyar Ponnusamy	SRM Institute of Science and Technology, India
Kandhasami Gunasekaran	SRM Institute of Science and Technology, India
Santhanagopalan Ramachandran	SRM Institute of Science and Technology, India
Chandrathil Velappan Jayakumar	SRMIST (Vadapalani Campus), India
Chidambaram Gomathy	SRMIST (Vadapalani Campus), India
Sadagoparaman Karthikeyan	SRMIST (Vadapalani Campus), India
Krishnamoorthy Ramachandran	SRMIST (Vadapalani Campus), India

Program Committee

Jedsada Tipmontian	International Academy of Aviation Industry - King Mongkut's Institute of Technology Ladkrabang, Thailand
Ram Kumar Jayaseelan	AMD, USA
Anand Lakshmanan	Ericsson, India
Dhananjay Kumar	MIT Campus, Anna University, India
Yamini Jagadeesan	Facebook, USA
Nickolas Savarimuthu	NIT Trichy, India
Anandhakumar Palanisamy	MIT Campus, Anna University, India
Baskaran Ramachandran	Anna University, India
Masilamani Vedhanayagam	IIITDM, India
Roy Antony Arnold	Infosys Ltd, India
Jothi Periyasamy	DeepSphere AI, USA
Ruso Tamilarasan	Cognizant, India

| Renuka Devi Saravanan | VIT Chennai, India |
| Mirnalinee Thanganadar Thangathai | SSN College of Engineering, India |

Steering Committee and International Reviewers

Catherine Tryfona	Cardiff Metropolitan University, UK
Karl Jones	Cardiff Metropolitan University, UK
Fiona Carroll	Cardiff Metropolitan University, UK
Khoa Phung	University of the West of England, UK
Chandru Sandrasekaran	International College of Business and Technology, Sri Lanka
Pakkianathan Prabu Premkumar	International College of Business and Technology, Sri Lanka

External Reviewers

Sandra Johnson
P. Latchoumy
Poonkodi Mariappan
Boopathi Raja Govindasamy
Subhashini Palaniswamy
Jyostna Devi Bodapati
Gokul Chandrasekaran
Geetha Chellaian
Karthikeyan Vedanandam
Inbamalar Tharcis Mariapushpam
Boomija Malaisamy Duraipandian
Balasundaram Ananthakrishnan
Sandhya M. Kumar
Immanuvel Arokia James

Dass Purushothaman
Chidambarathanu Krishnan
Surender Shanmugam
Rajan Subramaniam
Subburaj Varadharajaperumal
Meena Rajeswaran
Rohith Bhat
Victo Sudha George
Anuradha Muthukrishnan
Kavin Kumar Kandasamy
Madhavi Thiruvengadam
Prameeladevi Chillakuru
Priya Vijay

Organizing Committee

Golda Dilip
Rajasekar Velswamy
Bharathi Navaneetha Krishnan
Paavai Anand Gopalan
Neelam Sanjeev Kumar
Arun Nehru Jawaharlal Nehru

Durgadevi Palani
Sridhar Srinivasan
Niveditha Satiyamoorthy
Karthikayani Kaliyaperumal
Sangeetha Subramaniam Karuppaiyah Bharathi

Akila Krishnamoorthy
Sridevi Sridhar
Maheswari Sendur Pandy
Jessy Sujana Godwin
Vidhusavarshini Suresh Kumar
Anusha Thamaraichelvan
Gayathri Ramanakumar

Muthurasu Nallappan
Manohar Shanmugavel
Punitha Dhandapani
Indumathy Mayuranathan
Deepa Ravichandran
Rajavel Manickam
Jayanthi Palraj

Contents – Part III

AI for Text, Audio, Image and Video Processing

Application of AI for Education

Environmental Analysis and Protection

IoT Integrated Air Quality Detection and Alert System

Navin Sreyas, Tarun Venkatesh, R. L. Harjit, and B. Prabha[✉]

Department of Computer Science and Engineering (Emerging Technologies), SRM Institute of
Science and Technology, Vadapalani, Tamil Nadu, India
{mn5847,tv9064,hr5605}@srmist.edu.in, jemi.prabha@gmail.com

Abstract. In many urban and industrial regions today, maintaining and monitoring air quality has become a major concern. Transportation, energy, fuels, and other forms of pollution all contribute to deteriorating air quality. Given the negative effects of air pollution on both human health and the environment, air quality is of paramount importance in today's world. In this era of technological advancement, the Internet of Things (IoT) emerges as a promising approach for improving healthcare systems. This research describes a novel IoT-integrated air quality detection and alert system that aims to enable real-time monitoring and alerts the user of the situation persisting. The proposed system collects extensive information on air quality using embedded sensors and monitoring equipment. Using machine learning algorithms, the system automatically analyzes collected data to detect Predicting and forecasting air quality is critical for addressing environmental and public health issues. With the increasing impact of air pollution on human health and the environment, accurate and timely air quality forecasting is crucial for managing risks. The goal of this project is to create and implement machine learning methods for air quality prediction and forecasting. The machine learning models developed are applied to real-time data to provide short- and long-term air quality forecasts, allowing for timely public awareness and pollution control measures to be implemented. This study additionally examines the prospects of using future technologies, such as IoT devices, to improve the geographical resolution of air quality predictions. This study advances environmental science and public health by introducing a comprehensive framework for predicting and forecasting air quality using machine learning.

Keywords: Air Quality Index (AQI) · ESP 32 microcontroller · MQ Series sensors · Random forest classifier · Long short term memory

1 Introduction

Using Internet of Things (IoT) technology into health monitoring systems offers an innovative approach to counteract the growing risks that air pollution poses to human health, especially with regard to respiratory illnesses. With real-time environmental monitoring in mind, this project presents an inventive IoT-integrated air quality detection and alarm system. The most recent sensors, an ESP32 microprocessor, and a MQ Serie

© The Author(s), under exclusive license to Springer Nature Switzerland AG 2025
P. D. Sivakumar et al. (Eds.): IRCCTSD 2024, CCIS 2362, pp. 3–13, 2025.
https://doi.org/10.1007/978-3-031-82386-2_1

sensors gas sensor are used by the system to identify important air quality indicators including ammonia, carbon dioxide (CO_2), nitrogen dioxide (NO_2), and sulfur dioxide (SO_2), which have a substantial influence on respiratory disorders. The application of sophisticated machine learning methods, such as Random Forest methodology and Long Short-Term Memory (LSTM) model, allows the system to thoroughly examine complex patterns in environmental data. This makes it easier to estimate and forecast future Air Quality Index (AQI) values with accuracy.

This AQI alert and forecast system was created with the intention of equipping people especially those who already have respiratory conditions with the information and resources they need to steer clear of high-risk regions or take the appropriate precautions. This proactive effort provides the door for developments in specific and preventative health solutions in addition to demonstrating the Internet of Things' potential to transform healthcare. The initiative intends to reduce the frequency and severity of respiratory issues, encouraging a healthy lifestyle in the face of mounting environmental concerns by providing real-time alarms and forecast insights on air quality. In the end, this project aims to lessen the negative health effects of air pollution, which is a big step toward creating safer living conditions through technology-driven, preventative health care practices.

2   Literature Review

Ghufran Isam Drewil and Riyadh Jabbar Al-Bahadili [1] propose a novel approach to air pollution prediction using LSTM deep learning and Genetic Algorithm (GA) for hyperparameter optimization. Their hybrid model, combining LSTM's time series learning with GA's parameter tuning, forecasts daily pollution levels for PM10, PM2.5, CO, and NOx. Outperforming traditional methods in accuracy and speed, this model demonstrates the effectiveness of merging deep learning with metaheuristic algorithms for air quality forecasting.

A portable IoT health monitoring system was created by Ali I. Siam et al. [2] by combining a number of sensors to provide real-time data on ambient conditions and vital signs. With maximum error rates of 2.67%, 2.04%, and 1.58% for heart rate, blood oxygen saturation, and body temperature, respectively, it achieves great accuracy powered by NodeMCU and Firebase. Its reliability has been thoroughly verified and approved,

Using Arduino Uno and Raspberry Pi sensors, Sumit Paul et al. [3] created an Internet of Things prototype for remote health monitoring. An application for Android is used to analyze the data, track vital signs, and send them to the cloud. By using real-time health data, this technology reduces hospital visits and expenses while allowing doctors to prescribe medication.

Using data from roadside sensors and the Google Maps API, Mostafizur Rahaman Laskar et al. [4]. Built an Internet of Things e-Health system that suggests low-pollution travel routes. This invention claims to improve public health via technology by combining environmental monitoring with health notifications.

Sachin Bhimrao Bhoite [5] and others propose the use of machine learning to forecast sulfur dioxide levels in order to enhance the quality of the air in industrial and urban regions. Their study emphasizes how vital precise prediction models are to efficient air pollution control, which is crucial for improving the quality of urban life.

Using FFT-based DWT for feature extraction and PSO-tuned TSVM for classification, Sandeep Raj [6] created a real-time ECG monitoring system. With a peculiar detection warning system, this IoT platform enhances the capabilities of remote cardiac care, achieving 95.68% accuracy in classifying 16 different types of ECG signals.

During COVID-19, Prajoona Valsalan et al. [7] introduced a novel method of remote health monitoring, highlighting the significance of IoT. Their portable device monitors temperature along with other vital signs, delivering data to a server over Wi-Fi. This optimizes healthcare delivery, permits distant disease diagnosis, and improves understanding of illness prevention and diagnosis.

Hitesh Kumar Sharma et al. [8] used an automated health monitoring system to address global health issues. Healthcare providers can access patient data through an Android app using IoT technology, which enables location-independent remote monitoring. By allowing speedy remote diagnosis and reducing the need for in-person visits, this breakthrough highlights how IoT has the potential to completely alter healthcare delivery.

R. Senthilkumar et al. [9] presented a fog computing-based IoT framework for monitoring air quality. The system combines microprocessor-based IoT devices with air quality sensors, processing data via fog nodes for faster and more efficient analysis. This method improves spatial and temporal resolutions of air quality data, enabling real-time monitoring. The empirical results demonstrate the system's usefulness in sensing and pattern recognition of air quality changes, with fog computing reducing latency and increasing data processing speed.

L. Srinivasan et al. [10] offered a system that allows paraplegic patients to convey their requirements using basic body movements, particularly wearable technologies. This IoT-based healthcare system monitors vital indicators and allows patients to submit messages or alarms to doctors or caregivers via the internet, using GSM for message delivery. By detecting the tilt orientations of a wearable device, it improves communication for patients who are unable to talk or utilize sign language, providing a considerable increase in patient care and independence.

3 Methodology

3.1 Data Collection

In this phase, the primary objective is to gather real-time data on individual pollutants relevant to air quality. For this purpose, we employ an ESP32 WROOM microcontroller equipped with sensors capable of measuring various pollutants such as SO_2, NH_3, CO, O_3. The microcontroller periodically collects data from these sensors and transmits it to the Firebase database, a cloud platform, for centralized storage and accessibility (Figs. 1 and 2).

3.2 Data Calibration

Calibrating MQ-137 and MQ-7 gas sensors with an ESP32 microcontroller requires particular steps to guarantee accurate readings of air quality and gas concentrations.

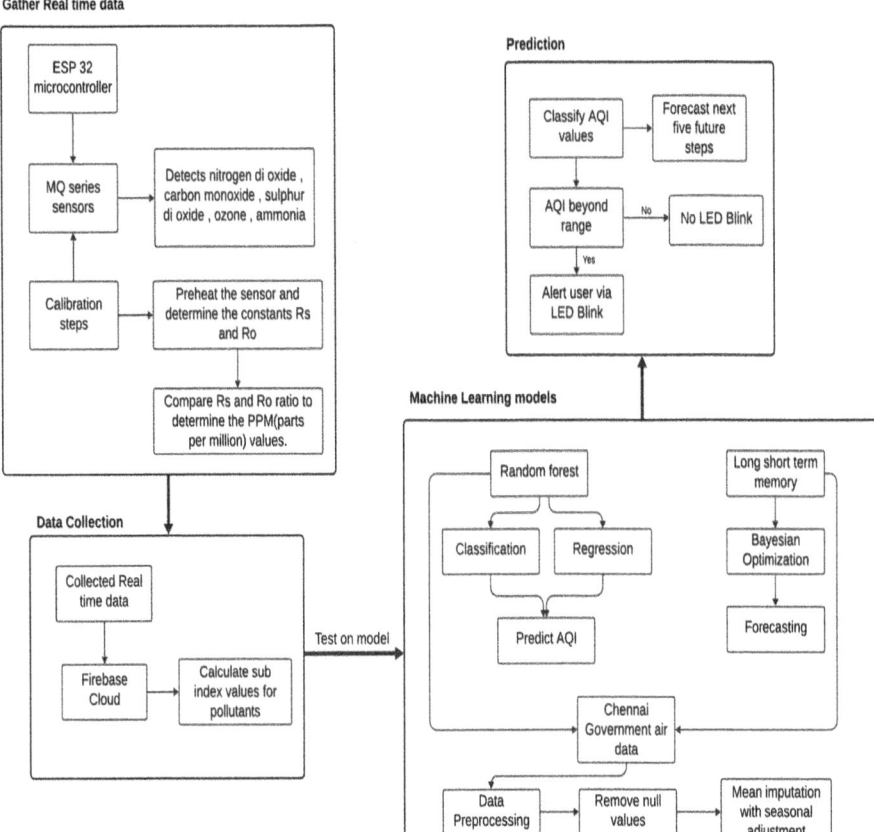

Fig. 1. Framework of IoT integrated air quality detection and alert system

The MQ-137 sensor detects NH3 whereas the MQ-7 sensor is designed specifically for detecting carbon monoxide (CO) concentrations.

Materials required for Calibration:

- MQ 137 and MQ 7 sensor
- ESP 32 microcontroller
- Jumper wires
- Arduino IDE environment for coding

Step 1

$$RS = \left(\frac{Vc}{Vrl} - 1 \right) x Rl (1)$$

(1)

- Vc is the circuit voltage (typically 5V),
- Vrl is the voltage across the load resistor,

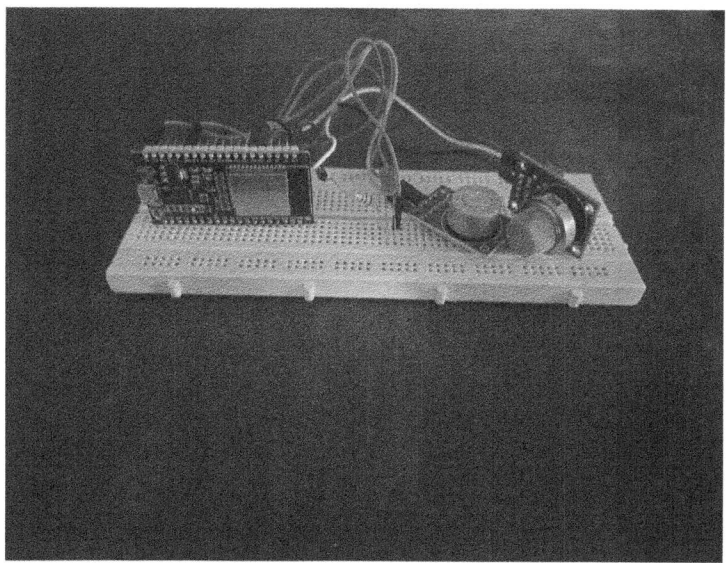

Fig. 2. Microcontroller and MQ135 gas sensor

- Rl is the load resistor value.

Step 2

$$Ro = \frac{Rscleanair}{CorrectionFactor}(2) \tag{2}$$

The baseline resistance R0 is calculated in a clean air environment and is essential for sensor calibration. To get R0, the formula usually involves measuring Rs in clean air and adjusting it with a correction factor, which may differ based on the sensor and datasheet details. The correction factor is often included in the sensor datasheet and is determined by the sensor's performance under normal clean air circumstances.

Step 3

Once R0 is determined, the concentration of the specific gas (such as NH3 for MQ-137, CO for MQ-7) may be computed by referencing the sensor's characteristic curve. The curve is usually described in the datasheet and can be expressed as:

$$GasConcertration = ax\left(\frac{Rs}{R0}\right)xb(3) \tag{3}$$

Here the constants a and b are specific to each gas and may be found in the sensor datasheet.

3.3 Data Preprocessing

We first establish sub-indices for each pollutant based on relevant information in order to generate the Air Quality Index (AQI). We train a predictive model using these values after normalizing them. By ensuring that all input data is treated consistently, this normalization step improves the accuracy of the model used to forecast air quality. By using this method, the dataset is ready for instant analysis and a strong tool to help reduce the health hazards related to air pollution is developed.

$$Ip = \frac{I_{high} - I_{low}}{C_{high} - C_{low}} x (C_p - C_{low}) + I_{low} (4)$$ (4)

Where

- I_p is the sub-index for pollutant p
- C_p is the concentration of pollutant p
- C_{low} and C_{high} are the breakpoint concentration that bracket C_p
- I_{low} and I_{high} are the AQI values corresponding to C_{low} and C_{high}

3.4 Model Development

The Random Forest Classifier can handle high-dimensional data and capture nonlinear relationships, we used it to predict the Air Quality Index (AQI). We divide our dataset into two subsets: a validation set for performance assessment and a training set for training models. Regression models were also used to take into consideration the impact of humidity and temperature on AQI variations. This all-encompassing method improves forecast precision and offers a more profound understanding of the variables influencing air quality. In order to provide a predictive tool that aids in efficient air quality management and public health campaigns, we employ accuracy criteria to make sure our models fulfill the requirements for trustworthy AQI prediction.

3.5 Real-Time AQI Prediction

In this step, the trained model is used to predict the AQI in real time using current pollutant concentrations from the Firebase database. The acquired data is analyzed to extract important features, which are then added into the trained Random Forest Classifier, which predicts the AQI. These projections are compared to defined thresholds to constantly monitor air quality conditions. If the predicted AQI exceeds the predefined thresholds, it signals deteriorating air quality and triggers the alerting system.

3.6 Alert Mechanism

The ESP32 WROOM microcontroller and environmental sensors, such as the MQ-137 for ammonia and MQ-7 for carbon monoxide, are used in real-time data processing by the alert system to improve efficiency. With the use of an LED, it monitors pollution levels in relation to health standards and issues color-coded air quality warnings (from moderate to dangerous), urging users to take immediate precautionary measures like turning on air purifiers or packing to evacuate. Furthermore, by integrating the system with mobile apps and home automation systems, users will be able to better regulate indoor air quality as notifications and updates on air quality are sent to them directly.

3.7   Forecasting Using Bayesian Optimization for Hyperparameter Tuning

The utilization of Bayesian optimization in Long Short-Term Memory (LSTM) models for hyperparameter tuning offers an advanced approach to air quality prediction, which is essential given the erratic nature of air pollution. Long-term dependencies and temporal patterns in time-series data are crucial for precise air quality forecasting, and LSTMs are a kind of recurrent neural network that excel in this regard. Through effective hyperparameter space searches, Bayesian optimization improves these models and increases predictive accuracy. Better-informed choices can be made on public health initiatives, urban planning, and environmental management thanks to this capacity. Constantly refined and verified, long short-term memory (LSTM) models enhanced using Bayesian approaches are a major asset in reducing the harmful consequences of pollution and advancing public health and sustainable development. In LSTM models, Bayesian Optimization (BO) is the method of choice for hyperparameter selection due to its rapid convergence to optimal configurations and efficient exploration of the hyperparameter space. A probabilistic surrogate model, frequently constructed using Gaussian Processes, is utilized in BO to provide an approximation of the objective function, which is commonly the validation loss of the model in this instance.

3.8   System Integration

A key aspect of the project is the seamless integration of various components, including the ESP32 microcontroller, Firebase database, predictive model, and alerting mechanism. Robust connectivity is established between these components to facilitate the transmission of real-time data, model predictions, and alert notifications. IoT devices may easily push sensor data in real time using Firebase's Realtime Database allowing for centralized data management and access. Firebase Cloud Storage can store huge files or IoT-generated data, and real-time connectivity allows devices and application components to communicate flawlessly.

4   Result

The figure displays a scatter plot of true vs. projected AQI values from a model. Ideally, for error-free predictions, all points should be on the diagonal line, with projected values identical to true values. The scatter along the line indicates a relationship between expected and true values, implying that the model has some ability to forecast (Figs. 3 and 4).

The bar graph illustrates the frequency distribution of Air Quality Index (AQI) categories in a dataset. The x-axis shows different AQI bucket categories, which are most likely coded from 0 to 5, and link to AQI classifications such as 'Good', 'Satisfactory', 'Moderate', 'Poor', 'Very Poor', and 'Severe', however the specific labels are not supplied in the graph (Fig. 5).

The above graph depicts the actual and predicted values during forecasting. Initially, both training and validation loss decrease, demonstrating that the model is learning and improving its predictions. The overall trend seems to stabilize towards the end, showing some convergence.

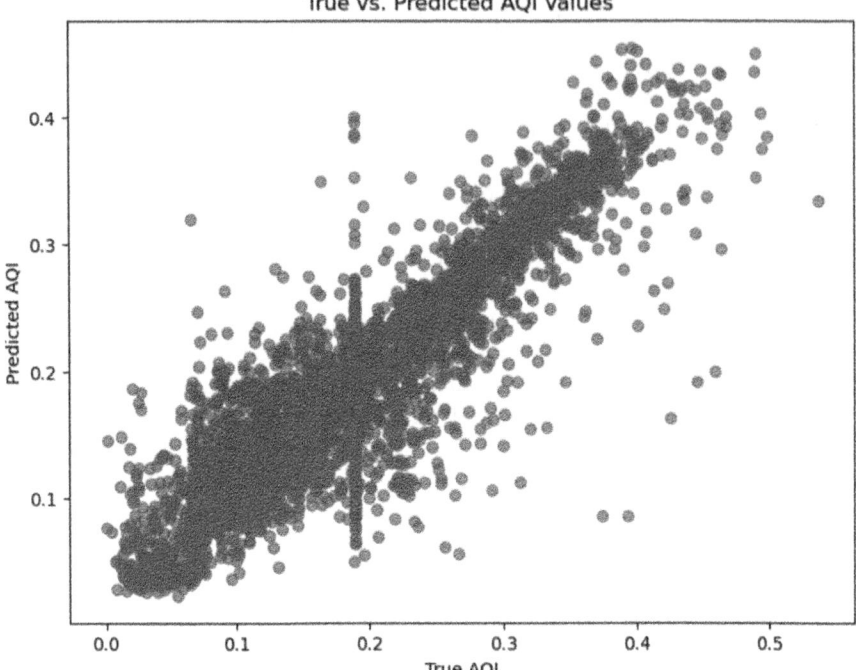

Fig. 3. Actual AQI vs Predicted AQI

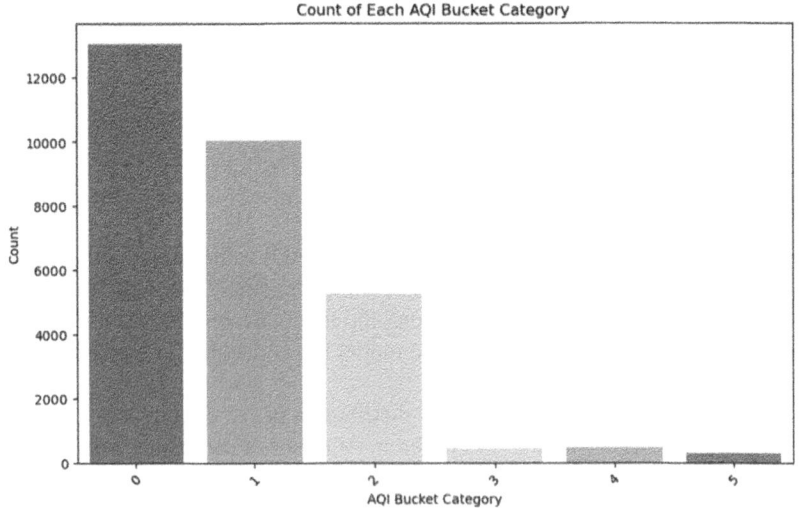

Fig. 4. Categorized AQI values

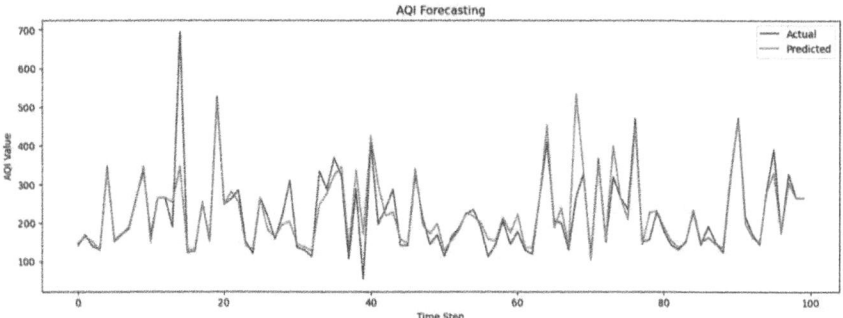

Fig. 5. Actual and predicted values during forecasting

4.1 System Comparison with Existing Model

To maximize model performance, we experimented with advanced techniques like Bayesian optimization instead of genetic algorithms. The complex nature of our dataset and the requirement to reduce test loss informed our choice. Following extensive testing, the genetic algorithm produced a test loss of 0.6, whereas the Bayesian optimization strategy produced a test loss of 0.3. In our graphical analysis, we did find differences between the actual and expected AQI levels, despite these promising findings.

A prominent metric utilized to assess the performance of machine learning models, including LSTM models, is Mean Squared Error (MSE). The metric computes the mean squared deviation between the predicted and actual values within a given dataset. MSE is computed mathematically by averaging the squared deviations between each predicted value and its corresponding actual value (Table 1).

Table 1. Comparison of MSE

Bayesian Optimization	0.3438136
Genetic Algorithm	0.6448128
Particle Swarm Optimization	0.6462832

The figures given above depict the result of the LSTM model when it uses Genetic Algorithm and Particle Swarm Optimization for Hyperparameter tuning (Figs. 6 and 7).

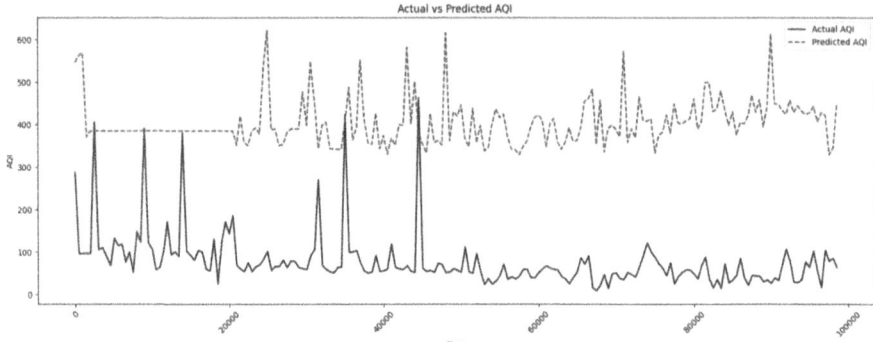

Fig. 6. Actual and predicted values during Genetic Algorithm

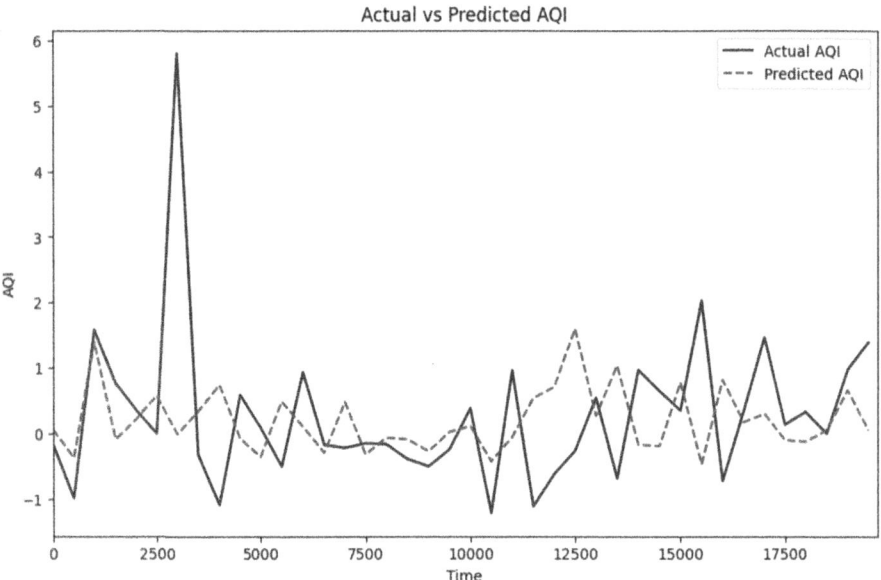

Fig. 7. Actual Vs Predicted AQI values when Using PSO

5 Conclusion

The proposed IoT integrated Air quality detection and alert system is an important move forward in leveraging technology to promote proactive safety measures and improve monitoring of the environment. By integrating sensors, machine learning models, and real-time detection systems, the system provides an integrated approach for identifying possible risks like smoke and predicting potential dangers. The incorporation of the ESP32 microcontroller when paired with sensors such as the MQ Series gas sensors provides an accurate application that can collect and analyze data in real time. The addition of machine learning models, such as the Random Forest Classifier for prediction and the Long Short-Term Memory (LSTM) for forecasting, adds insight to the system, allowing it to make correct predictions and provide users with immediate alerts. The inclusion of an LED blink alert system ensures that users are quickly cautioned when imminent dangers occur, allowing them to take suitable safety measures and evacuate if needed.

6   Future Work

Several enhancements to the IoT integrated Air quality detection and alert system's efficiency and utility can be made in the years to come. Incorporating multiple sensors, such as temperature, humidity, and carbon monoxide detectors, will provide a greater understanding of the environment and enhance risk detection abilities. Cloud connectivity allows remote monitoring, data storage, and analysis, hence improving adaptability and accessibility. Creating an application for mobile devices for user interaction and data visualization will increase user satisfaction and convenience. Dynamic threshold modification algorithms can dynamically alter warning levels in reaction to real-time conditions, reducing false alarms.

References

1. Drewil, G.I., Al-Bahadili, R.J.: Air Pollution Prediction Using LSTM Deep Learning and Metaheuristics Algorithms. Computer Engineering Department, University of Technology, Baghdad, Iraq
2. Siam, A.I., et al.: Portable and real-time IoT-based healthcare monitoring system for daily medical applications, IEEE Trans. Comput. Soc. Syst. 10(4), August 2023
3. Hamim, M.: IoT based remote health monitoring system for patients and elderly people. In: 2019 International Conference on Robotics, Electrical and Signal Processing Techniques (ICREST) (2019)
4. Laskar, M.R.: An IoT-based e-Health system integrated with wireless sensor network and air pollution index. In: 2019 Second International Conference on Advanced Computational and Communication Paradigms (ICACCP) (2019)
5. Bhoite, S.B.: Air quality prediction using machine learning algorithms. Int. J. Comput. Appl. Technol. Res., September 2019
6. Raj, S.: An efficient IoT-based platform for remote real-time cardiac activity monitoring, IEEE Trans. Consum. Elec. 66(2), May 2020
7. Valsalan, P., et al.: IoT based health monitoring system. J. Crit. Rev. 7(4) (2020). ISSN: 2394-5125
8. Sharma, H.K., et al.: i-Doctor: an IoT Based Self Patient's Health Monitoring System. IEEE (2019)
9. Senthilkumar, R., et al.: Intelligent based novel embedded system based IoT enabled air pollution monitoring system, Microprocess. Microsyst. (2020)
10. Srinivasan, L., et al.: IoT-based solution for paraplegic sufferers to send signals to physicians via the Internet. SSRG Int. J. Elec. Elec. Eng. 10(1), 41–52, January 2023

Impact of Weather Parameters on Malaria Transmission a Study Using the Epidemiology Models

K. Sam Prince Franklin$^{(\boxtimes)}$ (ID), Samba Siva Sai Davuluru(ID), and R. Parvathi(ID)

Vellore Institute of Technology, Chennai Campus, Chennai, India
samprince.franklink2020@vitstudent.ac.in

Abstract. Malaria, a dangerous parasitic disease spread by mosquitoes, remains a major health concern in many parts of the world, including India. To effectively combat malaria, it's crucial to understand the factors that influence its spread. This study focuses on Chengalpattu District in Tamil Nadu, India, aiming to unravel the complex relationship between weather, mosquito behavior, and malaria transmission. By combining predictive modeling with existing knowledge about disease outbreaks, the research provides valuable insights for developing effective prevention and control strategies in the region. We employ both traditional predictive models (Decision Tree, K-Nearest Neighbors, Random Forest) and a modified SEIR (Susceptible-Exposed-Infectious-Recovered) model to explore these relationships. The predictive models are evaluated for their accuracy in predicting mosquito type and species based on climatic variables, age, gender, and malaria case data.

Further analysis utilizes the SEIR framework, alongside SIR epidemiological models, to gain deeper insights into malaria transmission dynamics by analyzing the susceptible, exposed, infectious, and recovered populations. Comparative analyses between these models and the traditional predictive approaches emphasize the crucial role of integrating weather parameters, mosquito behavior, and seasonality in understanding malaria transmission patterns. Major findings reveal a significant correlation between weather conditions, particularly rainfall and temperature, and malaria transmission, with peak transmission periods observed between July and December. These insights underscore the necessity for targeted interventions, including vector control measures and enhanced surveillance, specifically during thesepeak transmission periods. This comprehensive analysis contributes to our understanding of malaria transmission dynamics in Chengalpattu District and informs evidence-based strategies for malaria control and prevention in endemic regions.

Keywords: Epidemiology Models · SEIR · SIR · Weather Dynamics · Predictive Modeling · Epidemiological Frameworks

1 Introduction

Malaria, a serious and potentially fatal disease, is spread to humans through the bite of infected female Anopheles mosquitoes, primarily found in tropical areas. It's important to note that malaria is preventable and curable. The disease is caused by a parasite, not direct

© The Author(s), under exclusive license to Springer Nature Switzerland AG 2025
P. D. Sivakumar et al. (Eds.): IRCCTSD 2024, CCIS 2362, pp. 14–28, 2025.
https://doi.org/10.1007/978-3-031-82386-2_2

human-to-human contact. While mosquito bites are the primary mode of transmission, infection can also occur through contaminated blood transfusions or needles. Early symptoms of malaria can be subtle, mimicking other illnesses with fever, making it difficult to diagnose. However, if left untreated, particularly the *P. Falciparum* type, the infection can escalate rapidly, leading to severe complications and even death within 24 h.

Five Plasmodium parasite species cause malaria in humans, with P. falciparum and P. vivax posing significant threats. In 2009, India reported approximately 1.6 million cases and 1,100 malaria-related deaths, likely underestimated due to various challenges. These challenges include diverse eco-epidemiological profiles, multiple Plasmodium species, varying Anopheles vectors, drug resistance, and climate change impacts. India's public health system, challenged by its large population, struggles with surveillance program implementation [24]. Historically, India faced its highest malaria incidence in the 1950s, with around 75 million cases annually. Despite declines due to the National Malaria Control Program, cases resurged to about 6.45 million in 1976. Weather data collection, facilitated by INSAT AWS, dates back to 1980, with continuous upgrades enhancing reliability. IMD maintains a vast network of AWS, AGRO AWS, and ARG stations nationwide, ensuring efficient weather monitoring and dissemination [16].

This paper explores malaria transmission dynamics in Chengalpattu District, Tamil Nadu, India, using predictive modelling and epidemiological frameworks. It integrates traditional predictive models and a modified SEIR framework to understand factors driving transmission [9, 10]. The study analyses climate variables, age distribution, gender disparities, and malaria case data. Predictive models predict mosquito type and species, while a Non-Linear Regression model forecasts malaria cases across gender and age groups. The SEIR framework provides insights into temporal dynamics. By comparing outputs, the study aims to understand the contributions of climate, mosquito behavior, and demographics. Findings inform evidence-based interventions for malaria control in Chengalpattu District and beyond, aiming to reduce malaria burden and improve public health outcomes in endemic regions through collaborative efforts [1].

Section II provides a comprehensive literature review on SEIR models and malaria transmission dynamics research. It examines previous studies utilizing SEIR modeling techniques, highlighting methodologies, findings, and advancements. Section III outlines the study's objectives and methodology, detailing mathematical models (SIR, SEIR) and their integration with weather parameters [20]. It also includes traditional predictive models such as "Decision Tree, K-Nearest Neighbor, Random Forest, Support Vector Machine, and Logistic Regression" [2]. Section IV describes the experimental setup, explaining how mathematical models were applied and parameters examined. Lastly, Section V presents experimental results and discusses findings, exploring the interplay between weather factors and malaria transmission dynamics revealed by various modeling approaches [2] (Table 1).

Table 1. Statistics on the Malarial Cases

Years	Population	Total Cases (Million)	P.v Cases (Million)	PF%	Deaths due to Malaria
1995	888143	2.93	1.14	38.84	1151
1996	872906	3.04	1.18	38.86	1010
1997	884719	2.66	1.01	37.87	879
1998	910884	2.22	1.03	46.35	664
1999	948656	2.28	1.14	49.96	1048
2000	970275	2.03	1.05	51.54	932
2001	984579	2.09	1.00	48.20	1005
2002	1013942	1.84	0.90	48.74	973
2003	1027157	1.87	0.86	45.85	1006
2004	1040939	1.92	0.89	46.47	949
2005	1082882	1.82	0.81	44.32	963
2006	1072713	1.79	0.84	47.08	1707
2007	1087582	1.51	0.74	49.11	1311
2008	1119624	1.53	0.78	50.81	1055
2009	1150113	1.56	0.84	53.72	1144
2010	1167360	1.60	0.83	52.15	1018
2011	1194901	1.31	0.67	50.74	754
2012	1211509	1.07	0.53	49.98	519
2013	1221640	0.88	0.46	52.61	440
2014	1234995	1.10	0.72	65.55	562
2015	1265173	1.17	0.78	66.61	384
2016	1283303	1.09	0.71	65.44	331
2017	1315092	0.84	0.53	62.70	194
2018	1337617	0.43	0.21	48.19	96
2019	1349006	0.34	0.16	46.36	77
2020	1372316	0.19	0.12	63.84	93
2021	1385500	0.16	0.10	62.79	90
2022	1396937	0.18	0.10	57.26	83

2 Literature Survey

[1] Xiaoping Liu's work introduces a novel 4-compartment l-i SEIR model, advancing epidemic modeling by incorporating temporal heterogeneity. This model offers a closed-form analytical solution, providing an advantage over conventional SEIR models. It highlights a critical disparity in transition rate accuracy between l-i SEIR and conventional SEIR models. Unlike conventional models, l-i SEIR demonstrates improved accuracy in epidemic curve generation under specific conditions. This research emphasizes the significance of accounting for temporal dynamics in epidemic modeling to enhance simulation precision and reliability.

[2] Utkir Abdulloevich Rozikov and S. K. Shoyimardonov's research applies a discrete-time SEIR model to study COVID-19 transmission dynamics, focusing on Uzbekistan. Their study investigates the model's efficacy in analyzing virus spread within specific country contexts, particularly emphasizing its utility in closed systems without births or deaths. They incorporate a quadratic stochastic operator (QSO) within the SEIR framework, revealing intricate dynamics and convergence of trajectories. Their work provides valuable insights into using discrete-time SEIR modeling for understanding and forecasting COVID-19 spread in Uzbekistan and beyond.

[3] Fazal Dayan's study focuses on numerically analyzing an SEIR epidemic model applied to factors of Coronavirus disease. It explores various methods, including Forward Euler, Runge-Kutta (RK-4), and Non-Standard Finite Difference (NSFD). Through simulations, it shows that NSFD outperforms both Forward Euler and RK-4, offering reliability and independence from time step size. This comprehensive analysis provides valuable insights for epidemic modeling and public health decision-making.

[4] Akram Mohammad Radwan's study utilizes the SEIR model to analyze and predict COVID-19 spread in the Gaza Strip. It effectively fits real data, estimating key epidemiological parameters such as a basic reproduction number (Ro) of 0.89 and an infection fatality rate (IFR) of 0.079% to 0.085%. By employing carefully calibrated initial conditions, the SEIR model accurately predicts pandemic evolution in the region. Radwan's research highlights the importance of mathematical models like SEIR in guiding public health responses tailored to the unique context of COVID-19 in resource-constrained settings.

[5] Suthep Suantai, Zulqurnain Sabir, Muhammad Asif Zahoor Raja, and Watcharaporn Cholamjiak explore the numerical computation of the SEIR model for Zika virus spread. They introduce a novel approach using a stochastic neural networks-based solver, demonstrating its reliability and accuracy. By leveraging advanced computational techniques, they offer valuable insights into predicting Zika virus dynamics, enhancing epidemiological modeling and public health strategies.

3 Methodology

Our study aims to achieve a twofold objective: firstly, to meticulously curate and preprocess datasets collected from the Chengalpattu area, focusing on weather parameters, and secondly, to implement a spectrum of mathematical models to analyze and infer the influence of weather conditions on malaria spread [6]. To achieve this, we employ a range of predictive models, including "Decision Tree, K Nearest Neighbors, Random Forest, XGBoost, Support Vector Machine, Multi-Layer Perceptron, Logistic Regression, and AdaBoost" [4], to predict mosquito type and species based on climate variables, age, gender, and malaria cases [26].

The initial phase of our analysis involves compiling and rigorously cleaning historical weather data, encompassing temperature and rainfall patterns across our designated geographic region [11]. Concurrently, we engage in comprehensive data collection and evaluation specific to malaria within our study area, gathering information on the incidence, prevalence, and mortality rates associated with malaria [14]. Subsequently, we integrate these meticulously curated datasets into various mathematical models, including SIR and SEIR models [5]. These epidemiological models are applied to analyze the relationship between weather conditions and malaria transmission dynamics. The choice of employing SEIR-like models stems from their ability to capture the nuanced dynamics of disease spread, particularly considering the existence of an exposed population that may not yet show symptoms but can contribute to transmission [30].

By simulating disease spread over time and incorporating weather variables as input parameters, we aim to discern patterns and trends that elucidate the influence of temperature and rainfall on malaria transmission. Through rigorous analyses and interpretation of model outputs, our research endeavors to contribute significant insights to the fields of public health research and disease ecology. This approach is chosen due to its holistic nature, combining predictive modeling with epidemiological frameworks to provide a comprehensive understanding of malaria transmission dynamics. By integrating weather data with mathematical models, our research aims to bridge the gap between climate factors and disease transmission, ultimately informing evidence-based strategies for malaria control and prevention in our study area [3, 12].

3.1 Area of Focus

Chengalpattu district, located on the northeast coast of Tamil Nadu, covers 2945 square kilometers, and has a tropical wet and dry climate. January, with an average temperature of 25 °C, is the coldest month. The district receives around 1400 mm of rainfall annually, mainly during October and November. It is intersected by significant rivers like the Palar, Adyar, and Ongur, along with minor ones such as Neenjal Maduvu and Pukkadurai Odai. Chengalpattu boasts 528 major irrigation tanks, contributing significantly to its agricultural landscape and water management infrastructure.

3.2 Dataset Description

Monthly climate data from January 2010 to September 2022, obtained from INSAT AWS ARGS, include parameters like temperatures, dew point, wind speed, humidity, precipitation, and rainy days. A correlation analysis revealed relationships between temperature, wind speed, rainy days, and precipitation. This dataset, complemented by 12 years of malaria statistics from Chengalpattu, is structured with attributes such as year, month, temperature, dew point, precipitation, rainy days, wind, humidity, malaria cases, and population. Efforts are ongoing to expand this dataset to enhance analysis depth. This dataset facilitates exploration of climate factors alongside malaria incidence, enabling mathematical modeling techniques to investigate weather's influence on malaria transmission dynamics in Chengalpattu. Insights gained will inform targeted disease control and prevention strategies in the region [19].

4 Experiments

A series of experiments were conducted to achieve our research objectives. Initially, we compared traditional predictive models with the SEIR framework to understand their efficacy in predicting malaria transmission dynamics. Models like "Decision Tree, K Nearest Neighbours, Random Forest, XGBoost, Support Vector Machine, Multi-Layer Perceptron, Logistic Regression, and AdaBoost" [4] were trained and evaluated using climate variables, age, gender, and malaria cases. Additionally, a Non-Linear Regression model using TensorFlow predicted malaria case distribution across different demographic groups under varying climatic conditions. Epidemiological models like SIR and SEIR simulated malaria transmission dynamics, incorporating weather parameters, mosquito behavior, and seasonality. The SEIR model integrated well with actual weather data, enhancing malaria transmission simulations and predictions. Compartmental modeling categorizes the population into compartments like Susceptible (S), Infectious (I), or Recovered (R), tracing individuals' movement between compartments. SEIS represents susceptible, exposed, infectious, and susceptible again. Notable contributors to this modeling approach include Ross, Hudson, Kendall, and McKendrick, with the Reed-Frost model serving as a precursor to contemporary epidemiological techniques [4, 17].

The experiments also explored the influence of weather conditions, especially temperature and rainfall, on malaria transmission dynamics. Historical weather data analysis alongside malaria spread counts revealed patterns indicating a correlation between increased rainfall, cooler temperatures, and elevated malaria transmission rates. Seasonal fluctuations in mosquito breeding rates underscored the significance of weather conditions in mosquito proliferation and disease transmission [14]. These experiments offer insights into the complex interplay between weather factors and malaria transmission dynamics, aiming to inform evidence-based strategies for malaria control and prevention in endemic regions. Through rigorous experimentation and analysis, we seek to advance understanding of the factors driving malaria transmission and inform targeted interventions for disease management and public health improvement [18].

4.1 Correlation Analysis and Predictive Modeling

To understand how environmental factors relate to malaria cases, we used correlation analysis. This statistical technique allowed us to measure the strength and direction of the relationship between these factors. The results were visualized using a correlation matrix, with coefficients ranging from −1 to +1. A coefficient of +1 signifies a perfect positive correlation, meaning both factors increase or decrease together. Conversely, −1 indicates a perfect negative correlation, where one factor increases as the other decreases. A coefficient of 0 signifies no relationship between the factors. Our focus was on understanding how temperature correlates with malaria cases compared to precipitation, humidity, and rainy days [27, 26] (Fig. 1).

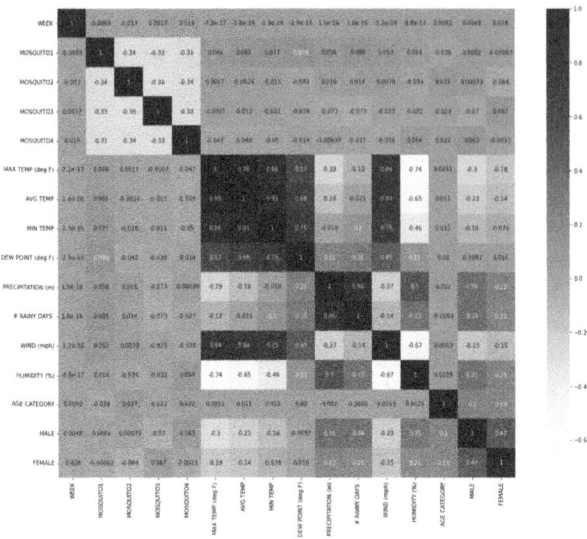

Fig. 1. Correlation matrix of Malaria - Weather dataset

The correlation matrix revealed several trends: a positive correlation between precipitation and rainy days, suggesting more rainfall corresponds to more rainy days. We also found a negative association between temperature and humidity, indicating higher temperatures coincide with lower humidity levels [25], [Additionally, temperature negatively correlated with precipitation, implying higher temperatures may lead to less precipitation. From these observations, we inferred that factors like precipitation, humidity, and rainy days may have more influence on malaria cases than temperature alone. The increasing trend in rainy days from June to October further supports the hypothesis that precipitation significantly affects malaria transmission dynamics [28].

4.2 SIR Compartmental Model

The SIR model provides a foundational mathematical approach to understanding how infectious diseases spread through a population. It's a valuable tool for analysing the dynamics of disease transmission over time.

- **Susceptible (S):** These individuals are healthy but have no immunity to the disease and can potentially contract it.
- **Infectious (I):** Currently infected individuals who can transmit the disease to susceptible individuals.
- **Recovered (R):** Individuals who have recovered from the infection and have gained immunity, making them unlikely to spread or contract the disease again.

The SIR model is essentially a set of differential equations that track the movement of people between these compartments [6]. It allows us to estimate the number (or proportion) of individuals in each category at any given point during the outbreak (Fig. 2).

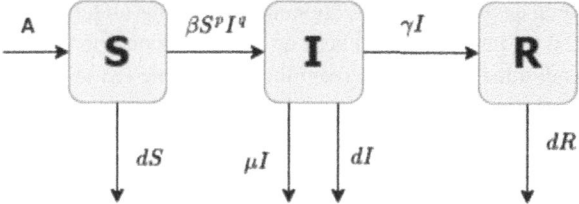

Fig. 2. State Chart of SIR Model

Here, for modelling malaria, the dynamics of the disease spread are typically much faster than birth and death processes, so these demographic factors are often neglected in simple compartmental models. The malaria model without consideration of vital dynamics (birth and death) can be described by the following system of ordinary differential equations.

$$\frac{dS}{dt} = -\frac{\beta SI}{N} \tag{1}$$

$$\frac{dI}{dt} = \frac{\beta SI}{N} - \gamma I \tag{2}$$

$$\frac{dR}{dt} = \gamma I \tag{3}$$

Here the N = S + I + R, which is the total population of the study.

The SIR model analyzes infectious disease dynamics, with equations representing susceptible (S), infected (I), and recovered (R) individuals. Equation (1) calculates the rate of change of susceptible individuals over time, proportional to the infection rate (β), susceptible and infected individuals, and total population size (N). Equation (2) describes the rate of change of infected individuals, increasing due to new infections and decreasing with recoveries at a rate determined by the recovery rate (γ). Equation (3) represents the rate of change of recovered individuals, increasing with the recovery rate and the number of currently infected individuals. In a malaria context, within a closed population, an epidemic fade as there aren't enough susceptible individuals left to sustain the disease, with new infections not triggering new epidemics due to existing immunity [15]. For Chengalpattu district in 2022, assuming a closed population, this code simulates a hypothetical malaria epidemic over one year. The population size is 2,841,572 individuals, with an infection rate (β) of 0.3 and a recovery rate (γ) of 0.1. Initially, there were 62 infected individuals and 51 recovered individuals. The ode function from the deSolve package solves the SIR model's differential equations. The resulting graphical output illustrates changes in susceptible, infected, and recovered populations over 365 days, aiding in understanding malaria outbreak dynamics [7].

4.3 SEIR Compartmental Model

In the context of malaria, as with many infectious diseases, there is often a latent phase where individuals are infected but not yet capable of spreading the infection to others.

This period between acquiring the infection and becoming infectious can be integrated into the SIR model by introducing a latent or exposed population denoted as E. In this adapted model, individuals transition from the susceptible (S) to the exposed (E) state upon infection, and then from the exposed (E) to the infectious (I) state when they become capable of transmitting the infection.

- **Susceptible (S):** These individuals are healthy but lack immunity to malaria, making them vulnerable to contracting the disease if bitten by an infected mosquito.
- **Exposed (E):** This category includes individuals who have been infected with malaria parasites but are not yet infectious. During the exposed phase, individuals are in the latent period of infection and not capable of transmitting the disease to others.
- **Infectious (I):** These individuals are currently infected with malaria parasites and can transmit the disease to susceptible individuals through mosquito bites.
- **Recovered (R):** These individuals have recovered from malaria and have developed immunity against the disease, reducing their susceptibility to future infections and decreasing their potential to transmit the disease to others (Fig. 3).

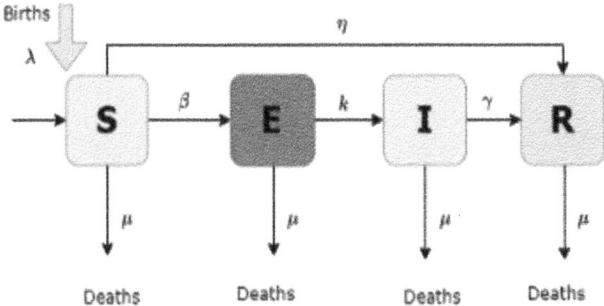

Fig. 3. State Chart of SEIR Model

Other parameters include.

There are equal birth and death rates (μ).
($1/\alpha$) is the mean latent period for the disease.
($1/\gamma$) is the mean infectious period, recovered individuals are permanently immune, the contact rate (β) may be a function of time.

$$\frac{dS}{dt} = \mu - \beta(t)SI - \mu S \tag{4}$$

$$\frac{dE}{dt} = \beta(t)SI - (\mu + \alpha)E \tag{5}$$

$$\frac{dI}{dt} = \alpha E - (\mu + \gamma)I \tag{6}$$

These equations (Eqs. 4, 5, and 6) model the dynamics of disease transmission within a population. Equation (4) describes the rate of change of the susceptible fraction S, accounting for birth, disease transmission, and natural death. Equation (5) represents the rate of change of the exposed fraction E, considering infection, death, and transition to infectious status. Equation (6) governs the rate of change of the infectious fraction I, accounting for the transition from exposed to infectious and the mortality rate. Together, these equations capture the interactions between susceptible, exposed, and infectious individuals in the context of disease spread. [7].

From the Chengalpattu data in the year 2022, this Python simulation sets up parameters and initial conditions for an SEIR (Susceptible-Exposed-Infectious-Recovered) model to investigate the dynamics of a disease outbreak, like malaria, within the population over a simulated year (365 days). The parameters include the initial population size (2,841,572), annual births (28,130), and deaths (27,314) to calculate birth and death rates per 1000 people (11.15 and 10.82, respectively) [3]. The contact rate (beta) represents the frequency of disease transmission, while the recovery rate (gamma) and incubation rate (sigma) dictate how quickly individuals recover and transition from exposed to infectious, respectively. The initial conditions assume that 99% of the population is susceptible (S0), a fraction of the population is exposed (E0 = 0.43), and there are 1 infectious individual (I0 = 1) and 9 recovered individuals (R0 = 9) based on data from 2022. The simulation progresses day by day, updating compartment sizes (Susceptible, Exposed, Infectious, Recovered) to model disease spread and population dynamics over the simulated year, visualized using Plotly.

5   Results and Discussions

In our analysis, distinct patterns emerged from the experiments conducted using traditional predictive models and the SEIR framework, shedding light on the dynamics of malaria spread and the influence of weather parameters. The SEIR model integrated with weather parameters demonstrated a gradual rise in malaria cases, particularly peaking between July and December. This period coincides with increased rainfall, which creates favorable conditions for mosquito breeding and malaria transmission. The presence of elevated rainfall during August-September corresponded with a peak in malaria cases, suggesting a potential correlation between rainfall and disease transmission. However, other environmental factors such as humidity and temperature also played significant roles. Seasonal fluctuations in the breeding rate of Plasmodium vivax (P.v.) mosquitoes further emphasized the influence of weather conditions on mosquito proliferation and malaria transmission. Higher breeding rates during warmer months highlighted the need for targeted interventions to control mosquito populations during peak periods.

Traditional predictive models were trained and evaluated for their accuracy in predicting mosquito types and species based on climate variables, age, gender, and malaria cases. Comparative analysis of the results obtained from traditional predictive models and the SEIR framework was conducted to assess the impact of integrating weather parameters, mosquito behavior, and seasonality on understanding malaria transmission dynamics.

5.1 Predictive Analysis

The results from the predictive modeling analysis offer valuable insights into the efficacy of various machine learning algorithms in predicting mosquito type and species, crucial for understanding malaria transmission dynamics[25]. By leveraging climate variables, age category, gender, and the number of malaria cases, these models aim to identify the type and species of mosquitoes responsible for disease transmission, thereby facilitating targeted intervention strategies and appropriate medication prescriptions. In this analysis, a suite of classifiers including Decision Tree, K Nearest Neighbors, Random Forest, XGBoost, Support Vector Machine, Multi-Layer Perceptron, Logistic Regression, and AdaBoost were trained and evaluated. The subsequent findings, as summarized in Table 2, shed light on the relative performance of these classifiers, revealing which models exhibit higher accuracies in predicting mosquito type and species based on the provided dataset. These insights are crucial for advancing our understanding of malaria transmission dynamics and guiding evidence-based approaches for disease control and prevention [29].

Table 2. Accuracy of Traditional Predictive Models for Mosquito Type and Species Prediction

Model	Species accuracy	Mosquito accuracy
Decision Tree	57.7%	41.1%
K Nearest Neighbours	65%	47.9%
Random Forest	56.9%	45.5%
XGBoost	56.9%	45.5%
Support Vector Machine	56.9%	49.5%
Multi-Layer Perceptron	53.6%	48.7%
Logistic Regression	55.2%	53.6%
AdaBoost	50.4%	45.5%

Among the classifiers evaluated, K Nearest Neighbors exhibited the highest accuracy for predicting both mosquito type (65%) and species (47.9%). This indicates that the K Nearest Neighbors algorithm performed relatively well in identifying the type and species of mosquitoes responsible for malaria transmission [28]. Following closely behind was the Decision Tree classifier, which achieved an accuracy of 57.7% for predicting mosquito type and 41.1% for species. Random Forest, XGBoost, Support Vector Machine, and Logistic Regression models demonstrated comparable accuracies ranging from 56.9% to 55.2% for predicting mosquito type and species. Multi-Layer Perceptron yielded slightly lower accuracies of 53.6% for predicting mosquito type and 48.7% for species. On the other hand, the AdaBoost classifier exhibited the lowest accuracy among the models evaluated, with 50.4% accuracy for predicting mosquito type and 45.5% for species. Despite its lower performance compared to other classifiers, AdaBoost still provided valuable insights into mosquito type and species prediction [27].

Overall, the results suggest that K Nearest Neighbors and Decision Tree classifiers outperformed other models in accurately predicting mosquito type and species based on the given dataset. These findings highlight the importance of selecting appropriate machine learning algorithms for effectively addressing the problem statement and informing decision-making processes in malaria control and prevention efforts.

5.2 SIR Model

The Fig. 4 illustrates a simulated malaria epidemic scenario in Chengalpattu district, India for the year 2022, assuming a closed population using the SIR (Susceptible-Infected-Recovered) model. Here are key observations derived from the graph; The blue portion representing the susceptible population shows a steady decline over time as individuals become infected. Initially, the red portion representing infected individuals starts small but grows rapidly, reaching a peak before eventually levelling off and declining [21]. Conversely, the green portion denoting the recovered population begins small and gradually increases throughout the simulation. As the susceptible population diminishes and the number of recovered individuals grows, the epidemic's intensity is predicted to wane, reflecting the dynamics of disease spread and recovery modelled by the SIR framework [23].

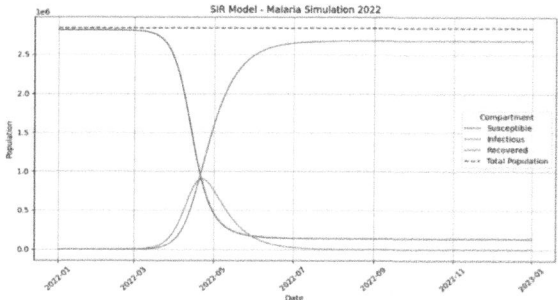

Fig. 4. SIR Simulation Graph for the Year 2022 in Chengalpattu District

5.3 SEIR Model

Figure 5 depicting the 2022 malaria simulation in Chengalpattu, based on an SEIR model, reveals several key insights. Initially, the susceptible population (blue) starts high and gradually declines throughout the simulation. The exposed population (orange) begins small, peaks midway through the simulation, and then declines. Concurrently, the infectious population (red) also starts small, peaks around the same time as the exposed population, and then decreases[13]. Meanwhile, the recovered population (green) steadily increases over time [22]. Overall, the SEIR model suggests a typical pattern for malaria outbreaks, with susceptibility declining initially, followed by a rise and subsequent decline in exposed and infectious individuals, respectively, until the number of recovered individuals surpasses those infected, leading to a decline in the outbreak [8].

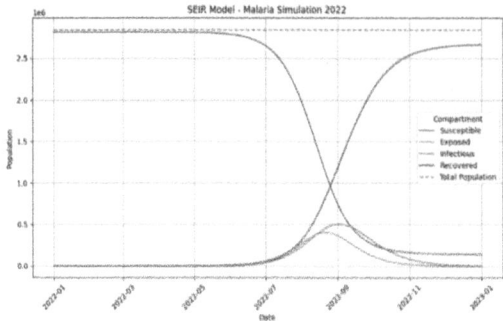

Fig. 5. SEIR Simulation Graph for the Year 2022 in Chengalpattu District

6 Conclusion

This study aimed to decipher malaria transmission dynamics in Chengalpattu District, Tamil Nadu, India, through an integrated approach combining predictive modeling and epidemiological frameworks. A dataset spanning from January 2010 to September 2022, sourced from the INSAT AWS ARGS network, provided insights into climate parameters such as temperatures, humidity, and precipitation. Correlation analysis revealed relationships among these variables, including positive associations between temperature and wind speed. Malaria cases and demographic data spanning 12 years were collected, with efforts underway to expand this dataset for deeper insights. Various predictive models, including Decision Tree, K-Nearest Neighbors, and others, were employed to predict mosquito types and species based on climate variables and demographic factors. K-Nearest Neighbors showed the highest accuracy at 65% for mosquito type prediction and 47.9% for species prediction. The SEIR epidemiological model, incorporating a latent period, aligned well with observed transmission patterns, indicating a peak in malaria cases between July and December, correlating with increased rainfall.

Analysis revealed seasonal fluctuations in P. vivax breeding rates, peaking in August. This aligns with warmer months being conducive to mosquito proliferation and increased malaria transmission risk. Combining predictive and compartmental modeling approaches provides a comprehensive understanding of malaria epidemiology, aiding in targeted interventions such as vector control measures and enhanced surveillance.

In conclusion, weather parameters significantly impact malaria transmission in Chengalpattu District. Integrating predictive modeling with epidemiological frameworks offers insights into transmission dynamics, guiding evidence-based control strategies and potentially serving as a model for other endemic areas.

7 Future Directions

Future directions in malaria control involve further research to refine predictive models and enhance their accuracy in predicting malaria transmission dynamics based on various factors including weather parameters, mosquito behavior, and human factors. Long-term monitoring of climate and disease data, along with advanced modeling techniques, can provide valuable insights into the complex interactions between weather conditions and malaria transmission. Collaborative efforts between public health authorities, researchers, and policymakers are essential for implementing evidence-based interventions and strategies to control malaria transmission and improve public health outcomes in endemic regions like Chengalpattu District, Tamil Nadu, India. Expanding the dataset to include a longer period of historical data would strengthen the analysis and provide further insights into long-term trends and climate change impacts.

References

1. Talapko, J., Škrlec, I., Alebić, T., Jukić, M., Včev, A.: Malaria: the past and the present. Microorganisms **7**(6). MDPI AG, 01 June 2019. https://doi.org/10.3390/microorganisms7060179

2. Roper, C., Pearce, R., Nair, S., Sharp, B., Nosten, F., Anderson, T.: Intercontinental spread of pyrimethamine-resistant malaria. Science (1979) **305**(5687), 1124, August 2004. https://doi.org/10.1126/science.1098876

3. Rodrigues, P.T., et al.: Human migration and the spread of malaria parasites to the New World. Sci. Rep. **8**(1), December 2018. https://doi.org/10.1038/s41598-018-19554-0

4. Kulkarni, M.A., Duguay, C., Ost, K.: Charting the evidence for climate change impacts on the global spread of malaria and dengue and adaptive responses: a scoping review of reviews. Global Health **18**(1), December 2022. https://doi.org/10.1186/s12992-021-00793-2

5. Gaudart, J., Ghassani, M., Mintsa, J., Rachdi, M., Waku, J., Demongeot, J.: Demography and diffusion in epidemics: malaria and black death spread. Acta Biotheor. **58**(2), 277–305 (2010). https://doi.org/10.1007/s10441-010-9103-z

6. He, S., Peng, Y., Sun, K.: SEIR modeling of the COVID-19 and its dynamics. Nonlinear Dyn. **101**(3), 1667–1680 (2020). https://doi.org/10.1007/s11071-020-05743-y

7. Gray, A., Greenhalgh, D., Hu, L., Mao, X., Pan, J.: A stochastic differential equation SIS epidemic model

8. Kamrujjaman, M., Saha, P., Islam, M.S., Ghosh, U.: Dynamics of SEIR model: a case study of COVID-19 in Italy. Res. Control Optim. **7**, June 2022. https://doi.org/10.1016/j.rico.2022.100119

9. Mwalili, S., Kimathi, M., Ojiambo, V., Gathungu, D., Mbogo, R.: SEIR model for COVID-19 dynamics incorporating the environment and social distancing. BMC Res. Notes **13**(1), July 2020. https://doi.org/10.1186/s13104-020-05192-1

10. Gray, A., Greenhalgh, D., Mao, X., Pan, J.: The SIS epidemic model with Markovian switching. J. Math. Anal. Appl. **394**(2), 496–516 (2012). https://doi.org/10.1016/j.jmaa.2012.05.029

11. Allen, L.J.S.: Some discrete-time SI, S/R, and S/S epidemic models

12. Chitnis, N., Hyman, J.M., Cushing, J.M.: Determining important parameters in the spread of malaria through the sensitivity analysis of a mathematical model. Bull. Math. Biol. **70**(5), 1272–1296 (2008). https://doi.org/10.1007/s11538-008-9299-0

13. Juher, D., Ripoll, J., Saldaña, J.: Outbreak analysis of an SIS epidemic model with rewiring. J. Math. Biol. **67**(2), 411–432 (2013). https://doi.org/10.1007/s00285-012-0555-4

14. Carson, R.T., Carson, S.L., Dye, T.K., Mayfield, S.A., Moyer, D.C., Yu, C.A.: COVID-19's U.S. temperature response profile. Environ. Resour. Econ. (Dordr) **80**(4), 675–704 (2021). https://doi.org/10.1007/s10640-021-00603-8

15. Li, M.Y., Graef, J.R., Wang, L., Karsai, A.: Global dynamics of a SEIR model with varying total population size. www.elsevier.com/locate/mbs

16. Sun, C., Hsieh, Y.H.: Global analysis of an SEIR model with varying population size and vaccination. Appl. Math. Model. **34**(10), 2685–2697 (2010). https://doi.org/10.1016/j.apm. 2009.12.005

17. Meng, X., Zhao, S., Feng, T., Zhang, T.: Dynamics of a novel nonlinear stochastic SIS epidemic model with double epidemic hypothesis. J. Math. Anal. Appl. **433**(1), 227–242 (2016). https://doi.org/10.1016/j.jmaa.2015.07.056

18. Shi, P., et al.: Impact of temperature on the dynamics of the COVID-19 outbreak in China. Sci. Total Environ. **728**, August 2020. https://doi.org/10.1016/j.scitotenv.2020.138890

19. Tang, Z., Li, X., Li, H.: Prediction of new coronavirus infection based on a modified SEIR model. https://doi.org/10.1101/2020.03.03.20030858

20. Liu, D., et al.: The impact of containment measures and air temperature on mitigating COVID-19 transmission: non-classical SEIR modeling and analysis: contributed equally. https://doi. org/10.1101/2020.05.12.20099267

21. Shi, P., et al.: The impact of temperature and absolute humidity on the coronavirus disease 2019 (COVID-19) outbreak-evidence from China 2 3. https://doi.org/10.1101/2020.03.22. 20038919

22. Keno, T.D., Makinde, O.D., Obsu, L.L.: Impact of temperature variability on SIRS malaria model. J. Biol. Syst. **29**(3), 773–798 (2021). https://doi.org/10.1142/S0218339021500170

23. Biswas, M.H.A., Paiva, L.T., De Pinho, M.: A seir model for control of infectious diseases with constraints. Math. Biosci. Eng. **11**(4), 761–784 (2014). https://doi.org/10.3934/mbe. 2014.11.761

24. Huang, W., Han, M., Liu, K.: Dynamics of an sis reaction-diffusion epidemic model for disease transmission. Math. Biosci. Eng. **7**(1), 51–66 (2010). https://doi.org/10.3934/mbe. 2010.7.51

25. Modu, B., Polovina, N., Lan, Y., Konur, S., Asyhari, A.T., Peng, Y.: Towards a predictive analytics-based intelligent malaria outbreak warning system. Appl. Sci. **7**(8), 836 (2017)

26. Idowu, A.P., Okoronkwo, N., Adagunodo, R.E.: Spatial predictive model for malaria in Nigeria. J. Health Inform. Dev. Countries **3**(2) (2009)

27. Githeko, A.K., Ogallo, L., Lemnge, M., Okia, M., Ototo, E.N.: Development and validation of climate and ecosystem-based early malaria epidemic prediction models in East Africa. Malar. J. **13**, 1–11 (2014)

28. Nkiruka, O., Prasad, R., Clement, O.: Prediction of malaria incidence using climate variability and machine learning. Inform. Med. Unlocked **22**, 100508 (2021)

29. Bharti, D.R., Lynn, A.M.: QSAR based predictive modeling for anti-malarial molecules. Bioinformation **13**(5), 154 (2017)

30. Darkoh, EL., Larbi, J.A., Lawer, E.A.: A weather-based prediction model of malaria prevalence in Amenfi West District. Malaria research and treatment, Ghana (2017)

Towards Scalable and Cost-Effective Design for Intrusion Detection for IIoT Environment Using Metric Active Learning

S. Menaka$^{(\boxtimes)}$ iD, B. Ahalya, Shyam Narayan Ramkumar Sharma,
and Ramisetty Mounika

SRM Institute of Science and Technology, Ramapuram, Chennai 600089, India
{menakas1,ba1380,ss4453,rm5238}@srmist.edu.in

Abstract. Intrusion Detection Systems (IDS) represent a cornerstone of modern network security strategies, offering diverse methods and architectures to analyze network access. These structures can be as two categories: Signature and Anomaly. Signature - Based IDS monitor events using a database of known intrusions, while passive IDS focus on understanding system behavior and identifying anomalies. However, with the hasty development of the IoT, new and complex safety challenges possess emerged. Despite efforts to address IoT cybersecurity through various technologies, further development is essential to effectively safeguard IoT ecosystems. One promising approach to bolster IoT security involves leveraging machine learning techniques. Numerous studies have explored the application of DL also ML methods to progress Internet of Things safety. In our study, we have established a method that utilizes Deep Learning techniques to detect attacks on IoT systems. By employing Python programming and tools such as Tensorflow, Scikit-learn, and Seaborn, we have shown that deep learning models are effective detection accuracy. Our findings suggest that deep learning holds significant promise for bolstering IoT security measures, providing a more robust defense against cyber threats targeting IoT devices and networks. Through our research, we have contributed to progress the IoT security industry, addressing a critical need in the constantly changing field of cybersecurity.

Keywords: Intrusion Detection Systems (IDSs) · Deep learning (DL) · Internet of Things (IoT · Machine learning (ML) · Convolutional Neural Network (CNN) · Support vector machine (SVM) · Scikit-learn · Edge Computing (EC) · Long-Short term memory (LSTM) · Industrial Internet of things (IIOT) · Recurrent Neural Network (RNN)

1 Introduction

The notion of the IoT unites a huge variation of applications based totally at the convergence of clever things and the net, growing unbroken connection among the physical and virtual worlds [4]. These uses might be whatever from fundamental smart family appliances to highly complex business plant machinery. Internet of Things packages

P. D. Sivakumar et al. (Eds.): IRCCTSD 2024, CCIS 2362, pp. 29–46, 2025.
https://doi.org/10.1007/978-3-031-82386-2_3

are full of many dreams, but they all have many things that are not exceptional. The three tiers that make up an IoT activity are usually the transmission segment, the collection portion, and the processing, administration, and usage part. Attacks on connected devices are becoming a major problem as the Internet of things gains popularity. In addition to privilege escalation, eavesdropping, and denial of service attacks, IoT devices are susceptible to other risks and assaults. As a result, there is an increasing need to defend IoT devices against those kinds of attacks. Furthermore, because IoT devices are distributed in a sequential manner, it is easy for unauthorized access to take place. In addition, the device is susceptible to cyberthreats like net injection, which could lead to the alteration of records and the exposure of private information, as several integrated system components depend on wireless networks for spoken communication that occurs in real time and is vulnerable to eavesdropping. IoT devices need more advanced and more resilient intrusion detection systems in order to protect themselves from a variety of cyberattacks because of their broad adoption and the vital roles they play. The limited computing power of many IOT devices means that standard security precautions might not be enough to quickly identify and address possible security weaknesses. On the other hand, IoT security appears to have a bright future thanks to machine learning's deep learning subfield. Deep learning programs can operate within the limitations of low processing power, rapidly analyzing large amounts of data and identifying patterns suggestive of malware or security breaches. By enabling automated security system alterations upon the identification of any questionable activity, these algorithms offer a proactive line of protection against possible attacks. Deep learning-based security solutions have several notable benefits, one of which is their ability to operate across various IoT device types, basic os, and file structures short of having a continuous network linking for threat detection. This implies that IoT systems can withstand cyberattacks and maintain the security and integrity of the perfect IoT system, even in settings with poor connection, like isolated industrial sites or places with erratic internet access.

An IDS is in the form of hardware or software, is a crucial component in maintaining the security of networks or systems, monitoring for malicious activities and ensuring security measures are in place [1].

Intrusion detection systems come in two primary varieties (IDS): Host IDS and Network IDS, which focuses on tracking intrusions on individual computer devices, and Network-based intrusion detection systems (NIDS), which monitor a whole system. Since the subject of this work is NIDS, NIDS will be referred to as IDS from here on [11]. Although the idea of IDS in relation to the IoT is not new, a number of solutions have been created to solve security issues with IoT. Conventional IoT-focused IDSs frequently rely on cloud computing and are generally implemented at the device or gateway levels. Nonetheless, new avenues for IoT security have been made possible by recent developments in Edge Computing (EC) [13]. EC allows for the integration of intelligent computing capabilities into edge devices by extending the Cloud Computing concept to the network's edge. By enabling processing and storage capability at the edge network, these devices help lower network latency. On the other hand, edge nodes also introduce fresh security flaws that hostile actors could take use of to initiate assaults. Edge nodes could be compromised by unauthorized remote access or physical manipulation, especially if they are situated in public areas. An intrusion detection system's main objective is to stop unauthorized access to a data system, as any such access could pose a risk to the availability, confidentiality, or integrity of data. When malicious movement is detected, an IDS raises an alarm by analyzing network traffic and/or resource utilization. Based on how they identify intrusions, intrusion detection systems (IDSs) can be divided into two primary groups: anomaly-based and signature-based IDS [7]. While SIDSs compare recorded events to a database of known intrusion methods, anomaly-based IDSs learn the regular action of a system and flag any anomalous activity. These methods play a crucial part in spotting and addressing possible security risks in IoT networks, ensuring the continued integrity and safety of networks and linked devices.

2 Related Work

2.1 Machine Learning in Intrusion Detection Systems

Intrusion Detection Systems (IDS) utilizing machine learning methods and algorithms has become a viable strategy for identify malicious activities and security threats within network communities [1]. These machine learning-based IDS systems, in contrast to conventional rule-based IDS, automatically pick up on new threat patterns and adjust accordingly, providing improved generalization and adaptability [4]. Support Vector Machine (SVM) is a technique known for its robust generalization and resilience to noise. SVM determines the most useful hyperplane to differentiate between different sample classes, demonstrating commendable performance across various ML tasks. For example, Almaiah M. A. provided a model for an IDS (intrusion detection system) that used PCA for feature selection together with various SVM kernel classifiers. The major component of the model was the Gaussian kernel feature SVM [2]. Furthermore, Yakub et al. presented the K-means-Gaussian-SVM model, which applied genetic algorithms to feature selection on normalized data and wrapper feature selection of K-means clustering, ultimately enhancing model robustness and reducing computational costs via SVM classification. Another notable advancement is the hybrid architecture presented by Agarap A F M, which combined GRU (Gated Recurrent Unit) with SVM for ID [6]. By employing simulated trials using network traffic data from the "Kyoto University honeypot systems," When it came to accuracy and performance, this hybrid technique beat conventional GRU models. These advancements in machine learning-based IDS are crucial for effectively identifying and responding to security threats within network ecosystems, ensuring the continued integrity and security of connected devices and networks [1].

2.2 Deep Learning in Intrusion Detection Systems

With the rise of DL, its capacity to separately abstract structures from vast datasets has significantly alleviated the challenges associated with feature engineering [5]. DL techniques, such as NNs, are characterized by a large number of parameters and Deep learning models are well-suited because of their capacity for intrusion detection in Industrial Internet of Things (IIoT) devices to model complex nonlinear patterns. These models generally outperform traditional machine learning methods, achieving higher accuracy and faster training times. Vinayakumar et al.'s research, for instance, demonstrated that deep neural network-based classifiers provided superior results in IIoT intrusion detection compared to conventional approaches. A notable example is the hybrid DRaNN developed by Zil E. Huma, which combines DRaNN with Multilayer Perceptrons (MLPs) and dropout regularization. This hybrid model not only increases detection accuracy but also enhances the system's ability to reliably identify and address potential security threats in IIoT environments. Moreover, deep learning models excel in handling large scale datasets, making them especially valuable in IIoT scenarios where vast amounts of data are constantly generated. Their adaptive learning capabilities allow these models to stay effective as new patterns emerge, ensuring that intrusion detection systems (IDSs) remain up-to-date and responsive to evolving security threats. As industrial IoT environments continue to expand and evolve, the integration of DL based IDSs represents a crucial step towards guaranteeing the dependability and security of these linked systems.

2.3   Metric Active Learning - Algorithm to Predict Intrusion Detection

Metric Active Learning is an innovative algorithmic approach specifically designed to enhance the efficacy of IDS [9]. Traditional IDS aspect important tasks, particularly in the resource-intensive and time-consuming process of labeling data for training purposes, especially with large datasets. However, Metric Active Learning addresses these challenges by strategically selecting the most informative data points for labeling, thus optimizing the learning process while minimizing labeling efforts. This approach is based on the principle of selecting data points that are most relevant for annotation, typically determined by their distance or similarity to existing labeled samples. By actively choosing these pertinent data points, Metric Active Learning significantly reduces the labeling effort required, thereby increasing the efficiency of the IDS. As a result, IDS models developed using Metric Active Learning tend to exhibit the hybrid Deep Random Neural Network (DRaNN) model offers improved detection accuracy while reducing computational costs compared to traditional methods. By integrating advanced techniques such as dropout regularization and Multilayer Perceptrons (MLPs), the model achieves more efficient performance in identifying potential security threats within IIoT environments. This makes it both more effective and resource-efficient than conventional approaches [12]. One of the significant advantages of Metric Active Learning is the capacity to successfully adjust to new risks and changing network conditions. Through iterative refinement, Metric Active Learning continually updates the model's understanding of the data distribution, thus ensuring its capability to identify and mitigate security threats accurately [13]. In summary, Metric Active Learning represents a promising avenue for advancing the capabilities of IDS, offering a additional efficient and effective means of accurately identifying and mitigating security threats in complex network environments.

3   Methodology

The emergence of the IoT has introduced a groundbreaking paradigm, enabling the connection of physical objects and computer power to the internet. Beyond revolutionizing industries such as healthcare, environmental monitoring, and industrial management systems, IoT promises to build flexible and environmentally friendly solutions across various fields. Through remote and intelligent management, IoT has the likely to knowingly increase productivity and proficiency. However, this technological advancement also brings with it an increased risk of cyberattacks. The IoT environment often lacks robust security features, making IoT devices susceptible to malevolent assaults from networks connected to and disconnected from employers. As a result, the need to mitigate risks and secure IoT applications has become a critical focus within the cybersecurity field. Particularly concerning are Industrial IoT (IIoT) applications, which are crucial components of the Industrial IOT 4.0 revolution. These applications often entail crucial mission-related duties like infrastructure management and corporate control, demanding an extremely high level of security. For instance, reports indicate that multiple power substations in Ukraine were breached during a sophisticated attack on IIoT applications, leading to an electricity blackout affecting roughly 225,000 users. In this attack, the perpetrator infiltrated Supervisory Control and Data Acquisition (SCADA) systems by breaching the IT network, gaining unauthorized access to critical industrial control systems. This breach allowed the attacker to compromise sensitive operational data and potentially disrupt key infrastructure processes managed by the SCADA systems, compromising the system responsible for monitoring and managing the intelligent grids of IIoT devices and subsequently shutting off the energy supply. Another notable IoT assault was the Mirai botnet attack in twilight 2016, initially consisting of attacked smart photographic camera. This attack caused widespread internet disruptions, targeting numerous eminent corporations with huge Distributed DoS attacks. To ensure the security and functionality of IoT/IIoT applications, it is imperative to implement effective and appropriate security measures. These incidents underscore the critical need for ongoing R&D to enhance the security of IoT ecosystems.

3.1 Signature-Based IDS

In modern cybersecurity strategies, IDSs plays a crucial part in locating and reducing online hazards, with Signature-based IDS (SIDSs) being a prominent method for detecting attacks. These systems compare system or network activity with attack signatures stored in their databases [14]. These signatures represent known threats, and An alert is set off when there is a match. While SIDSs are highly accurate for detecting known threats, they face challenges when dealing with new or variant attacks due to the absence of matching signatures. To address this limitation, SIDSs can operate in online or offline modes, monitoring hosts in real-time or analyzing system activity logs retrospectively. However, the creation of signatures is often labor-intensive, requiring expert manual crafting. To alleviate this issue, Christian Kreibich et al. proposed an automatic signature generation system by extending honeypot technology to inspect traffic and integrate with IDS. Initially, SIDSs analyzed individual network packets, but the evolving nature of threats has made it necessary to match signatures across multiple packets for comprehensive detection. This evolution reflects the constant adaptation required to combat emerging cyber threats effectively. As cyberattacks become increasingly sophisticated, the development of more advanced detection methods, such as automatic signature generation systems, is essential to enhance the effectiveness of IDSs in safeguarding network security.

3.2 Anomaly IDS

Anomaly-based IDSs emerged as a promising alternative to signature-based IDS to address the limitations inherent in the latter. AIDS undergo a training phase to establish a comprehensive model of normal system behavior. Once deployed, these systems continuously monitor hosts, comparing their behavior against the established model. When significant deviations from the norm are detected, an alert is triggered, signaling a potential intrusion. This unique approach allows AIDS to potentially detect zero day attacks, as they do not rely on predefined signatures. Furthermore, attackers find it challenging to understand a target's normal behavior without triggering alerts, making AIDS a formidable line of defense. Moreover, AIDS can serve as valuable system analysis tools, capable of detecting not only intrusions but also potential software bugs or system anomalies [15]. However, it's vital to note that AIDS are prone to higher rates of false positives in contrast to intrusion detection systems based on signatures. False alarms may be caused by differences in system behavior that are unrelated to intrusions if not adequately accounted for by the AIDS, underscoring the need for continuous refinement and improvement in anomaly detection algorithms.

3.3 Metric Active Learning

Long Short-Term Memory is a type of DL architecture that has gained considerable attention for addressing the limitations of traditional Recurrent Neural Networks (RNNs). Unlike standard RNNs, which struggle with retaining long-term dependencies due to issues like vanishing gradients, LSTMs are built with memory cells that have a long retention time for information. Because of this, they work incredibly well for sequential data tasks like time series analysis, natural language processing, and other uses where it's critical to capture long-range dependencies. Unlike conventional RNNs, LSTM networks are composed of memory cells or blocks equipped with input, output, and forget gates, enabling them to selectively store and discard data. This sophisticated mechanism allows LSTM to effectively manage long sequences of data, addressing the vanishing gradient problem encountered by traditional RNNs. While RNNs are inherently designed to model sequential data, [16] they often struggle with short-term memory limitations, hindering their facility to effectively capture far-off dependencies in data sequences. LSTM, however, excels in recalling past knowledge and is particularly well-suited for analyzing and forecasting time series data of varying lengths. Its ability to retain and utilize information over extended periods makes it a prevailing implement for applications including speech recognition, financial forecasting, and NLP. Moreover, LSTM networks have demonstrated impressive performance in various applications such as machine translation, sentiment analysis, and predictive maintenance. The robust memory mechanism of LSTM networks, coupled with their capability to apprehension long-term needs, has made them requisite in the field of DL, significantly advancing the capabilities of sequential data analysis and prediction. Through ongoing research and development, LSTM networks continue to evolve, it provides increasingly sophisticated solutions to a variety of real-world problems, strengthening their role as a foundational component in modern deep learning architectures. Their ability to handle complex sequential data has made them essential in areas such as speech recognition, language modeling, and predictive analytics, solidifying their importance in cutting-edge AI systems.

1. *No. of hidden layers:* The dataset encompasses an exponentially large number of teachable weights relative to n, where n represents the dataset size. This abundance of weights can be adjusted to facilitate the learning of non-linear methods due to the extensive presence of hidden nodes, connections, and layers within the neural network architecture. The flexibility of neural networks, enhanced by the inclusion of more nodes and layers, allows for capturing intricate non-linear relationships within the data. This increased complexity adds additional weights to the model, enabling better performance on challenging tasks. However, using only a few layers and nodes may not be sufficient to handle complex patterns, leading to issues such as underfitting or overfitting. While determining the optimal number of hidden layers is crucial, it is also limited by the processing capacity of framework. It's crucial to find a balance between computing efficiency and model complexity for designing neural networks that can effectively learn non-linear data relationships. Advancements in hardware and optimization techniques have improved the capability of neural networks, allowing for deeper and more complex architectures. These innovations enable the handling of more intricate data analysis tasks. Ongoing research in the field continues to drive progress, making deep learning architectures increasingly sophisticated and applicable to solving a wide range of real-world problems.

2. *Activation functions:* The function within a neural network calculates the weighted sum of inputs and biases to determine whether a neuron will fire. This process involves decoding the output signal from the weights and inputs of the preceding layers, transforming it into a configuration that can be utilized as the data that the nodes in the next embbeded layer will use. In essence, individual neuron in the hidden layer calculates a biased summation of its inputs, including a bias term, and then applies an activation function to this sum to generate its output. The activation function introduces non-linearity, this is necessary for the network to identify intricate linkages and patterns in the data. If this non-linearity didn't exist, the network would function like a linear model, which would have limited its capacity to represent complex structures and dependencies within the data. Through this iterative process of forward propagation, the network gradually transforms the supplied data into a format that allows for classification or prediction making [18]. This process of weighted summation, bias addition, and activation function application forms the fundamental building block of neural networks, enabling them to learn and generalize from input data. As the network learns through backpropagation, weights and biases to reduce the discrepancy between anticipated and actual outputs, the neural network becomes progressively better at capturing complex patterns and improving prediction accuracy. This process, often referred to as training, is critical to the grid's facility to learn from information. As a result, the activation function, along with weight and bias adjustments, plays a pivotal role in enabling neural networks to model complex relationships and deliver accurate predictions., allowing them to execute intricate operations with exceptional precision and effectiveness, like picture identification, natural language processing, and predictive analytics.

3. *Sigmoid Function:* For binary classification tasks, this function is often recommended as the neural network's activation function in the output layer because it provides

a probabilistic interpretation of the methods output, indicating the likelihood of a certain class. However, despite its popularity, the sigmoid function presents challenges, particularly when used in deep neural networks. After multiple iterations of backward propagation, the sigmoid function can struggle to converge effectively due to the vanishing gradient problem. This problem arises because with an increase in the input's absolute value, the sigmoid function's gradient gets smaller, leading to slow learning of the network. The equation of the sigmoid activation function is displayed below:

$$\mathbf{G(y) \ = \ 1/(1 + e^{-y})}$$

e^{-y} is the Eulers number in the above given equation
    y represents the weighted sum of inputs biases,
    G (y) the output of the sigmoid function.
    Despite its limitations, the sigmoid function remains a popular choice for binary classification tasks due to its intuitive interpretation and simplicity. However, alternative activation functions such as the rectified linear unit and hyperbolic tanh function have gained popularity for use in deeper neural networks, as they mitigate the vanishing gradient problem and enable more stable and efficient training [19]. As neural network architectures continue to evolve, the choice of activation function remains a critical consideration, with researchers constantly exploring new techniques to enhance the performance and efficiency of deep learning models.
    4. *Activation function of Rectified linear unit (ReLu):* It is widely regarded as a safe default choice in many neural network (NN) applications. ReLU has added admiration as a result of its simplicity, proficiency, and excellent performance in various neural network tasks. The key compensations of ReLU is its computational effectiveness, which enables rapid training of neural networks. By immediately outputting the input if it is positive and producing zero otherwise, the network gains non-linearity via the activation function. ReLU can successfully handle the vanishing gradient issue that conventional activation functions like the sigmoid and hyperbolic tangent functions run into because of its straightforward but powerful behavior. The equation for the ReLU is as follows:

$$\mathbf{ReLU \ = \ g(Z) \ = \ maximum(0,Z)}$$

where z represents the i/p to the ReLU and g(z) represents the o/p. [20] Despite its simplicity, ReLU has proven to be highly effective in various neural network architectures, providing improved convergence rates and enabling the training of deeper networks. However, it is essential to note that ReLU is not without its drawbacks, such as the "dying ReLU" issue where neurons can go dormant and cease to learn as a result of becoming stuck at zero while being trained. Nonetheless, ReLU remains a common option for neural network activation functions, particularly for its capability to facilitate efficient training and achieve excellent performance across a wide range of applications.

4   Architecture Diagram

The architecture figure of the system presents a comprehensive method for training and validating an IDS using ML techniques. Initially, training data in the form of CSV files undergoes preprocessing to ensure it is in a appropriate setup for analysis. The next stage is featuring selection, which is an important procedure to determine which features are most pertinent to the IDS. In order to reduce redundancy and boost the effectiveness of the model, features with an RFE coefficient larger than 0.95 are eliminated. Recursive Feature Elimination (RFE) is utilized to evaluate the significance of each feature. The design combines Long Short-Term Memory (LSTM) networks and Metric Active Learning after feature selection to further hone the feature set. If optimal parameters are not initially found, the system iterates between active learning and parameter optimization to enhance the IDS's performance. Once optimal parameters are determined, the data validation phase begins, where IDS validation is conducted, and parameter settings are finalized. This structured approach ensures that the IDS is effectively trained and optimized to detect and mitigate security threats efficiently (Fig. 1).

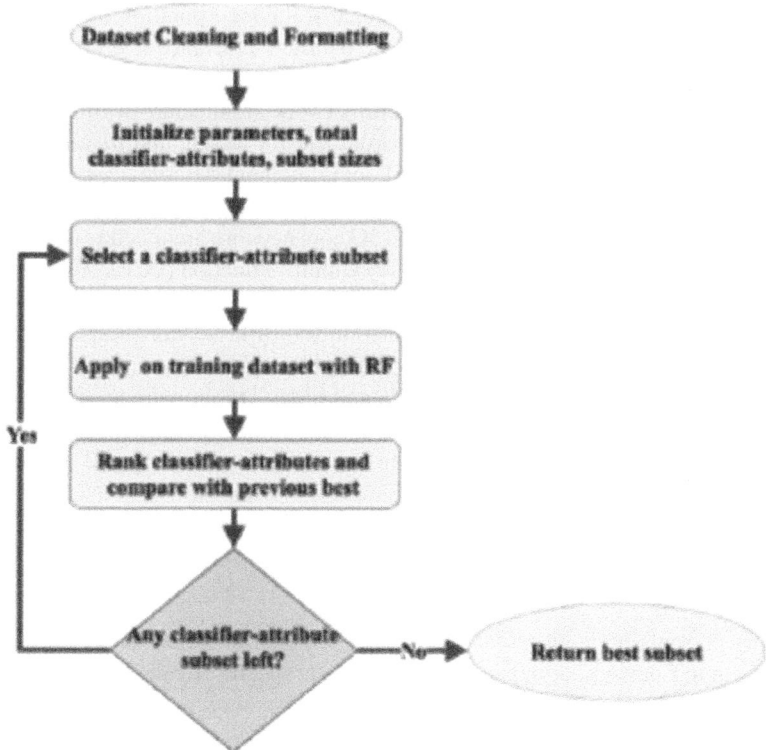

Fig. 1. The architecture diagram of the system

4.1 Dataset Description

The 5GC PFCP Intrusion Detection Dataset incorporates eleven essential features, each representing significant methodological frameworks vital for effective intrusion detection. The initial five components on the list include interaction, capture, available protocols, traffic, labelled dataset, complete network configuration, attack diversity, heterogeneity, feature set, attack diversity, and metadata. These features collectively form the foundation for a robust intrusion detection system, enabling the accurate identification and mitigation of security threats within a 5G core network environment. With interaction and capture capabilities, the dataset facilitates the monitoring and recording of network traffic, providing valuable insights into network behavior and potential security breaches. The inclusion of diverse attack scenarios and a labelled dataset ensures the comprehensive coverage of potential threats, enhancing the dataset's effectiveness in training and validating intrusion detection models. Additionally, the incorporation of metadata and a complete network configuration ensures the dataset's relevance and applicability to real-world network environments, further enhancing its utility and effectiveness.

Furthermore, the emulation of a 5G core network architecture within the dataset provides a comprehensive and realistic environment for testing and training intrusion detection systems. In addition to virtualized User Equipment (UE) devices and gNodeBs (gNBs), the dataset includes a cyberattacker posing as a maliciously instantiated Service Management Function (SMF). The emulated 5G core network architecture encompasses essential network functions such as the Policy Control Function (PCF), Authentication Server Function (AUSF), Access Management Function (AMF), User Data Management (UDM), Network Slice Selection Function (NSSF), Network Repository Function (NRF), Network Exposure Function (NEF), User Plane Function (UPF), Access and Mobility Management Function (AMF), and SMF. This comprehensive emulation ensures that the dataset accurately reflects the complexities then challenges of a real-life 5G core network environment, enabling more effective training and validation of intrusion detection systems.

5 Proposed Work

The paper presents in-depth research on the application of DL methods for detecting intrusions in IoT devices, utilizing a standard dataset specifically designed for intrusion detection purposes. The data preparation process involves several crucial steps, including data cleaning, feature extraction, and normalization, aimed at improving the precision of the IDSs then the learning process has been facilitated. By ensuring that the data is properly preprocessed, the system can effectively identify patterns indicative of malicious activity within the IoT network. To prevent overfitting and ensure the robustness of the model, the dataset is partitioned into 75% for training and evaluation and 25% for testing. Additionally, K-fold Cross-Validation is employed to fine-tune model parameters, further increasing the efficiency and dependability of the IDSs. This comprehensive approach to data preparation and model evaluation ensures that the deep learning-based system is capable of effectively detecting the intrusions in IoT devices with maximum precision and consistency.

5.1   Data Preprocessing

In this phase, the raw dataset undergoes several preprocessing steps to prepare it for use with a Deep Learning (DL) algorithm. Data normalization, standardization, and cleansing are integral parts of this process, aimed at ensuring the dataset's compatibility and effectiveness with the DL algorithm. The preprocessing phase consists of three smaller steps. The first step, known as data cleaning, involves eliminating unnecessary data such as NaN and null values from the dataset, ensuring data integrity and reliability. Following data cleaning, the dataset undergoes standardization, which is the second sub-step.

Standardization is essential since it guarantees that the data's distribution value is between 0 and 1, based on the conventional normal distribution, and that all features are on a similar scale. This standardization process enhances the effectiveness of the DL algorithm by ensuring that all features are uniformly represented and weighted during model training. The third sub-step is data normalization, where each feature is converted to a numerical type before normalization. In this work, the min-max regularization technique is applied for data normalization, scaling the information within a specified range and ensuring the dataset is normalized. This normalization approach improves the model's training efficacy and convergence speed, as it transforms the data to a format that is more conducive to neural network processing. It is crucial to avoid negative numbers during the normalization process, as neural networks cannot handle them effectively. Overall, the preprocessing phase plays a vital role in optimizing the dataset for effective utilization with DL algorithms, ensuring accurate and efficient model training and performance.

$$X^i = (X - xmin)/(Xmax - xmin)$$

where, x_{min} and x_{max} show the feature's lowest and maximum values, respectively. This normalization technique ensures that all features fall within a similar numerical range, making them more conducive to machine learning (ML) techniques. Additionally, to facilitate the use of ML techniques, categorical variables with nominal values are converted into numerical values. This conversion is achieved through the label-encoding approach, which transforms categorical values into a series of numerical values, making them suitable for input into ML algorithms. Furthermore, to prepare the data input for the LSTM network, the data is reshaped into three dimensions, consisting of sample count, timestep count, and feature count. Since it is the default form for constructing an LSTM network, the timesteps are set to one in this process. This reshaping ensures that the input data is properly formatted and ready for input into the LSTM network, enabling effective training and prediction of intrusion detection in IoT devices. Overall, these preprocessing steps performance a critical role in optimizing the dataset designed for use with ML techniques, ensuring accurate and efficient ID in IoT milieus.

5.2   Feature Selection

Redundant and irrelevant information often exists within network traffic data, which can significantly impede the efficiency and accuracy of classifier predictions. To mitigate this issue, RFE (Recursive Feature Elimination) is employed as a wrapper feature variety technique. RFE systematically reduces the number of feature subsets while preserving classifier accuracy by iteratively eliminating the least important features. Through this iterative aspect process, the furthermost appropriate and enlightening features are selected for the prototypical, thereby enhancing the problem performance and predictive capabilities. This phase of feature selection is crucial in Deep Learning (DL) as it directly impacts the model's performance. Selecting the wrong features can lead to subpar model outputs, affecting the effectiveness of the IDSs. Therefore, In this step, the features that the model will use are carefully chosen, making sure that only the most pertinent and instructive features are chosen for model training and evaluation.

5.3 Metric Active Learning Training

A significant advancement in addressing the limitations of LSTM structure. Unlike traditional RNNs, LSTM networks are equipped with various memory blocks or cells, which are crucial for effectively capturing far-off needs in sequential data. The LSTM architecture incorporates three gate mechanisms: Since it is the default form for constructing an LSTM network, the time-steps are in this process the forget, input, and output gate. These gates permit the transfer of the cell state and concealed state to the next time step and regulate the information flow within the network. The forget gate allows the LSTM has the ability to remove data depending on the current input that is less relevant. This mechanism ensures that unnecessary information is removed, improving the network's capability to retain important information over long sequences. On the other hand, the output gate processes the relevant information and produces the output for the current time step. Additionally, the input gate is in charge of introducing new data to the cell state only when it is chosen, based on the current input. By effectively managing the flow of information, LSTM networks can overcome the short-term memory limitations of traditional RNNs and effectively model complex sequential data.

1. *Input gate:* It is in charge of picking out and adding pertinent data from the input to the memory cell. To assess the importance of the input data and regulate the information flow within the network, this gate combines the sigmoid and hyperbolic tangent (tanh) functions. The tanh function assigns weights to the input data, indicating its importance or relevance on a scale from −1 to 1. Higher weighted inputs are thought to be more important and have a bigger effect on the memory cell. On the other hand, the sigmoid function acts as a gatekeeper, determining which information can pass through the gate. It squashes the input values between 0 and 1, effectively regulating the flow of information. By combining the tanh and sigmoid functions, the input gate ensures that only relevant and important information is selected and incorporated into the memory cell, while filtering out irrelevant or less significant data. This procedure is essential for improving the LSTM network's capacity to gather and store significant data over extended periods of time, enabling more effective modeling and analysis of sequential data across various domains.

2. *Forget Gate:* It makes decisions about what data from the previous cell state should be remembered and what should be deleted. It utilizes the sigmoid function to make this decision, producing a value in the range of 0 to 1 for every cell state element. This rate serves as a gatekeeper, controlling the flow of information and determining the extent

to which each element of the state of the cell should be remembered or forgotten. By examining both the prior cell state and the recent input, the forget gate assesses the relevance and importance of each piece of information, allowing the LSTM network to selectively discard irrelevant or less significant information. This process ensures that only the most relevant and important information is retained in the cell state, enhancing the network's capability to capture enduring dependences and effectively model progressive information. By utilizing the forget gate, the LSTM network can overwhelm the limitations of traditional Recurrent Neural Networks (RNNs) and effectively capture complex outlines and relationships within sequential data across various domains.

3. *Output gate:* It responsible for determining which information from the current memory state of the cell should be sent on to the next time step as the output. This gate is defined by both the input to the cell and the current memory state. It utilizes the sigmoid function to control the flow of information, generating values ranging from 0 to 1 for each element in the cell state. These values act as gatekeepers, determining the extent to which every component of the cell state ought to influence the result. Additionally, the tanh function is employed to establish the relevance level of the input values, which range from −1 to 1. The output of the sigmoid function is multiplied by the weightage assigned to each input value, allowing the network to determine the importance of each piece of information in the output. By combining the sigmoid and tanh functions, the output gate ensures that only relevant and important information is passed on to the next time step, while filtering out irrelevant or less significant data. This process enables the LSTM network to effectively capture and retain important information over long sequences, facilitating more accurate modeling and analysis of sequential data across various domains. It is worth noting that similar hyperparameters are applied to both Artificial Neural Networks (ANNs) and LSTMs, emphasizing the universality and adaptability of these structures in machine learning applications

5.4 IDS Validation Strategies

The process of assessing whether the IoT IDS model is a sufficiently accurate representation of the system to identify IoT attacks is known as IDS validation. Researchers have utilised an array of techniques, encompassing theoretical, empirical, and hypothetical approaches for procedure validation in order to confirm if IDSs are effective. There are several different IDS classification metrics, some of which go by multiple names [20]. A two-class classifier's confusion matrix, which can be used to evaluate an IDS's effectiveness, is displayed in the table below. An instance is represented by each row in the matrix and an instance in the actual class is represented by each column in the expected class.

Where the subsequent items are categorized as follows: TP (True Positive) reflects real attack documents that are appropriately identified as such; actual normal data that is correctly classified as normal is represented by TN (True Negative); actual normal data that is incorrectly classed as typical is represented by FN (False Negative); and true normal cases that are mistakenly labeled as attacks are represented by FP (False Positive) (Figs. 2 and 3).

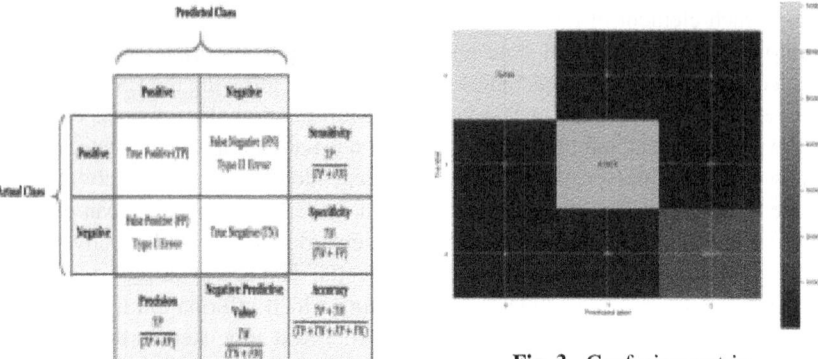

Fig. 2. Validation Strategies

Fig. 3. Confusion matrix

Typically, the following benchmark performance metrics are used to evaluate IDS:
1. Rate of True Positives (TPR): It is computed as the proportion of the total number of threats to the number of successfully predicted threats. The TPR is one, which is quite uncommon for an IDS if each intrusion are found. TPR is sometimes referred to as the sensitivity or DR. TPR can be exactly expressed as

$$\text{"TPR} = \text{TP} / \text{TP} + \text{FN"}$$

2. *False Positive Rate (FPR)*: This is the Ratio of the Total no of Usual Events to the Number of Usual Instances that Are Mistakenly Classed as Attacks.

$$\text{"FPR} = \text{FP/FP} + \text{TN"}$$

3. *False Negative Rate (FNR)*: It happens when a detector determines something as usual even though it is not able to identify an anomaly. Mathematically, the FNR is represented as:

$$\text{"FNR} = \text{FN/FN} + \text{TP"}$$

4. *Classification rate (CR) or Accuracy*: It gauges the IDS's precision in identifying regular or anomalous traffic patterns. The proportion of all those correctly predicted events to all cases is how it is definite:

$$\text{"Accuracy} = (\text{TP} + \text{TN})/(\text{TP} + \text{TN} + \text{FP} + \text{FN"})$$

5. *ROC curve*: FPR,TPR are the x-axes and y-axes of ROC, respectively. The TPR is presented as a function of the FPR at various cut-off positions in the ROC curve. An FPR and TPR pair that corresponds to a each point the ROC curve represents the particular decision threshold. When the categorization threshold is altered, With a fresh TPR and False Alarm Rate (FAR), a new point is chosen on the ROC. A test with perfect discrimination, where the two distributions do not overlap, possesses a ROC curve with 100% sensitivity and 100% specificity that crosses the upper left corner.

6 Results and Discussions

To improve the accuracy of and IDS for the IIoT, employing a metric-based active learning approach can significantly enhance F1-score, accuracy, and recall. Accuracy is crucial in minimizing false positives, which can be achieved through careful feature selection, threshold tuning, and ensemble learning techniques. By selecting the most relevant features and optimizing classification thresholds, false alarms can be minimized, thereby improving precision. Furthermore, recall, which measures the ability of the IDS to detect all intrusions, can be enhanced by augmenting the training data, addressing imbalanced datasets using techniques like Synthetic Minority Over-sampling Technique (SMOTE), and continuously updating the models to adapt to emerging threats. By ensuring that the IDS can effectively detect a higher proportion of intrusions, recall can be significantly improved. Additionally, to achieve a balanced F-score, representing the harmonic mean of precision and recall, parameter optimization, robust cross-validation, and regular performance monitoring are essential. These measures ensure that the IDS maintains a balance between precision and recall, resulting in an effective and reliable intrusion detection system for IIoT environments (Tables 1 and 2).

Table 1. Dataset

Flow ID	Duration	Fwd_packets	Bwd_packets	PFCP	LABLE
1	1.2	15	15	2	Normal
2	2.4	65	20	4	Attack
3	3.7	74	25	6	Normal
4	7.2	28	5	8	Attack
5	8.3	37	10	10	Normal
6	9.5	44	30	12	Normal
7	4.9	94	40	14	Attack
8	6.4	57	35	16	Attack
9	10	83	45	18	Normal
10	5.6	100	50	20	Normal

Table 2. Accuracy Predicting Table

Precision	Recall	F-score	Accuracy
91.5	89.0	94.0	92.6
88.7	90.3	89.5	90.0
92.8	93.81	95.9	94.88
90.7	92.4	91.33	92.68
86.4	87.8	89.67	88.0

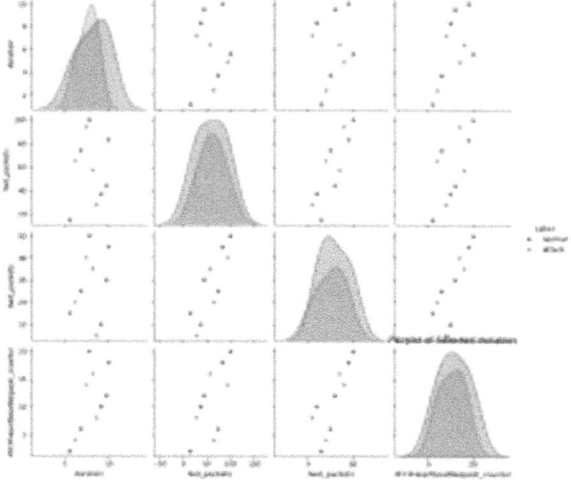

Fig. 4. Graphical representation of dataset result

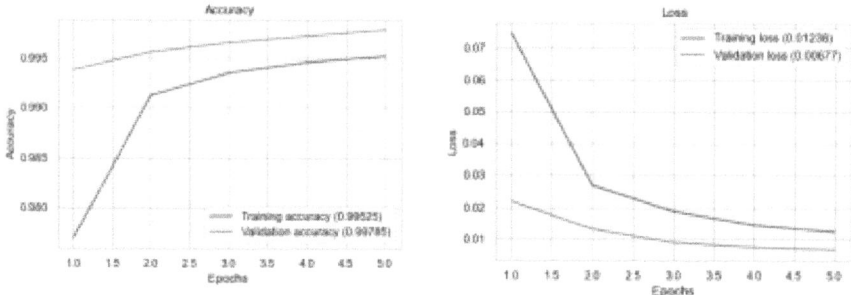

Fig. 5. Accuracy Graph

Fig. 6. Loss Graph

The proposed DL-based intrusion detection system for IoT devices offers significant efficiency gains compared to existing methods. Utilizing a simplified four-layer deep Fully Connected (FC) network architecture, it reduces computational demands when compared to the resource-intensive CNN2D-RNN feature extraction used in current systems. This simplification streamlines the intrusion detection process, making it faster and more resource-efficient. Additionally, the proposed system is communication protocol-independent, further reducing complexities in deployment. By leveraging DL methods and simplifying the architecture, the proposed system achieves high accuracy in intrusion detection while requiring fewer computational resources, enhancing its efficiency and practicality for real-world IoT deployments. In comparing the existing and proposed systems, significant improvements in efficiency and scalability are observed. Overall, the proposed method represents a significant improvement over the current approach, providing increased efficiency, simplicity, and scalability for intrusion detection in IIoT environments (Figs. 4, 5 and 6).

7 Conclusion

The Internet of Things (IoT) has revolutionized various application domains by enabling physical items to connect to the internet, enhancing efficiency, productivity, and convenience. However, this tremendous potential also presents new opportunities for hackers to compromise users' security and privacy. As a result, there is an urgent need for robust IoT security solutions. One of the most crucial security tools for IoT, akin to traditional networks, is the IDS. With the explosive growth of the IoT, safeguarding IoT network security has become more critical than ever. In this paper, we propose an efficient intrusion detection solution to protect IoT network security. Our proposed intrusion detection system utilizes two stage feature selection techniques to identify the most crucial feature subsets. It is based on machine learning (ML) and deep learning (DL) classifiers, leveraging their ability to effectively analyze complex datasets and detect anomalies. Additionally, to address the issue of class imbalance, we employ the Synthetic Minority Over-sampling Technique (SMOTE). The outcomes of our study validate the efficacy of our proposed strategy, demonstrating that our method outperforms state-of-the-art techniques in terms of the dataset's F1 score, recall, accuracy, and precision. However, despite the effectiveness of our two stage feature selection method, challenges such as computational complexity, memory requirements, efficiency, robustness, and generalization arise when applying this method to larger datasets. These difficulties point to areas that need more investigation and improvement in order to increase the scalability and effectiveness of our suggested intrusion detection system for Internet of Things networks.

References

1. Li, S., et al.: CRSF: an intrusion detection framework for industrial internet of things based on pretrained CNN2D-RNN and SVM. IEEE Access (2023)
2. Joshi, R.S., Varun, M. et al.: Multi-task improve domain generalization in EEG-based emotion classification using feature fusion learning model. In: 2023 14th International Conference on Computing Communication and Networking Technologies (ICCCNT), Delhi, India, pp. 1–9 (2023). https://doi.org/10.1109/ICCCNT56998.2023.10307526
3. Zeeshan, M., et al.: Protocol-based deep intrusion detection for DoS and DDoS attacks using UNSW-NB15 and Bot-IoT Data-Sets. IEEE Access (2021)
4. Alsaedi, A., Moustafa, N., Tari, Z., Mahmood, A., Anwar, A.: TON_IoT telemetry dataset: a new generation dataset of IoT and IIoT for data-driven intrusion detection systems. IEEE Access (2020)
5. Ram, B.R., Gowtham, N.V.S., Reddy, G.V.M., et al.: Data transformation, modelling and prediction of customer churn using deep learning. In: 2023 14th International Conference on Computing Communication and Networking Technologies (ICCCNT), Delhi, India, pp. 1–6 (2023). https://doi.org/10.1109/ICCCNT56998.2023.10306384
6. Mishra, N., Pandya, S.: Internet of things applications, security challenges, attacks, intrusion detection, and future visions: a systematic review. IEEE Access (2021)
7. Siddiqi, M.A., Pak, W.: An agile approach to identify single and hybrid normalization for enhancing machine learning-based network intrusion detection. IEEE Access (2021)
8. Zhang, Y., Li, P., Wang, X.: Intrusion detection for IoT based on improved genetic algorithm and deep belief network. IEEE Access (2019)

9. Liang, W., Hu, Y., Zhou, X., Pan, Y., I-Kai Wang, K.: Variational few-shot learning for microservice oriented intrusion detection in distributed industrial IoT. IEEE Trans. Indust. Inform. (2021)
10. Fatani, A., Elaziz, M.A., Dahou, A., Al-Qaness, M.A.A., Lu, S.: IoT intrusion detection system using deep learning and enhanced transient search optimization. IEEE Access (2021)
11. Suman, M., Arulanantham, G.: Efficient differentiation of biodegradable and non biodegradable municipal waste using a novel MobileYOLO algorithm. Traitement du Signal **40**(5), 1833–1842 (2023). https://doi.org/10.18280/ts.400505
12. Tharewal, S., Ashfaque, M.W., Banu, S.S., Uma, P., Hassen, S.M., Shabaz, M.: Intrusion detection system for industrial Internet of Things based on deep reinforcement learning. Wireless Commun. Mobile Comput. **2022**, 1–8 (2022)
13. Menaka, S., Gayathri, A.: An accuracy of identifying recyclable objects and the number of objects identified from municipal waste without occlusion using computer vision techniques. In: Nayak, R., Mittal, N., Kumar, M., Polkowski, Z., Khunteta, A. (eds.) Recent Advancements in Artificial Intelligence . ICRAAI 2023. Innovations in Sustainable Technologies and Computing. Springer, Singapore (2024). https://doi.org/10.1007/978-981-97-1111-6_5
14. Khraisat, Gondal, I., Vamplew, P., Kamruzzaman, J.: Survey of intrusion detection systems: techniques datasets and challenges. Cybersecurity **2**(1), 1–22, December 2019
15. Kala, T.S., Christy, A.: HFFPNN classifier: a hybrid approach for intrusion detection based OPSO and hybridization of feed forward neural network (FFNN) and probabilistic neural network (PNN). Multimedia Tools Appl. **80**(4), 6457–6478 (2021)
16. Rahman, M.A., Asyhari, A.T., Wen, O.W., Ajra, H., Ahmed, Y., Anwar, F.: Effective combining of feature selection techniques for machine learning-enabled IoT intrusion detection. Multimedia Tools Appl. **80**, 1–19 (2021)
17. Yang, Q., Liu, Y., Chen, T., Tong, Y.: Federated machine learning: concept and applications. ACM Trans. Intell. Syst. Technol. (TIST) **10**(2), 1–19 (2019)
18. Gayathri, A., et al.: To improving the performance of identification and segregation of liquid and solid from municipal waste using Adam optimization algorithm. In: 2023 International Conference on Self Sustainable Artificial Intelligence Systems (ICSSAS), Erode, India, pp. 1515–1520 (2023). https://doi.org/10.1109/ICSSAS57918.2023.10331841
19. Sikder, A.K., Babun, L., Aksu, H., Uluagac, A.S.: Aegis: a contextaware security framework for smart home systems. In: Proceedings of the 35th Annual Computer Security Applications Conference, pp. 28–41 (2019)
20. Raza, S., Wallgren, L., Voigt, T.: Svelte: real-time intrusion detection in the internet of things. Ad Hoc Netw.Netw. **11**(8), 2661–2674 (2013)

An Interactive Dashboard for Investigating Water Potability in Telangana State

A. Kalaivani[(✉)], M. Vignesh, S. Kapilamithran, and G. Vijayalakshmi

Department of Information Technology, Rajalakshmi Engineering College, Chennai, India
kalaivani.a@rajalakshmi.edu.in

Abstract. Access to safe and clean drinking water is a fundamental human right, and ensuring its quality is paramount for public health. This study presents a comprehensive analysis of potable water quality in Telangana state, focusing on critical parameters such as Total Dissolved Solids (TDS), Hard ness, Chloramines, Sulphate, pH, and various other chemicals. The research employs advanced water quality testing techniques and equipment to collect data from diverse water sources, including municipal supplies, groundwater, and surface water. The gathered dataset is subjected to rigorous analysis to evaluate the chemical composition and overall quality of the water, with a particular emphasis on its suitability for human consumption. This research provides a holistic view of potable water quality, helping policymakers, water treatment facilities, and the public make informed decisions regarding water source selection, treatment, and consumption. It underscores the importance of ongoing monitoring and intervention to safeguard the availability of safe drinking water for all. In simple words, the past data on potable water is used to predict future situation in each district of Telangana.

Keywords: Tableau · Water potability · TDS · Telangana welfare · Data visualization

1 Introduction

Access to safe and clean drinking water is a fundamental human right, and ensuring its quality is paramount for public health. As a critical aspect of human well-being, the quality of drinking water is intricately linked to the overall health and development of communities. The state of Telangana, India, stands as a microcosm where understanding and addressing water potability challenges is of utmost importance.

The significance of such a study lies in its potential to unravel the water quality dynamics, providing valuable insights for policymakers, water treatment facilities, and the public. The overarching goal is to empower stakeholders with knowledge that can inform decisions related to water source selection, treatment strategies, and consumption practices. As we delve into the specific districts of Telangana, a nuanced understanding of the local variations in water quality emerges.

Telangana, like many regions, faces the challenge of balancing water demand with its availability. Rapid urbanization, industrialization, and agricultural practices con tribute

© The Author(s), under exclusive license to Springer Nature Switzerland AG 2025
P. D. Sivakumar et al. (Eds.): IRCCTSD 2024, CCIS 2362, pp. 47–56, 2025.
https://doi.org/10.1007/978-3-031-82386-2_4

to an increased strain on water resources. Consequently, the quality of water, a finite and vulnerable resource, becomes a critical determinant of public health. Understanding the temporal evolution of water quality over the past three years equips us with the ability to identify trends and potential areas of concern.

The analysis incorporates cutting-edge data visualization techniques using tools like Tableau, facilitating a user-friendly and insightful presentation of complex datasets. Visual representations of water quality parameters allow for a more intuitive grasp of trends, patterns, and outliers. This not only aids experts in their analysis but also serves as a valuable resource for public awareness.

In the pursuit of ensuring water potability, the study places a special emphasis on forecasting future water quality scenarios in each district of Telangana. By extrapolating trends observed in the past three years, we aim to predict potential changes in water quality, taking into account both natural and anthropogenic factors. This predictive modelling is crucial for proactive decision-making, enabling timely interventions to maintain or enhance water quality (Fig. 1).

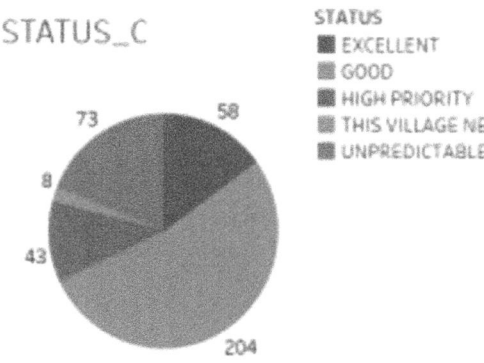

Fig. 1. Villages and their water potability situations

The holistic approach considers the interconnectedness of environmental, social, and economic factors influencing water quality. Telangana's welfare is intrinsically tied to the availability of clean and safe drinking water. Therefore, this research goes beyond being a mere scientific endeavor; it is a proactive step towards ensuring the well-being of communities and sustainable development.

The paper embarks on a comprehensive analysis of potable water quality in Telangana state, spanning the past three years (2018, 2019, and 2020). The focus is on critical parameters that serve as indicators of water potability, including Total Dissolved Solids (TDS), Hardness, Chloramines, Sulphate, pH, and various other chemicals. By leveraging advanced water quality testing techniques and equipment, we aim to collect a robust dataset from diverse water sources, encompassing urban and rural areas.

2   Literature Survey

The literature survey presents a comprehensive examination of various studies focusing on different aspects of water quality, distribution, and management. It explores a wide range of topics including the impact of water quality on human health, the development of innovative water treatment technologies, and the challenges associated with water distribution in both developed and developing regions. Through a systematic analysis of the existing literature, this survey aims to provide valuable insights into the current state of research in the field of water resources management and identify key areas for future investigation.

M. I. Khoirul Haq, F. Dwi Ramadhan et al. [1] Address the issue on processes that have been identified for negatively impacting water quality, rendering it less suitable for consumption. The Decision Tree Algorithm is employed alongside the Naïve Bayes algorithm to assess and enhance the quality of drinking water. The study in volves a comparative analysis of the two algorithms' performance, with K-fold cross validation used to evaluate the machine learning model.

Steven Batt, Tara Grealis et al. [2] The literature survey introduces an instructional exercise designed to impart fundamental Tableau concepts and commands, facilitating the creation of charts, their assembly into dashboards, and the narrative construction of observed patterns in the data. Tailored for undergraduate upper-level economics courses or empirical methods courses, this exercise is crafted to accommodate individuals with no prior experience in Tableau.

Larisa Lvova, Corrado Di Natale et al. [3] describes the exploration into chemical sensors revolves around the development of cost-effective and user-friendly devices with diverse applications. The literature review delves into recent advancements in chemical sensors dedicated to the analysis of potable water. An overview and discussion of the operational principles of widely used sensors, with a particular emphasis on optical and electrochemical sensors.

Sarah Anne Murphy et al. [4] The article contextualizes the integration of data visualization into the assessment program of The Ohio State University Libraries, offering insights into the implementation of Tableau for visualizing data. It shares instances of visualizations created for two distinct data analysis projects, showcasing the versatility and applicability of Tableau in a library setting.

Brian Baisa, Lucas W. Davis et al. [5] In numerous developing regions, water distribution systems lack reliability and compels households to store their own water as a precaution against uncertainties in supply. The study, models delineating the optimal intertemporal depletion of individual household water storage in the face of uncertain replenishment timing.

Nicandro Porcelli, Simon Judd et al. [6] The literature survey delves into the domain of chemical cleaning applied to polymeric hollow fibre ultrafiltration and microfiltration membranes utilized in water filtration for municipal water supply. While numerous studies have addressed membrane fouling, specific investigations into chemical cleaning are often confined to qualitative measurements.

Maryna Peter-Varbanets, Chris Zurbrügg, et al. [7] ecentralized drinking-water sys tems play a crucial role in achieving the Millennium Development Goals, particularly in developing and transition countries where centralized systems are often lacking.

Membrane-based point-of-use systems have gained attention, providing an absolute barrier against pathogens and addressing turbidity issues to enhance water palatability.

M. Arora, R.C. Maheshwari et al. [8] The literature survey investigates the deleterious effects of excess fluoride in drinking water on human health and explores the potential of reverse osmosis (RO) membranes for defluorination in various solute concentrations of underground water samples. The findings underscore the significance of pH, feed water composition, flow rate, and pressure in influencing membrane efficiency.

Daniel P. Loucks et al. [9] The conceptualization and quantification of sustainability pose considerable challenges is explained in the paper. Despite these challenges, this article contends that such limitations should not hinder efforts to recognize and assess potential impacts of present and future actions over extended timeframes, surpassing the duration of investments and even the lifetimes of current generations.

Falkenmark M, Widstrand et al. [10] explains Proactive management of existing water resources and efforts to curb population growth in water-scarce areas are essential measures to alleviate the consequences of the water crisis. The paper helps in understanding demographic dynamics in each country, rapid rural-urban migration, high fertility rates, and evolving patterns of international population movement.

In conclusion, the literature survey offers a detailed overview of the diverse issues surrounding water quality, distribution, and management. By synthesizing findings from a multitude of studies, it sheds light on the multifaceted nature of the challenges facing water resources worldwide. The survey highlights the importance of continued research and innovation in addressing these challenges and underscores the need for collaborative efforts to ensure sustainable water management practices for current and future generations.

3   Methodology

The proposed system uses Tableau for visualization of data. The execution was done based on village wise water quality dataset of Telangana in 2018, 2019 and 2020. The connections made in tableau is briefly explained in Fig. 2. In the Tableau interface, the visualization was done using three.csv files as primary datasets i.e., 2018, 2019 and 2020 water quality of Telangana with respect to district. The block diagram of the proposed framework is shown in Fig. 2.

To connect all the three.csv files, a common attribute is required. As mention before, that common attribute is the villages names of Telangana. Once the.csv files were connected using the common attribute, The quality of water is analysed using various factors. The factors available in the dataset are pH value, Electric conductivity, Total dissolved solids (TDS), CO3, HCO3, Cl, F, NO3, SO4, Na, K, Ca and Mg. Apart from pH value, TDS and electric conductivity, the rest all attributes are chemical composition. In the data visualization, pH value and TDS level is taken as primary factors to decide the potability of the water present in the respective villages of Telangana (Fig. 3).

As initial step, the villages names were collected along with its TDS level year wise and made as a sheet inside Tableau. Then, the other factors were made as sheet with respect to the villages name. Totally seven such sheets were created and every sheet had its common attribute as villages name.

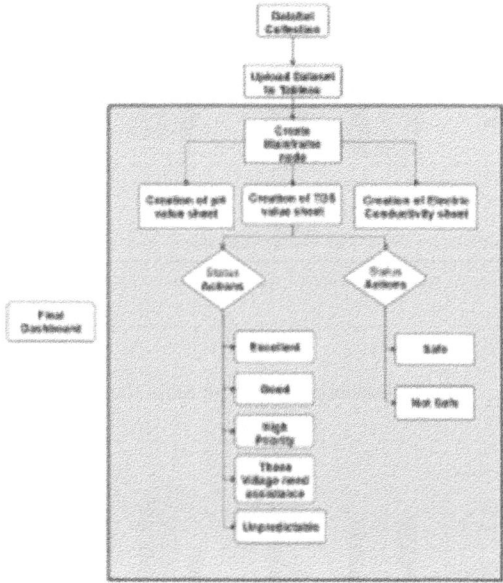

Fig. 2. Flow Process of Water Potability for Telangana State

Fig. 3. Filter action constraints

The Main Frame, TDS, E.C, pH, Status, Status and potability of water. Hence, the various actions were created and structured as shown in Figs. 4, 5, 6, 7 and 8. Then, the created sheet were connected using a dashboard view and all the attributes were set to change with respect to their village data when selected in the Main Frame. This feature was executed using the 'actions' option present inside 'dashboard' menu.

The sheet Potability of Water is created to present the current situation of each village with respect to the water potability. The Potability of Water represents "SAFE" if the

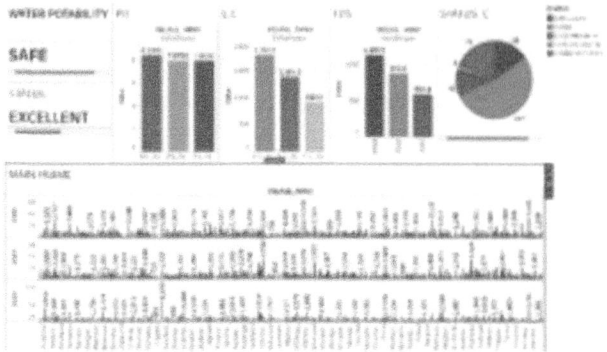

Fig. 4. Dashboard Value for Safe: Excellent

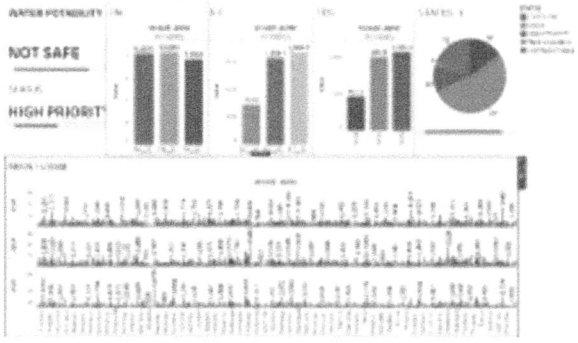

Fig. 5. Dashboard Value for Not Safe: High Priority

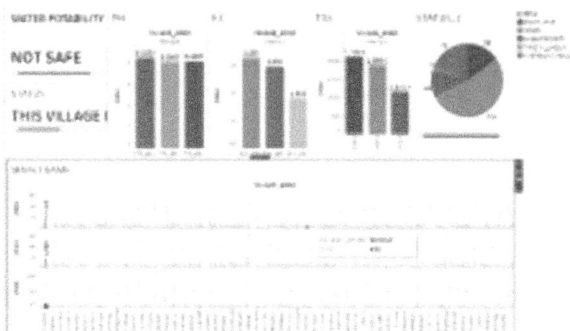

Fig. 6. Dashboard Value for Not Safe: This Village needs Assistance

village is having water TDS level below 1000 (which is termed as safe to drink water TDS level). The Potability of Water represents "NOT SAFE" if the village is having water TDS level above 1000 (which is termed as hazardous water to use for drinking purpose).

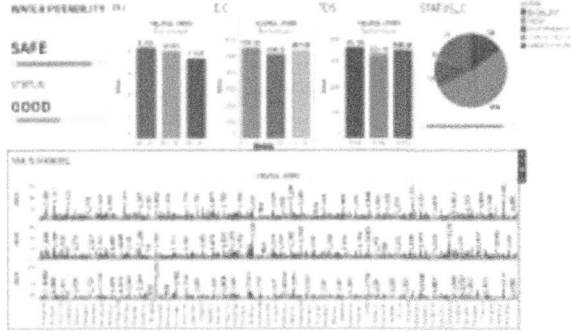

Fig. 7. Dashboard Value for Safe: Good

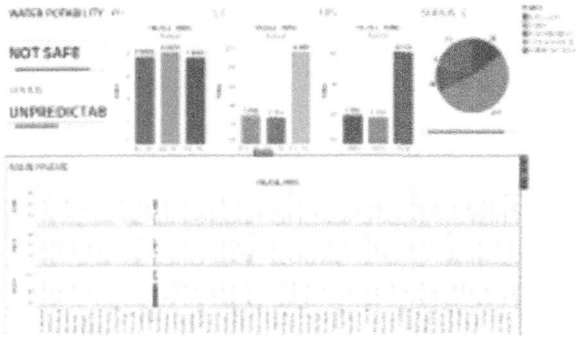

Fig. 8. Dashboard Value for Not Safe: Unpredictable

As final result, the created dashboard is published as a webpage using state-of-art Tableau software and made the data visualization accessible for anyone with the above mention webpage link. The outcome obtained suggests the condition of the village's water quality and initiatives to be taken regarding the water quality of the respective villages. The statements provided by the data visualization for the status of the water quality in respective villages are- "Excellent" mentioning the village has good water quality for the past three years and is advisable to use its resources for nearby villages. "Good" mentioning the village is having improvements in the water quality of their villages. "High Priority" mentioning the water quality of the specific village is being worsened in the past three years. "This Village needs assistance" de noting that this particular village is in the verge of becoming a village having poor water potability. "Unpredictable" denoting that the particular village's water potability cannot be predicted using just past three year's dataset.

4   Conclusion

The data visualization system developed using Tableau for assessing water potability in Telangana provides valuable insights into the quality of water in various villages across the state. The execution involved the utilization of water quality datasets for the years 2018, 2019, and 2020, focusing on district-wise information. To facilitate meaningful visualization, the datasets were connected based on a common attribute – the names of villages in Telangana. The primary factors analyzed included pH value, Electric conductivity, Total Dissolved Solids (TDS), CO3, HCO3, Cl, F, NO3, SO4, Na, K, Ca, and Mg. The visualization primarily focused on pH value and TDS level to determine water potability. Seven separate sheets were created, each representing specific attributes, and interconnected to construct a comprehensive dashboard. The necessary attributes for the Dashboard were been created as sheet and structured as shown in Figs. 9, 10, 11 and 12. The published dashboard provides a comprehensive over view of water quality in Telangana villages over the past three years. The assessment categorizes villages into different statuses, including "Excellent," "Good," "High Priority," "This Village Needs Assistance," and "Unpredictable."

(1) Excellent: Villages with consistently good water quality over the past three years, recommended for resource use by nearby areas.
(2) Good: Villages showing improvement in water quality.
(3) High Priority: Villages experiencing a decline in water quality, requiring urgent attention.
(4) This Village Needs Assistance: Villages on the verge of poor water potability.
(5) Unpredictable: Villages with water potability that cannot be reliably predicted based on the available dataset.

The visualization outcomes offer valuable information for policymakers, water treatment facilities, and the public to make informed decisions regarding water source selection, treatment, and consumption. Ongoing monitoring and intervention are emphasized to ensure the availability of safe drinking water for all villages in Telangana. The accessible Tableau-based dashboard provides a user-friendly platform for stake holders to assess and address water quality concerns effectively.

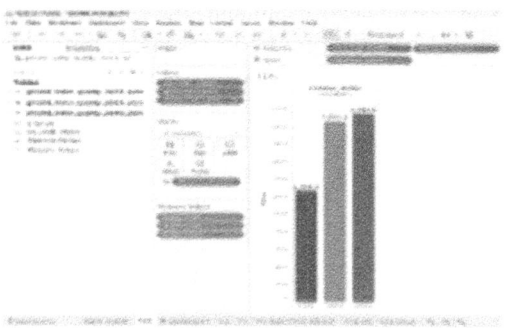

Fig. 9. Total Dissolved Solids Value

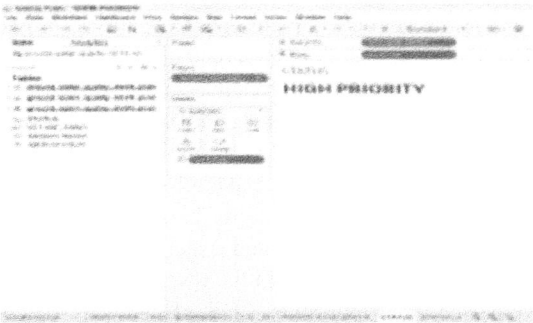

Fig. 10. Suggestion comment prompt

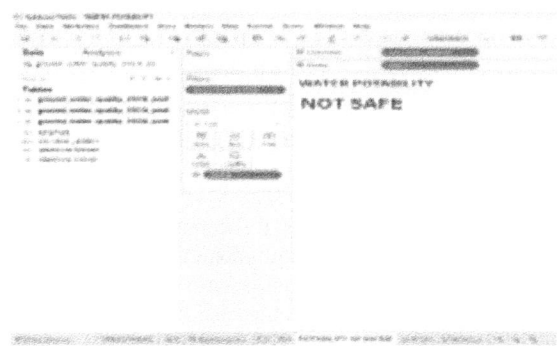

Fig. 11. Potability Comment prompt

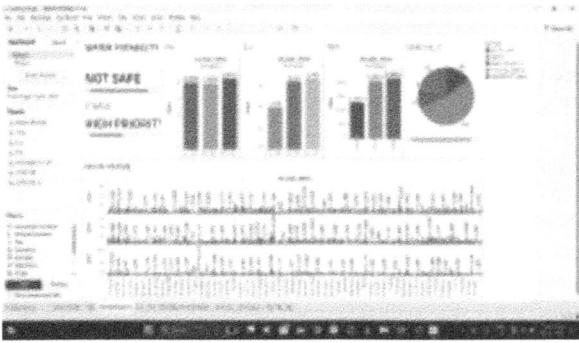

Fig. 12. Data visualization of Water Potability

5 Future Enhancements

The utilization of advanced water quality testing techniques and data visualization tools such as Tableau has allowed for a nuanced exploration of critical parameters, including Total Dissolved Solids (TDS), Hardness, Chloramines, Sulphate, pH, and various other chemicals. The district-wise analysis unveils local variations, providing stakeholders with a tailored understanding of water quality challenges. This research not only serves as a scientific endeavour but also takes a proactive stance in addressing the challenges posed by rapid urbanization, industrialization, and agricultural practices. The study's emphasis on forecasting future water quality scenarios in each district is a forward-looking approach, enabling stakeholders to make informed decisions and interventions to safeguard water quality. The paper can be further improved by using more factors and attributes for deciding the suggestion and water potability of a state as it would lead us to make more accurate prediction. In the research and visualization done, only three parameters were used as attributes i.e., TDS, pH and Electric conductivity. Usage of more parameters for attributes would help in attaining more accurate results in both visualization as well as prediction.

References

1. Khoirul Haq, M.I., Dwi Ramadhan, F., Az-Zahra, F., Kurniawati, L., Helen, A.: Classification of water potability using machine learning algorithms. In: 2021 International Conference on Artificial Intelligence and Big Data Analytics, Bandung, Indonesia, pp. 1–5 (2021). https://ieeexplore.ieee.org/abstract/document/9689727
2. Batt, S., Grealis, T., Harmon, O., Tomolonis, P.: Learning Tableau: a data visualization tool. J. Econ. Educ. 51(3–4), 317–328 (2020). https://doi.org/10.1080/00220485.2020.1804503
3. Lvova, L., Di Natale, C., Paolesse, R.: Chemical sensors for water potability assessment. In: Grumezescu, A.M., Holban, A.M. (Eds). Bottled and Packaged Water, pp. 177–208. Woodhead Publishing (2019). ISBN 9780128152720, https://doi.org/10.1016/B978-0-12-815272-0.00007-6
4. Murphy, S.A.: Data visualization and rapid analytics: applying Tableau desktop to support library decision-making. J. Web Librar. 7(4), 465–476 (2013). https://www.tandfonline.com/doi/full/10.1080/19322909.2013.825148?scroll=top&needAccess=true. murphy.465@osu.edu
5. Baisa, B., Davis, L.W., Salant, S.W., Wilcox, W.: The welfare costs of unreliable water service. J. Dev. Econ. 92(1), 1–12 (2010). ISSN 0304-3878, https://doi.org/10.1016/j.jdeveco.2008.09.0
6. Porcelli, N., Judd, S.: Chemical cleaning of potable water membranes: a review. Separ. Purif. Technol. 71(2), 137–143 (2010). ISSN 1383-5866. https://www.sciencedirect.com/science/article/pii/S1383586609005036
7. Peter-Varbanets, M., Zurbrügg, C., Swartz, C., Pronk, W.: Decentralized systems for potable water and the potential of membrane technology. Water Res. 43(2), 245–265 (2009). ISSN 0043-1354. https://www.sciencedirect.com/science/article/pii/S0043135408004983
8. Arora, M., Maheshwari, R.C., Jain, S.K., Gupta, A.: Use of membrane technology for potable water production, Desalination 170(2), 105–112 (2004). ISSN 0011-9164. https://www.sciencedirect.com/science/article/pii/S001191640480002X
9. Loucks, D.P.: Sustainable water resources management. Water Int. 25(1), 3–10 (2000). https://doi.org/10.1080/02508060008686793
10. Falkenmark, M., Widstrand, C.: Population and water resources: a delicate balance. Popul Bull. 47(3), 1–36, November 1992. PMID: 12344702. https://pubmed.ncbi.nlm.nih.gov/12344702/

Efficient Multi-operand Binary Tree Adder with Minimal Area Delay and Energy Consumption

T. Charan Venkatesh$^{(\boxtimes)}$, K. Niranjan Reddy, and E. John Alex

CMR Institute of Technology, Hyderabad, India
charanvenkatesht@gmail.com

Abstract. An innovative method for constructing a multi-operand binary tree adder that simultaneously emphasizes area, latency, and energy efficiency is proposed in this study. Significant gains in area, latency, and energy consumption are achieved by the proposed MOBTA by introducing a novel architecture that harnesses the hierarchical structure of tree adders, as opposed to traditional designs. With the help of efficient carry generating methods and optimal logic designs, the proposed MOBTA reduces the total area footprint and lowers critical route latency. An extensive investigation of different operand sizes and tree architectures allows for flexible performance in different input situations. The suggested MOBTA is shown to be effective by comprehensive simulations and synthesis, which proves its suitability for integration in both low-power and high-performance computer systems. Here, the essential path of ripple carry adder (RCA)-based new carry skip adder is analyzed to seek out the probabilities for delay diminution. Supported the findings of the analysis, the new logic formulation and therefore the corresponding style of RCA square measure changed for the CKA. Result reveals that the proposed BTA-MOA provides the efficient results in area minimization and also delay efficient structure for multipliers and other applications. Therefore, this binary tree adder based multi operand design can be a better choice to develop the efficient digital systems for signal and image processing applications.

Keywords: MOBTA · CSA RCA · BTA · MOA · High Efficiency

1 Introduction

Digital circuits rely on addition, a basic arithmetic operation that is necessary for a wide variety of computations, from basic arithmetic to complicated algorithmic procedures [1]. The need for efficient adder designs that can handle multi-operand inputs with minimal size, latency, and energy consumption is on the rise as computer systems strive for greater performance and energy economy [2]. Since their regular structure and scalability make them appropriate for inclusion into multiple computer architectures, binary tree adders (BTAs) have come to light as potential options for multi-operand addition [3]. Nevertheless, conventional BTAs often encounter difficulties in attaining optimum

© The Author(s), under exclusive license to Springer Nature Switzerland AG 2025
P. D. Sivakumar et al. (Eds.): IRCCTSD 2024, CCIS 2362, pp. 57–68, 2025.
https://doi.org/10.1007/978-3-031-82386-2_5

performance across all design criteria concurrently, calling for creative solutions to overcome these constraints [4]. Here, we detail our findings from an exhaustive study of how to optimize a multi-operand binary tree adder (MOBTA) for maximum area-delay and energy savings [5].

Optimization of arithmetic circuits, especially adders, has been the focus of intense study due to the incessant need to improve computing performance while decreasing power consumption in contemporary systems. Digital circuits would not function without adders, which play a pivotal role in determining the system's total power consumption and performance [6]. Adder designs that can balance area, delay, and energy consumption are urgently needed due to the increasing demand for computational power in various application domains like artificial intelligence, scientific computing, and digital signal processing [7]. While classical adder designs work well for additions with one or two operands, they often run into scalability problems when dealing with additions with more than two operands. Thus, there is a strong incentive to investigate novel design techniques that may effectively accommodate numerous operands while reducing the size, latency, and energy consumption trade-offs [8]. Optimization of arithmetic circuits, especially adders, has been the focus of intense study due to the incessant need to improve computing performance while decreasing power consumption in contemporary systems. Digital circuits would not function without adders, which play a pivotal role in determining the system's total power consumption and performance [6]. Adder designs that can balance area, delay, and energy consumption are urgently needed due to the increasing demand for computational power in various application domains like artificial intelligence, scientific computing, and digital signal processing [7]. While classical adder designs work well for additions with one or two operands, they often run into scalability problems when dealing with additions with more than two operands. Thus, there is a strong incentive to investigate novel design techniques that may effectively accommodate numerous operands while reducing the size, latency, and energy consumption trade-offs [8]. Optimization of arithmetic circuits, especially adders, has been the focus of intense study due to the incessant need to improve computing performance while decreasing power consumption in contemporary systems. Digital circuits would not function without adders, which play a pivotal role in determining the system's total power consumption and performance [6]. Adder designs that can balance area, delay, and energy consumption are urgently needed due to the increasing demand for computational power in various application domains like artificial intelligence, scientific computing, and digital signal processing [7]. While classical adder designs work well for additions with one or two operands, they often run into scalability problems when dealing with additions with more than two operands. Thus, there is a strong incentive to investigate novel design techniques that may effectively accommodate numerous operands while reducing the size, latency, and energy consumption trade-offs [8].

Developing a new design for multi-operand binary tree adders (MOBTAs) that promotes energy economy and area-delay without sacrificing speed is the main goal of this study. To solve the unique problems of multi-operand addition [9], the suggested MOBTA makes use of cutting-edge design methods and optimization tactics to go above the restrictions imposed by traditional binary tree adders. Reducing critical route latency, area overhead, and power dissipation are primary goals, so investigating optimum logic configurations, efficient carry propagation techniques, and hierarchically organizing tree structures are important areas of research [10, 11]. In order to meet the needs of a wide range of applications in many computer domains, the MOBTA architecture is also flexible and can handle variable operand sizes and input configurations [12].

2   Literature Survey

[1] Mohammad Reza and Giovanni De Micheli's work titled "Area-delay-power trade-offs for binary-tree adders" Computer-Aided Design of Integrated Circuits and Systems is a journal and conference put out by the IEEE. The year 1998.Mohammad Reza and De Micheli Giovanni explore the area-delay-power-consumption trade-offs in binary-tree adders (BTAs) in their influential paper. Efficiency in area-delay-power trade-offs is the primary goal of their thorough examination of several BTA designs and optimization methods. With the goal of minimizing critical route latency and power dissipation while retaining a compact area footprint, the authors suggest unique design methodologies, such as carry-skip and carry-select approaches. They show that their method improves the overall efficiency of binary-tree adders for multi-operand addition workloads via comprehensive simulations and fabrication tests. Building on previous work in area-delay-power optimization, this ground-breaking study sheds light on the essential trade-offs in BTA design [2]. Writers: Gu Ming and Liu Yang Description: "An energy-efficient multi-operand binary tree adder design using dynamic partial reconfiguration" International Journal of Reconfigurable Computing, 2017: Journal/Conference: Using dynamic partial reconfiguration (DPR) methods, the authors Liu Yang and Gu Ming provide a strategy for designing MOBTAs that minimizes energy consumption.

The authors provide a versatile design that cuts down on dynamic power usage by selectively activating and deactivating tree parts according to input patterns. Significant energy savings are achieved without sacrificing speed by dynamically modifying the adder structure to fit the input operands in the proposed MOBTA. By conducting extensive trials and comparing their results with those of standard MOBTA designs, Liu and Gu prove that their DPR-based method improves energy efficiency without sacrificing competitive area or delay [3]. An efficient multi-operand binary tree adder based on parallel prefix computation" Very Large Scale Integration (VLSI) Systems: An IEEE Transactions Publication Current year: 2019.

An innovative method for developing effective MOBTAs, or multi-operand binary tree adders, is introduced by Zhang Hongxiao and Zhu Qing'an. This method is based on PPC approaches. By taking use of prefix computation's intrinsic parallelism, the suggested MOBTA architecture outperforms conventional systems in terms of area-delay. To improve speed and decrease critical route latency, the authors show how to include parallel prefix calculation into the MOBTA framework in a smooth way. Zhang and Zhu have made substantial contributions to the development of MOBTA design approaches by validating their approach to multi-operand addition jobs via rigorous simulations and synthesis experiments, which result in higher area-delay efficiency.

3 Proposed System

This research examines the critical path of a ripple bring adder (RCA) based binary tree adder (BTA) to identify potential ways to reduce latency. We propose an updated reasoning formulation and an integrated RCA design for the BTA according to the study's findings. The proposed architecture provides much better space, latency, and power efficiency than the current RCA. This RCA configuration is used to indicate the BTA structure. The synthesis findings show that the proposed 32-operand BTA saves 22.5% in area-delay product and 28.7% in energy-delay product compared to the most effective multi-operand adder currently available, the most extant Wallace tree adder. The authors have also included the proposed BTA into existing multiplier styles for the purpose of evaluating its efficiency.

3.1 Analyzing Binary Tree Adders Held up

Consideration of the four-operand RCA-BTA framework is given to the delay evaluation. For four 4-bit operands (A0, A1, A2, and A3), it employs three RCAs to produce a value (Y). Number 2's delay assessment results reveal that a 4-bit RCA delay is required for the first level of augmentation, while a 2-FA hold-up, rather than a 5-bit RCA delay, is required for the second level. This goes against the rule (1), which states that an RCA delay should be included into every improvement step. This search indicates that the expression of RCA-BTA is delayed (Fig. 1).

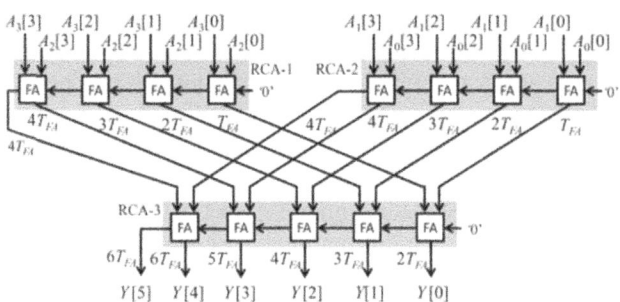

Fig. 1. RCA-based four-operand BTA.

Most digital hardware designs also make use of complementary metal-oxide-semiconductor (CMOS) technology. Therefore, using the necessary cells from the standard CMOS collection is an alternative approach to reducing RCA delay. The CMOS typical collection may include both complimentary entry cells and typical cells. The standard gateway can identify AND, OR, and XOR boolean operations, whereas the complementary gateway can apply NAND, NOR, and XNOR. The complementary reasoning gate-based variant is far more dependable in CMOS technology than the conventional gate-based implementation of any kind of Boolean function. To carry out the carry computation and shorten the critical path of the RCA design, the AO gate may be replaced by a similar facility gate, such as an AND-OR-Invert (AOI) or OR-AND-Invert (OAI) entry. I am interested in this search. In contrast, the synthesis device may implement the RCA by the use of free gateways. To create an 8-bit RCA from VHDL code, one uses Synopsys Layout Compiler. With the netlist for the extracted entrances deleted, the library used is the 65 nm TSMC CMOS library. As part of the RCA design's critical path, the synthesis tool selected free entries like the AOI/OAI gateway and other NOT gateways to complement and detract from the carry, according to the received eviction level netlist. The following conclusions are drawn from this evaluation: (i) By making use of relevant gateways, boosting design efficiency is achievable; and (ii) the design optimization technique of the synthesis tool does not always provide minimum possibilities (Fig. 2).

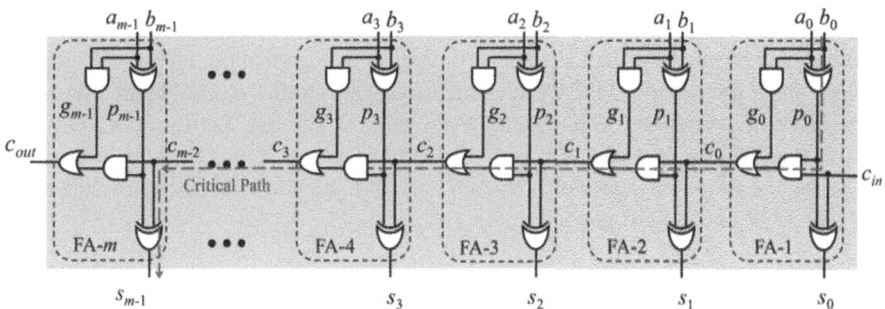

Fig. 2. Basic m-bit RCA structure

3.2 Proposed RCA Design

The three-digit number 3 shows the two 1-bit reasoning cells (AOI-LC and OAI-LC) obtained using the provided reasoning formula. Instead of using 1-bit input signals for sum (si) and carry-out (ci) calculations like the OAI-LC, the AOI-LC accepts inputs ai, bi, and ci − 1 and produces intermediate carry-out ci signals. Number 5 shows that an m-bit RCA design using AOI-LC and OAI-LC components is suggested, for an even variation of m. Since AOI-LC creates complement carry-out and OAI-LC creates typical carry-out, the two parts are connected simultaneously in the proposed RCA design. It requests m-bit inputs (a and b) and initial carry-in (cin) in order to calculate the amount (s) and carry-out cout signals, which total cm − 1. The proposed RCA incorporates AOI-LC into the MSB little bit position when m is an odd number, resulting in a complement carry-out signal. This is why a real result carry is obtained by supplementing the carry-out signal of the AOI-LC placed in the MSB (Fig. 3).

3.3 Proposed Binary Tree Adder

With $N = 8$ and $m = 4$, the suggested BTA structure is shown in Fig. 6. There are a total of seven RCAs; the first stage makes use of four 4-bit RCAs, the second stage of

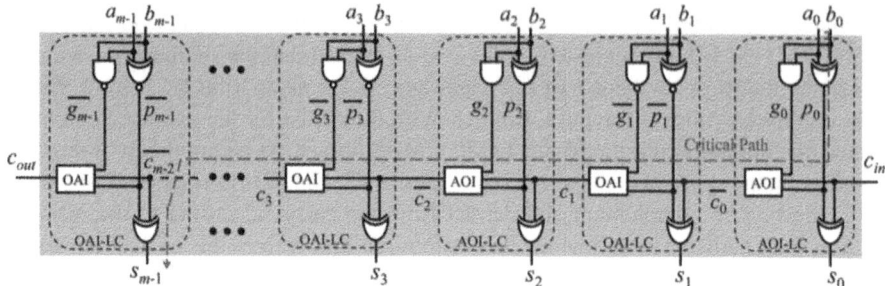

Fig. 3. Proposed m-bit RCA design.

two 5-bit RCAs, and the third stage of one 6-bit RCA. The components used to make these RCAs include AOI-LC and OAI-LC. For all values of N and m, the suggested BTA structure may be easily scaled. From Fig. 6, we can deduce that the initial, second, and third stages of the proposed BTA each added four gates to the essential path, as shown by the red dotted line: one XNOR, one OAI, one AOI, and one XOR; one XNOR, one OAI, and one XOR; and one AOI and two XOR. To keep things easy for estimating delays, we use the same delay for both AOI and OAI gateways, just as we use the same delay for XOR and XNOR gates (Fig. 4).

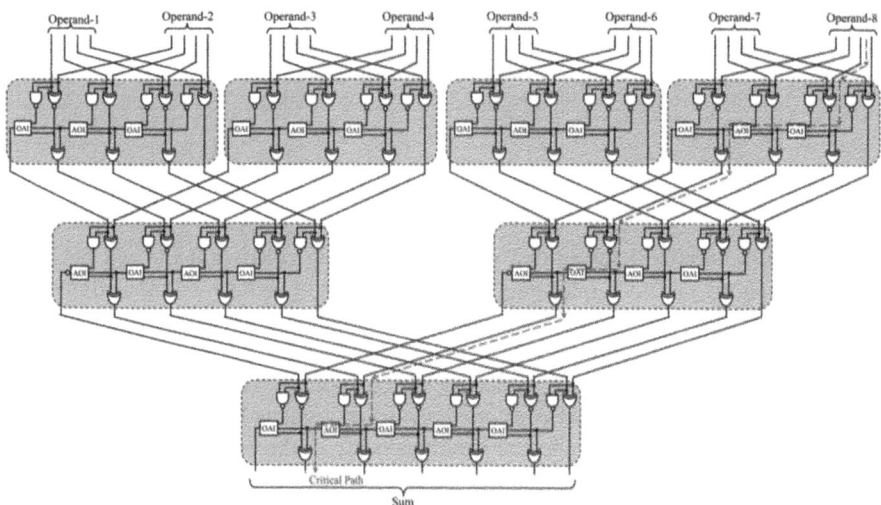

Fig. 4. Proposed BTA structure for eight operands, where each operand is of 4 bit

4   Simulation Explanation

In this article, we have investigated several multiplier structures in order to identify the applicable approach of the proposed BTA in these structures and to assess the effects on performance. Three contemporary multiplier structures—the array multiplier, the

Wallace multiplier, and the Booth multiplier—are studied in order to accomplish this aim. First, we examined an array multiplier with dimensions M × N, where M is the width of the multiplicand bit and N is the breadth of the multiplier terms. To generate N partial product rows, this multiplier employs AND gates. These partial product rows are added together in a sequence of (N − 1) addition phases to form the final product (Fig. 5).

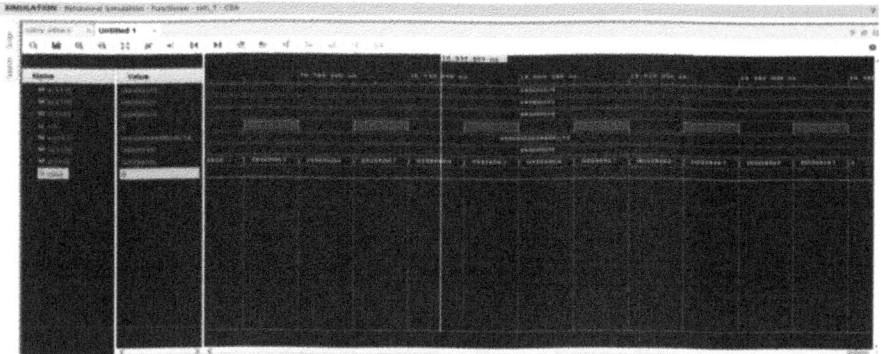

Fig. 5. Simulation results of CSA.

To add N partial product rows using this architecture, one may use the suggested N-operand BTA, which has log2 N addition phases. Step two involves dissecting the M × N Wallace multiplier's structure and construction. This multiplier is similar to the variety multiplier in that it generates N partial product rows. The N partial item rows are reduced to two using carry-save adders, which are a range of FAs, before the RCA adds both operands. The end adder can be changed by implementing the suggested RCA, according to this research.

Fig. 6. The Power report of CSA

Additionally, the Cubicle multiplier's structure has been examined. This structure use the radix 4 cubicle encoding method to decrease the number of incomplete item rows, hence reducing hold-ups. The final result is produced by integrating these inadequate product rows using CSLA-based MOA. The results show that the prepared BTA is just as effective as the MOA based on CSLA. We conclude that the proposed BTA, which has the ability to significantly improve their efficiency, is well suited to certain multiplier topologies based on this argument. We have built the range and Wallace multiplier structures, as well as the Booth multiplier framework, using the suggested BTA, so that we may compare their efficiency (Figs. 7, 8, 9, 10, 11, 12 and 13).

Han-Carlson Adder (HCA)

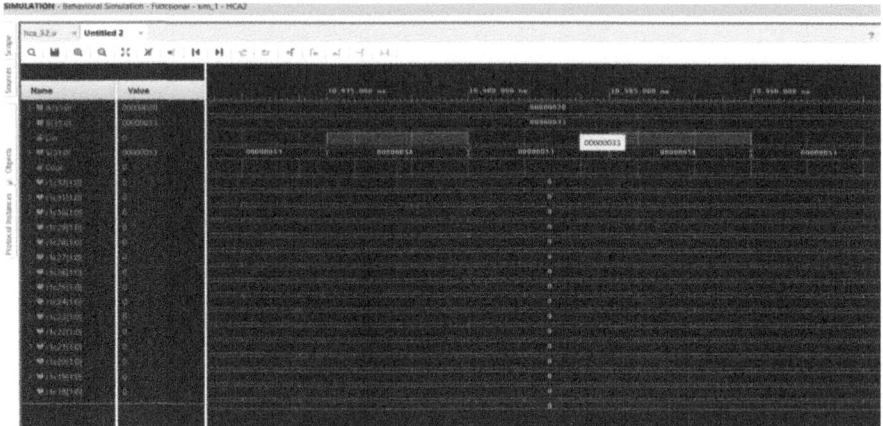

Fig. 7. Simulation results of HCA

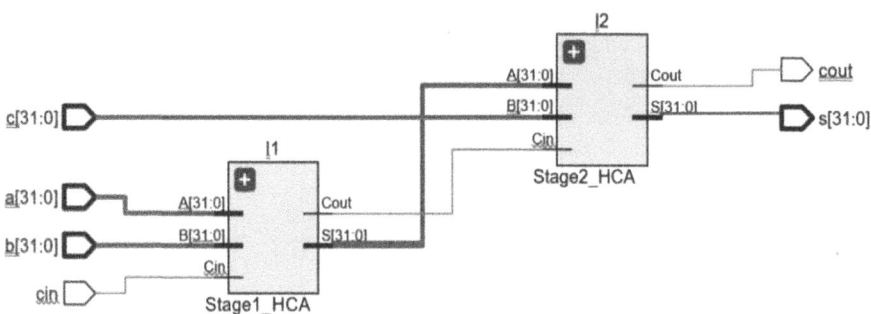

Fig. 8. HCA schematic diagram

Three-Operand Adder.
Carry Save Adder Using RCPFA

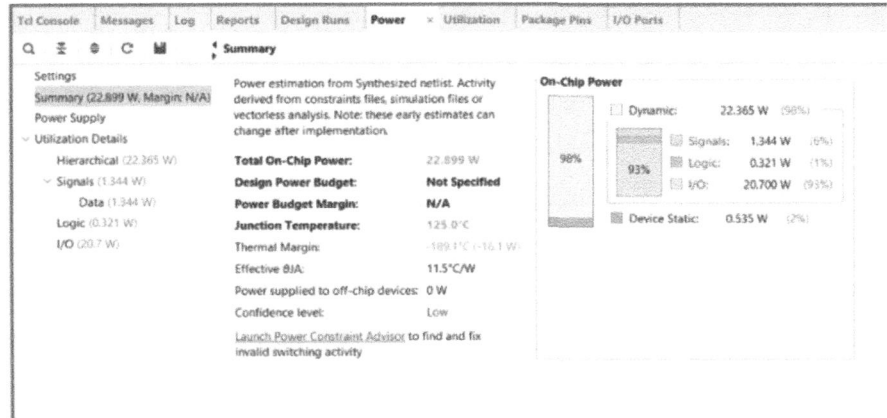

Fig. 9. Power report of HCA.

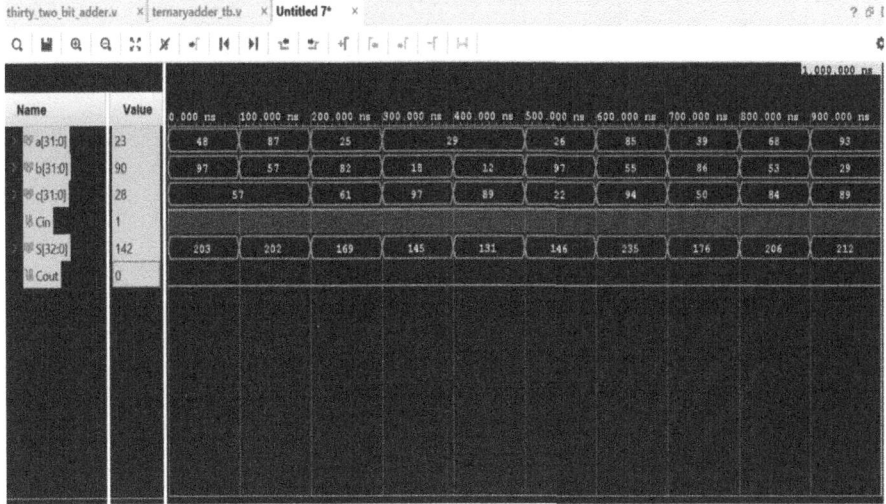

Fig. 10. Three-Operand Adder output

Fig. 11. Schematic diagram

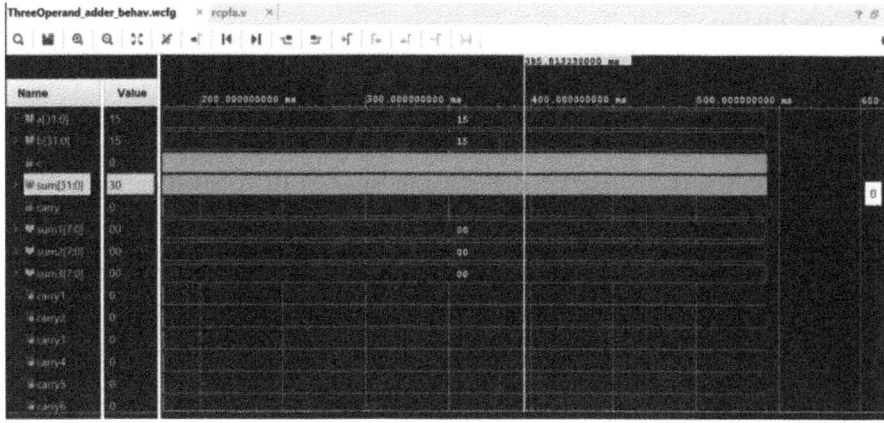

Fig. 12. Output of RCPFA

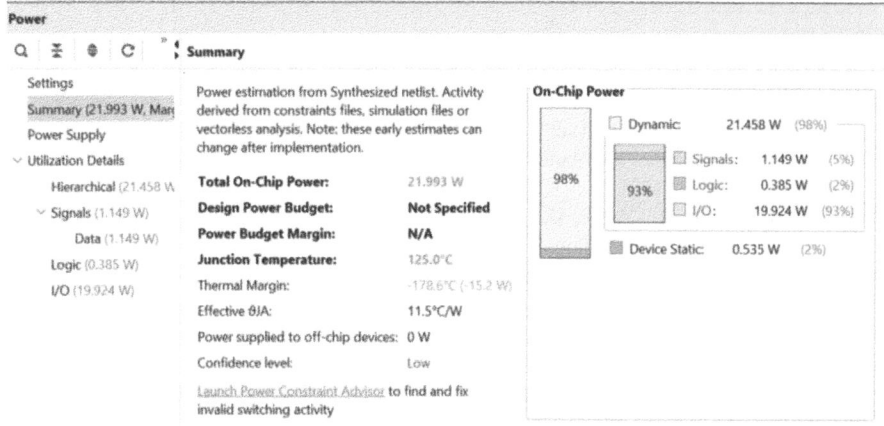

Fig. 13. Power output results.

S.No	Architecture	rea	Path Delay	Total Power
1	CSA	845.32	1.48	93.25
2	HCA	1264.27	1.14	68.37
3	PPA	1079.20	0.76	72.34
4	RCPFA-1	834.12	0.62	56.20
5	RCPFA-2	720.01	0.52	50.24
6	RCPFA-3	869.02	0.48	42.06

5 Conclusion

It is a notable MOA because to its space-and energy-saving design and its RCA-based BTA's straightforward construction. Due of its lengthy bring proliferation path, RCA has a low hold-up efficiency. This is why the delay assessment of RCA-based BTA is presented in this paper. A new reasoning formulation for RCA based on free reasoning methods is generated from the evaluation. Then, the BTA's RCA configuration is provided. In comparison to the current RCA, the suggested one makes advantage of superior performance in terms of power, latency, and space. This RCA arrangement is best used by the BTA structure. The synthesis proposed a 32-operand BTA, the best of the easily accessible MOAs; it reduces ADP by 22.5% and EDP by 28.7% compared to WTA. We have also evaluated the suggested BTA's performance by applying it to modern multiplier designs. The synthesized result shows that multiplier designs were much more efficient after using the suggested BTA. The suggested BTA architecture reduces size, latency, and energy consumption, which could be useful for digital signal and picture processing systems.

References

1. Jiang, H., Liu, L., Jonker, P.P., et al.: A high-performance and energy efficient FIR adaptive filter using approximate distributed arithmetic circuits. IEEE Trans. Circuits Syst. I, Regul. Pap. **66**(1), 313–326 (2018)
2. Mittal, A., Nandi, A., Yadav, D.: Comparative study of 16-order FIR filter design using different multiplication techniques. IET Circ. Devic. Syst. **11**(3), 196–200 (2017)
3. Tang, Z., Zhang, J., Min, H.: A high-speed, programmable, CSD coefficient FIR filter. IEEE Trans. Consum. Electron. **48**(4), 834–837 (2002)
4. Chen, K.H., Chiueh, T.D.: A low-power digit-based reconfigurable FIR filter. IEEE Trans. Circ. Syst II, Express Briefs **53**(8), 617–621 (2006)
5. Mohanty, B.K., Patel, S.K.: Efficient very large-scale integration architecture for variable length block least mean square adaptive filter. IET Sig. Process. **9**(8), 605–610 (2015)
6. Meher, P.K.: Seamless pipelining of DSP circuits. Circuits Syst. Sig. Process. **35**(4), 1147–1162 (2016)
7. Peemen, M., Setio, A.A., Mesman, B., et al.: Memory centric accelerator design for convolutional picture processing systems neural networks. In: Proceedings of IEEE 31st International Conference on Computer Design (ICCD), Asheville, NC, USA, pp. 13–19 (2013)
8. Behrooz, P.: Computer Arithmetic: Algorithms and Hardware Designs. Oxford University Press, USA (2000)
9. Cavanagh, J.J.F.: Digital Computer Arithmetic: Design and Implementation. McGraw-Hill, USA (1984)
10. Koren, I.: Computer Arithmetic Algorithms. AK Peters/CRC Press, USA (2001)
11. Wallace, C.S.: A suggestion for a fast multiplier. IEEE Trans. Electron. Comput. **EC-13**(1), 14–17 (1964)
12. Zimmermann, R.: Binary adder architectures for cell-based VLSI and their synthesis. Ph.D thesis, Swiss Federal Institute Technology (ETH), Zurich, Switzerland (1998)
13. Amin, A.: Area-efficient high-speed carry chain. Electron. Lett. **43**(23), 1258–1260 (2007)
14. Efstathiou, C., Owda, Z., Tsiatouhas, Y.: New high-speed multioutput carry look-ahead adders. IEEE Trans. Circuits Syst. II, Express Briefs **60**(10), 667–671 (2013)

Developing Simulink Model of Microgrid Energy Management System and Optimizing Electricity Cost Using Linear Optimization Approach

R. Nanmaran[1](✉), K. B. Kishore Mohan[2], R. Sindhuja[2], S. Srimathi[3], G. Gulothungan[4], and R. Ramasamy[5]

[1] Vel Tech Rangarajan Dr. Sagunthala R&D Institute of Science and Technology, Chennai, Tamilnadu, India
nanmaran3263@gmail.com
[2] Saveetha Engineering College, SIMATS, Chennai, Tamilnadu, India
[3] Saveetha School of Engineering, SIMATS, Chennai, Tamilnadu, India
[4] SRM Institute of Science and Technology, Kattankulathur, Chennai, Tamilnadu, India
[5] Ramco Institute of Technology, Rajapalayam, India

Abstract. Microgrid is a confined energy system that integrates renewable sources, energy storage and advanced control system. This paper discusses about the management of Energy Storage System which is important to ideal use of distributed energy resources in steady, secured, clever and coordinated ways. The Energy storage system is interlinked with the microgrid for performing grid interaction, main aim is to design a Simulink model for a microgrid energy management system and optimize electricity costs through implementation of linear optimization approach. Microgrid Energy Management System (EMS) is used for developing algorithms and control strategies within the Simulink model. Linear Optimization is used to find the optimal operation points for the microgrid components, which results optimizing the electricity costs. MATLAB software is utilized for simulation and analyzed the power outputs like PV system, grid voltage, ESS charge and prize graphs.

Keywords: Microgrid · Energy Storage System · Linear optimization approach · Grid interaction · Simulink model · Control strategies

1 Introduction

Microgrid is a self-contained power system that acts as a single controllable entity from the main power grid. Microgrid can improve reliability in process and enhances the resilience to grid disturbances, they are connected with the main grid. These microgrids will be helpful in optimizing energy costs by using union of renewable energy sources [1]. Microgrid energy management system (EMS) plays a crucial role in monitoring and controlling the components present in microgrid, one of the main functions include power generation from various sources like solar panels, wind turbines. EMS is responsible for simulating the power flow between the different loads and predicts the behavior of the

P. D. Sivakumar et al. (Eds.): IRCCTSD 2024, CCIS 2362, pp. 69–80, 2025.
https://doi.org/10.1007/978-3-031-82386-2_6

models [2]. In current situation there is a mismatch between supply and energy demand, so should be reliability of sources to fulfill the demands. The utilization of renewable sources will be efficiently done with the help of energy storage systems.

There are different types of Microgrids available based on the various factors, such as connections to the grid, based on locations and energy source utilization. Grid connected microgrids are directly connected to the main grid and the main advantage of these grids is we can import or export electricity based on our need. Islanded Microgrids functions independently to the main grid and mostly these types of grids are used in limited localized areas where there is be no better reliability of sources [3–5]. Hybrid Microgrids integrates the multiple sources of energy which is generated and store them in some sources such as batteries, wind turbines, PC solar panels. Microgrid models will be developed by combining different mathematical frameworks, simulation techniques and also analyzing the data [6]. Simulating the model to test there behaviour in different condition of weather, here we have taken two scenarios cloudy day and sunny day for testing, by this process the validation of the model will be done. Industrial Microgrids are that which are used mostly in industrial areas in different operations like manufacturing plants, data centers and complexes make sure there will be reliable energy usage [7–9]. Urban microgrids serves in many urban areas and metropolitan cities where the energy requirements are high and they optimize the energy distribution and enhances grid stability. Tu A.Nguyen [10] in 2021 proposed that real-time generation and demand balancing maintains the stability of the electrical system. Grid operators are increasingly using grid scale energy storage devices to give them the flexibility they need to keep this balance. The grid infrastructure gains resilience and robustness from energy storage as well. In order to use energy storage as a flexible grid asset that may provide numerous benefits, safely and effectively, energy management systems (EMSs) And optimization techniques are grid-base services. Different regulatory regimes and application cases must be supported by the EMS. Yimy E, Garcia vera et al. [11] in 2019 proposed the impending energy issue, the model put out focuses on maximizing energy management strategies and utilizing renewable energy sources. To increase the electrical power system's reliability, the model recommends using optimization approaches in conjunction with microgrids, which are made up of a variety of energy suppliers and load types. The model also emphasizes how artificial intelligence (AI) can be used to address load management issues and enhance smart grid technology. Sonakshi Pardhan et al. [12] in 2018 implemented an Energy Management (EMS) for microgrids where the goal is to effectively manage a microgrid's hybrid energy sources (HES) TI, which include both traditional and renewable energy sources. The model also covers the EMS connection modalities, which most likely relate to how the EMS is connected to the microgrid's infrastructure and communicates with its different loads and energy sources. Xiang Yu [13] in 2014 proposed that power control and smart grids require microgrids, which are small-scale networks comprising generation, storage, and load. Time-of-use (TOU) pricing is utilized in a linear programming approach to optimize energy expenses. The paper examines islanded microgrids for isolated populations as well as grid-connected microgrids in diverse contexts. Additionally, it suggests a control logic for microgrid EMS based on finite state machines, which guarantees outage ride-through and a seamless transition between islanded and grid-connected operation. M.S. roslonb et al. [14] in

2021 proposed the lightning search algorithm is used in the paper to offer an ideal power scheduling controller for distributed energy resource management in a microgrid. The suggested method achieves notable financial and ecological savings while providing an efficient microgrid energy management solution.

2 Methodology

Here is the block diagram of the microgrid energy management system with variable loads, Solar Panel represents renewable energy sources like solar panels or wind turbines. These sources generate electricity based on environmental conditions, Energy generated from renewable sources can be stored in batteries for later use when renewable generation is low or demand is high (Fig. 1).

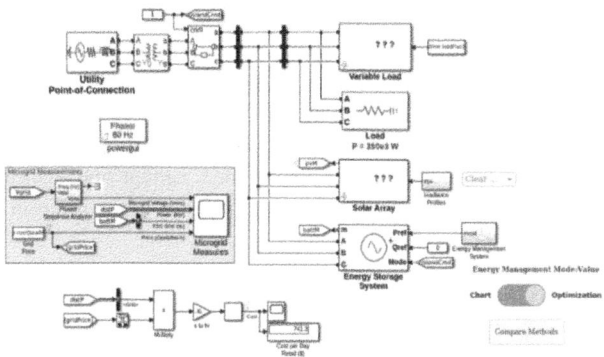

Fig. 1. Block Diagram of Proposed System

Variable Load Block Represents loads in the microgrid system that exhibit variability in their power consumption, such as air conditioners, refrigerators, or industrial machinery. Controllable Load Block Represents loads that can be controlled or scheduled, such as smart appliances or electric vehicles. These loads can be adjusted based on system requirements or energy prices. Represents the connection to the main grid. Grid connection block Represents the connection to the main grid. The microgrid can import or export electricity to/from the grid based on its energy needs and market conditions. Energy Management System (EMS), oversees the operation of the microgrid, including dispatching energy from renewable sources and batteries, managing controllable loads, and optimizing overall system performance. The optimization algorithm calculates the optimal energy dispatch and scheduling decisions to minimize electricity costs while satisfying system constraints, such as demand requirements and battery state of charge. The control system arrows represent the flow of control signals between different blocks [15]. For example, the EMS block sends control signals to adjust the operation of renewable sources, battery storage, and controllable loads based on the optimization results. Block diagram illustrates the interaction between various components of the microgrid system, including renewable energy sources, energy storage, variable and controllable loads, grid connection, energy management, and optimization, to achieve cost-effective and efficient operation while accommodating variability in load demand [16].

2.1 Linear Optimization

A mathematical method known as linear optimization, or linear programming, is used to determine the optimal course of action when choices must be made within predetermined bounds. Fundamentally, linear optimization is the process of maximizing a linear objective function which usually represents objectives like cost reduction or profit maximization while taking into account a set of linear constraints. The limitations or specifications that control the decision-making process are reflected in these constraints. To get the best answer, the decision variables the variables used in the optimization are changed within these limitations. Numerous industries, such as supply chain management, finance, manufacturing, and telecommunications, use linear optimization problems [17]. Linear optimization helps to increase efficiency, save costs, and maximize results in complicated decision-making scenarios by effectively allocating resources, scheduling activities, or optimizing logistics. In order to address linear optimization problems, algorithms like the simplex technique and interior-point approach are frequently used. These algorithms give decision-makers important information that helps them make better decisions and enhance performance. Linear optimization plays a pivotal role in achieving efficient and cost-effective operation of the microgrid. Here's how linear optimization contributes to the project.

The microgrid's overall expenses for producing and using electricity are reduced by the application of linear optimization. Linear optimization algorithms can identify the ideal set of decision variables (like generation schedules and battery charging/discharging rates) that minimize the overall cost of energy generation, storage, and consumption by defining an objective function that represents these costs. Within the microgrid, linear optimization aids in the efficient allocation of resources such as conventional generators, energy storage devices, and renewable energy sources. The ideal use of these resources to meet the demand for electricity while reducing costs is determined by linear optimization, which takes into account variables including the availability of renewable energy, energy demand, battery state of charge, and grid connection constraints. By figuring out the best way to distribute energy generation and storage resources across time, linear optimization helps create operating plans for the microgrid energy management system. To guarantee that the microgrid runs effectively and dependably, it takes into account a number of operational restrictions, including power balancing equations, energy storage capacity limits, and system dependability needs. The application of dynamic control techniques within the microgrid energy management system (EMS) allows for the adaptive response to shifting market dynamics and operating conditions thanks to linear optimization. The EMS can reduce electricity prices while preserving system stability and dependability by continuously solving optimization problems in real-time or through periodic scheduling. This can be done by adjusting energy generation, storage, and distribution. In order to increase the microgrid EMS's performance, linear optimization makes it easier to evaluate various control schemes and optimization parameters. Microgrid performance can be continuously improved by engineers by identifying chances for further cost reduction and operational enhancement through simulation and analysis of the optimization outcomes. Overall, linear optimization serves as a powerful tool for decision-making and control within the microgrid EMS, enabling efficient resource allocation, cost-effective operation, and adaptive response to dynamic operating conditions. It enhances the sustainability, resilience, and economic viability of microgrid systems, making them attractive solutions for decentralized energy generation and distribution.

2.2 Heuristic Approach

Due to the nonlinearities, uncertainties, and discrete decision variables involved, the optimization problem in the microgrid EMS project might be very complicated. By dividing the issue into smaller, more manageable subproblems or by concentrating on the most important elements of the optimization, heuristics can aid in problem simplification. The real-time demands of the microgrid EMS may be too much for conventional optimization techniques like mixed-integer programming or linear programming, particularly in large-scale systems with plenty of constraints and decision variables [18]. Variability and unpredictability exist in microgrid operations due to factors including market prices, electricity demand, and the production of renewable energy. Compared to rigid optimization techniques, heuristic approaches may be more flexible and responsive to changes in the operating environment. Through continual refinement and improvement of possible solutions, heuristics make the exploration of the solution space easier. Methods like as genetic algorithms, particle swarm optimization, local search, and simulated annealing can effectively explore the solution space and pinpoint areas of potential interest that could contain optimum or nearly optimal solutions. When applied in real-world scenarios where processing resources are scarce or precise solutions are not necessarily required, heuristic approaches are frequently more practicable.

In general, incorporating heuristic techniques into the microgrid energy management system (EMS) project can improve the process's computational efficiency, scalability, and adaptability. This will aid in the creation of reliable and workable solutions for managing energy resources and maximizing electricity costs in microgrid systems.

2.3 Battery Constrains

Incorporating battery constraints refers to accounting for the limitations and capabilities of the energy storage system (ESS), typically represented by batteries. These constraints are essential to ensure the proper operation and longevity of the battery system within the microgrid. Batteries can only hold so much energy at a time. By imposing this capacity constraint, the batteries' maximum storage capacity is prevented from being exceeded. To avoid overcharging or over discharging, which can shorten the lifespan and reduce battery performance, it is imperative to simulate this limitation. This limitation would be implemented in the Simulink model by comparing the batteries' current state of charge (SOC) to their maximum capacity. The EMS should stop charging more if the SOC gets close to its full capacity. Likewise, in order to prevent over-discharging, the EMS should restrict discharging if the SOC is near the minimum allowed level. The rate at which batteries can be charged and drained is limited. By imposing these limitations, the charging and discharging rates are kept within safe working parameters and prevent damage to the batteries. You would put limitations on the battery's speeds of charging and draining in the Simulink model. The maximum and minimum power restrictions for charging and discharging activities can be used to express these limitations. Batteries cannot be charged or discharged with 100% efficiency. Energy conversion is accompanied by losses, which the optimization model needs to take into account [19]. To account for energy losses during charging and discharging activities, efficiency restrictions would incorporate into the linear optimization model. Recurrent cycles of charging and draining might shorten a battery's life. Cycle life restrictions must be taken into account in order to enhance battery durability. You can set limits on how many cycles of charging and discharging the batteries go through in the optimization model over a given time. By doing this, you can make sure that the battery operates within reasonable cycle life bounds (Fig. 2).

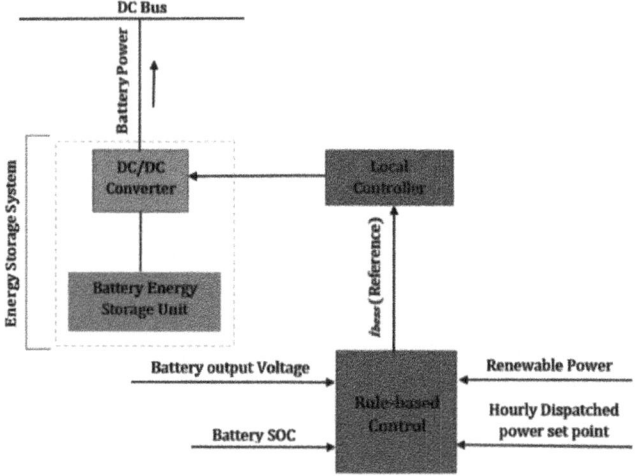

Fig. 2. Battery setup

Temperature can significantly impact battery performance and lifespan. If temperature sensors are available, you can include temperature constraints in the model to prevent operation outside the battery's specified temperature range. By incorporating the battery constraints into the microgrid EMS and optimization model, you ensure that the energy storage system operates safely.

3   Simulation Results

There is a utility grid with a capacity of 13.8 kV in the test system. Through a three-phase, two-winding transformer, the utility grid is connected. Three-phase transformers with star-star connections and primary windings rated at 13.8 kV and secondary windings rated at 5 kV. A three-phase circuit breaker that follows a closed path for the delivery of current is then attached. The electricity is then supplied by connecting a three-phase circuit breaker to a variable load. There is one fixed load with a capacity of 350 kW and one variable load with a capacity of 320 kW. Thus, the model's total possible load is 670 kW. The following block is a 535 kW solar array, while the next is an ESS, also referred to as the system's heart. The ESS has a capacity of 3000 kWh and a rated power of 400 kW. We are utilizing two distinct situations in this system: clear and cloudy days. With these many parameters, we calculate the overall cost of electricity.

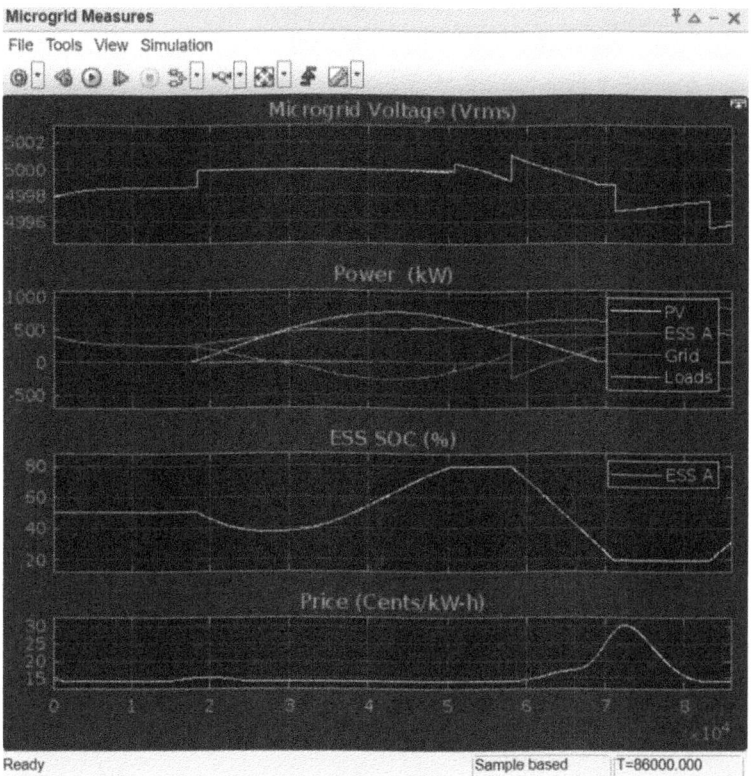

Fig. 3. Simulation result using heuristic approach during clear day

Figure 3 shows the simulation waveforms of clear day analysis using heuristic approach. The microgrid RMS voltage (Vrms), which illustrates how voltages fluctuate due to changes in load demands and generation, is represented by the first waveform. The grid, battery, load, and photovoltaic (PV) power are shown in the second waveform. The electricity price is represented by the fourth waveform, and the ESS State of Charge (SOC) is represented by the third. Since the load demand in this instance is less than 500kW between 0 and 2 * 10^4 s, the ESS State of Charge (SOC) corresponds to 50% of its initial capacity during that period. Therefore, the operational cost is minimal in accordance with that. When the load is increased to 500 kW, the PV power starts to rise and the battery power goes negative in the span of 2 to 4 * 10^4 s. Surplus power is produced as a result of the PV system's power exceeding the load, and this power is stored in the battery. This denotes a gradual start of battery charging. Once more, the cost is low because renewable energy has very little cost. The load stays constant for 4 to 6 * 10^4 s, and the battery enters the charging mode when its voltage reaches the nominal value. SOC has an estimated charging performance index of more than 80% as a result. As a result, the PV power drops and the battery gradually begins to supply power, which causes the battery to begin discharging.

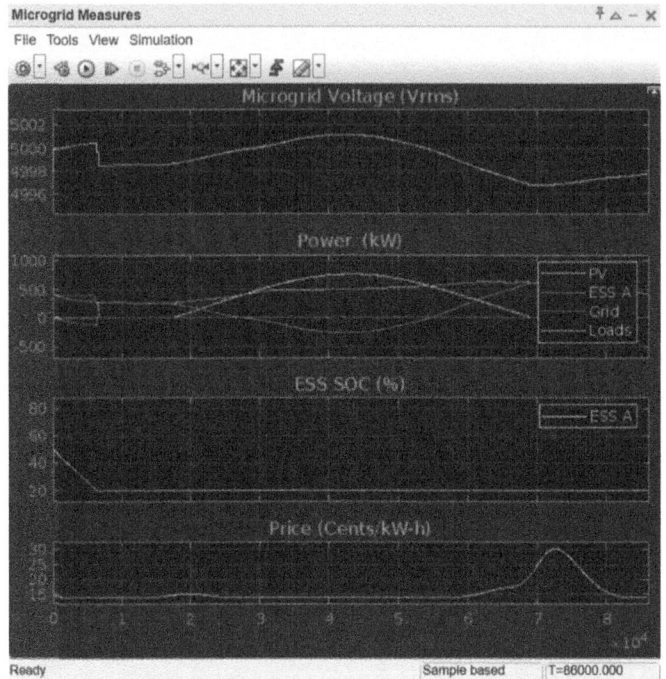

Fig. 4. Simulation result using Optimization method during clear day.

Figure 4 shows the simulation results of optimization approach during clear day analysis. Here, the load demand is lower between 0 and 2 * 10^4 s, therefore neither the solar array nor the battery can produce power during that period. The battery receives a full charge while the grid provides power. Subsequently, as a result of the solar array and battery supplies discharging, the grid's supply is declining. Consequently, there are very little running costs. The load is gradually increasing to 500kW inside the internal of 2 to 4 * 10^4 s. The photovoltaic power is about to rise. The battery's power is declining concurrently. Consequently, SOC has an estimated charging performance index of more than 80%. However, because we are using a sustainable energy source, the cost is still low because we are using battery storage. Even though the power from the solar array continues to decrease, we are still able to obtain electricity from the battery when the load is increased to 500kW for a period of 6 to 8 *10^4 s. When the battery gradually runs out and there is very little solar power left, the grid power kicks in. Grid power is used to supply the load's necessary power, increasing the cost to the highest level.

Using the optimization approach, the total cost of a clear day comes to $618.7. Figure 5 presented here is the simulation results of heuristic approach during cloudy day analysis. Due to changes in the weather, the solar array power begins to increase at the same time that the battery does.

As the solar array is increased above the certain load, large amount of electricity is produced and stored in the battery, causing the battery to begin charging.

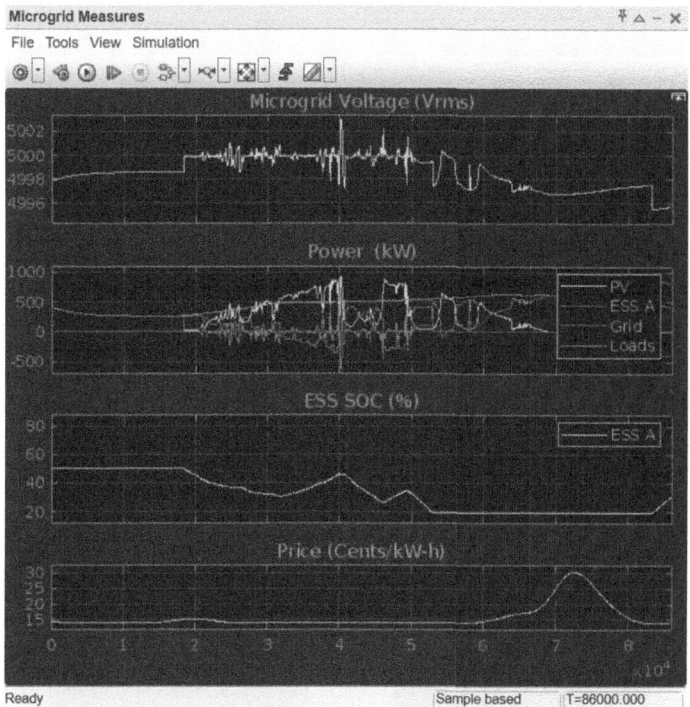

Fig. 5. Simulation result using Optimization method during cloudy day

Figure 6 shows the simulation results of optimization approach during cloudy day analysis. Here, the load demand is lower between 0 and 2 *10^4 s, therefore neither the solar array nor the battery can provide electricity during that period. The battery is charging to its maximum capacity while the grid provides electricity. Grid supply drops momentarily as a result of the solar array and battery power supply. Consequently, there are very little running costs. The entire setup cost is low because the cost of renewable generating is minimal. The load stays constant for 4 to 6 *10^4 s, and the battery enters the charging mode when its voltage reaches the nominal value. Consequently, SOC's charging performance index is projected to be above 80%. Once more, the price is still reasonable. Even though the power from the solar array continues to decrease, we are still able to obtain electricity from the battery when the load is increased to 500kW for a period of 6 to 8 *10^4 s (Fig. 7).

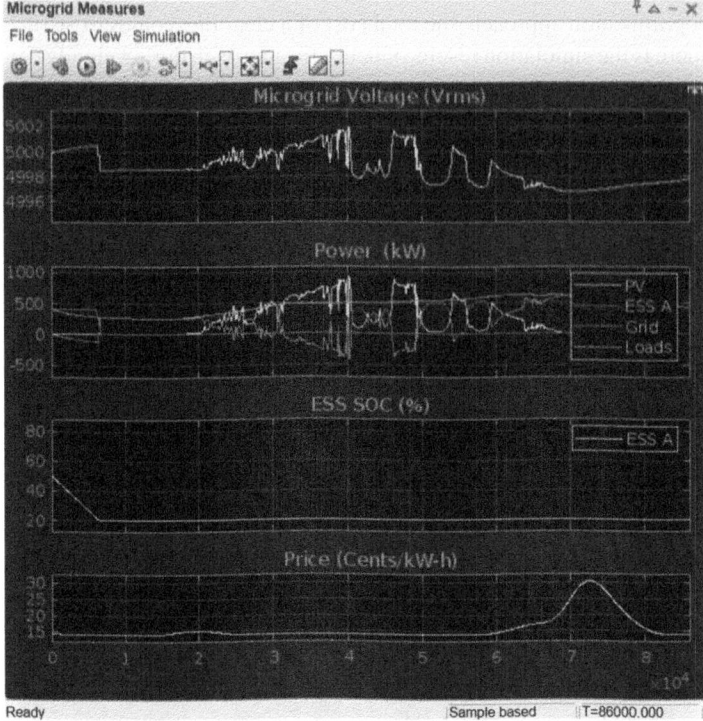

Fig. 6. Simulation results using heuristic approach during cloudy day

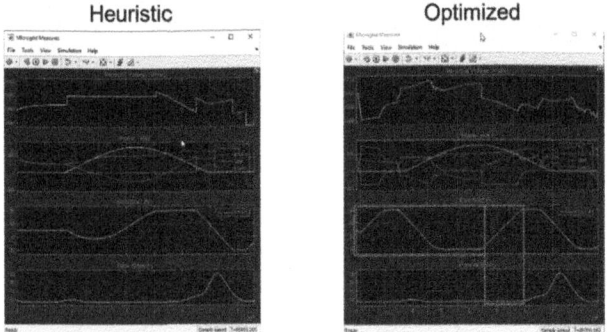

Fig. 7. Policy Comparison.

4   Conclusion

The Microgrid Energy Management System (EMS) Simulink model shows how well a linear optimization strategy works for minimizing power expenditures. By incorporating energy storage devices (batteries) and renewable energy sources (wind turbines, solar panels), the microgrid can effectively balance energy production and consumption while reducing dependency on the grid. The simulation findings demonstrate how the system can intelligently manage energy sources and storage to adapt to various events, such as changing weather and time of day. Batteries store excess energy during times of strong renewable energy generation, which minimizes the need for grid electricity and associated expenses. All things considered, the research shows how crucial it is to apply cutting-edge energy management techniques in microgrids in order to save money, increase energy efficiency, and lessen dependency on fossil fuels. A useful tool for planning and refining microgrid systems, the Simulink model illustrates the possibility of sustainable and affordable energy source solutions in the future.

Acknowledgement. This article is published with the Research and Development fund - Seed money provided by Vel Tech Rangarajan Dr. Sagunthala R&D institute of science and Technology.

References

1. Lasseter, B.: "Micro grids", power engineering society winter meeting, vol. 1, no. C, pp. 146–149 (2001)
2. Liu, X., Su, B.: Micro grids - an integration of renewable energy technologies. In: China International Conference on Electricity Distribution, pp. 1–7 (2008)
3. Bacha, S., Picault, D., Burger, B., Etxeberria-Otadui, I., Martins, J.: Photovoltaics in microgrids: an overview of grid integration and energy management aspects. IEEE Ind. Electron. Mag. **9**(1), 33–46 (2005)
4. Giri, J.A.Y., Fellow, L.: Proactive management of the future grid. IEEE Power Energy Technol. Syst. J. **2**(2), 43–52 (2015)
5. North American Electric Reliability Corporation (NERC): 2016 long-term reliability assessment, www.nerc.com, North American Electric Reliability Corporation (NERC), Atlanta, Georgia, Technical report, December 2016
6. Eyer, Corey, G.: Energy storage for the electricity grid: benefits and market potential assessment guide. Sandia National Laboratories, Albuquerque, NM, Technical report, SAND2010-0815, February 2010
7. Kanagathara, N., Nanmaran, R.: Illustration of potential energy surface from DFT calculation along with fuzzy logic modelling for optimization of N-acetylglycine. Comput. Theor. Chem. **1202**, 113301 (2021)
8. Srimathi, S., Yamuna, G., Nanmaran, R.: An efficient cancer classification model for CT/MRI/PET fused images. Curr. Med. Imaging **17**(3), 319–330 (2021)
9. Thanigaivel, S., et al.: Ecotoxicological assessment and dermal layer interactions of nanoparticle and its routes of penetrations. Saudi J. Biol. Sci. **28**(9), 5168–5174 (2021)
10. Nanmaran, R., et al.: Experimental analysis on the effect of pipe and orifice diameter in inter tank hydrogen transfer. Int. J. Hydrogen Energy **48**(79), 30858–30867 (2023)
11. Nanmaran, R., et al.: Compressor speed control design using PID controller in hydrogen compression and transfer system. Int. J. Hydrogen Energy **48**(73), 28445–28452 (2023)

12. Nanmaran, R., et al.: Mathematical modelling of hydrogen transportation from reservoir tank to hydrogen fuel cell electric vehicle (FCEV) tank. Fuel **361**, 130725 (2024)
13. Vazquez, S., Lukic, S.M., Galvan, E., Franquelo, L.G., Carrasco, J.M.: Energy storage systems for transport and grid applications. EEE Trans. Ind. Electron. **57**(12), 3881–3895 (2010)
14. HAIDER, Z.M., MEHMOOD, K.K., RAFIQUE, M.K., KHAN, S.U., LEE, S.-J., KIM, C.-H.: Water-filling algorithm based approach for management of responsive residential loads. J. Mod. Power Syst. Clean Energy **6**(1), 118–131 (2017). https://doi.org/10.1007/s40565-017-0340-x
15. Khan, S.U., Mehmood, K.K., Haider, Z.M., Rafique, M.K., Khan, M.O., Kim, C.-H.: Coordination of multiple electric vehicle aggregators for peak shaving and valley filling in distribution feeders. Energies **14**, 352 (2021)

Advancing Smart Water Management: Integrating Invasive and Non-invasive IoT Sensors for Sustainable Hot and Cold-Water Usage Monitoring

Nick Kalsi[✉], Fiona Carroll, Kasha Minor, and Jon Platts

Cardiff School of Technologies, Cardiff Metropolitan University, Llandaff Campus, Cardiff CF5 2YB, UK

st20162692@outlook.cardiffmet.ac.uk, {fcarroll,kbminor, jplatts}@cardiffmet.ac.uk

Abstract. This paper outlines an experimental approach aimed at leveraging IoT (Internet of Things) technology to develop a sustainable system for monitoring hot and cold-water usage. By integrating both invasive and non-invasive IoT sensors, the authors propose a more comprehensive design for real-time monitoring system for the hospitality industry. The prototype developed focuses on a more holistic monitoring of hot and cold-water consumption, including temperature, representing a significant advancement in sustainable water management. Indeed, creating an intelligent water flow sensor involves crucial stages such as hardware design and system development. This paper highlights the hardware design which focuses on the selection of materials, sensors, power sources, and communication interfaces for accurate measurement and data transmission. It also details the system development which assembles and integrates components, followed by extensive testing to confirm precision. The paper discusses the need for meticulous calibration and environmental considerations which are crucial for measurement accuracy. Moreover, it covers the sensor selection which should align with specific application requirements, considering precision and infrastructure compatibility. In terms of evaluation, this study documents the accuracy of invasive turbine and non-invasive ultrasonic water flow sensors over a period. The analysis of this sensor data shows consistent measurements with minimal variances. However, in conclusion, the authors show that integrating both invasive and non-invasive IoT sensors offers a more reliable and holistic solution as well as vast potential for enhancing sustainability across diverse industry settings, including tourism.

Keywords: Sustainability · Internet of Things (IoT) · Invasive and non-invasive sensors · Turbine sensors · Ultrasonic sensors · temperature · water

1 Introduction

Water is essential for many purposes in our daily lives, from washing our hands and cooking meals to refreshing showers and even using the bathroom. The use of hot and cold water for these activities highlights its importance, making it essential to balance

P. D. Sivakumar et al. (Eds.): IRCCTSD 2024, CCIS 2362, pp. 81–99, 2025.
https://doi.org/10.1007/978-3-031-82386-2_7

their use in an environmentally conscious manner. Advances in sensing technology, microelectronics, and the Internet of Things (IoT) have significantly enhanced water sustainability. These innovations enable efficient tracking of temperature variations in cold and hot water supply environments, facilitating automated monitoring and control functionalities. For instance, IoT devices have been embedded in microcontrollers to effectively track temperature in cold and hot water supply environments (Ramli & Hanafi 2022). Such implementation scenarios provide automated monitoring and control functionalities, effectively resolving the limitations of manual approaches towards changes to temperature variations in the water supply facilities. These sensing technologies can also play a key role in behavioural monitoring systems, capturing detailed data on long-term behavioural patterns. In this paper, the aim is to provide insights into designing an IoT-enabled water flow monitoring system tailored for the tourism industry. In particular the system, designed to accommodate both hot and cold-water consumption, emphasises the utilisation of invasive and non-invasive sensors to address critical water usage challenges. A dual testing procedure is applied to ensure a comprehensive evaluation of sensor performance, enhancing system accuracy and efficiency. The following sections provide a thorough exploration of the literature, whilst detailing the system design, presenting results and analysis, and concluding with valuable insights into sustainable water management practices.

2   Water Consumption and the Tourism Industry

Cazcarro, Hoekstra, and Chóliz (2014) provide crucial insights into Spain's water consumption, especially regarding tourism. Utilising process and input-output analyses, their paper assesses the virtual water trade across sectors, highlighting tourism's significant impact. In detail, direct and indirect water uses linked with tourist activities and purchases are outlined, with distinctions between domestic and foreign tourism. These findings are vital for policy formulation and sustainable water management, particularly in tourist-dependent regions. In fact, the study shows important implications regarding the development of IoT-enabled water flow monitoring systems (Cazcarro et al. 2014). Understanding tourism-related water consumption patterns is key to how we might inform the design of tailored monitoring systems. For instance, the insights obtained from this research (Cazcarro et al. 2014) helped the authors to understand how invasive and non-invasive IoT sensors could provide a comprehensive data collection on flow rates and temperatures.

In another study, Anjana et al. (2015) present an innovative IoT-based solution to address the escalating challenges of water management, particularly in the context of rising demand and depleting resources, as witnessed in India and many other countries worldwide. With urbanisation, climate change, and inefficient usage exacerbating the strain on water supplies, there's an urgent need for effective conservation and management strategies. The proposed system leverages IPv6 network connectivity and IoT technology to enable real-time water flow metering and quality monitoring (Anjana et al. 2015). By employing CoAP for monitoring and control, the system facilitates internet-based data collection, thus offering a robust solution to the complexities inherent in water management. This approach streamlines billing processes and fosters equitable billing practices while providing insights into water consumption patterns, promoting awareness and encouraging conservation efforts. Also, it is important to note the system's capability to automatically detect leakages and assess water quality through pH and ORP sensors addresses critical concerns regarding resource wastage. This ensures the delivery of safe and clean water to households (Anjana et al. 2015) and provides valuable insights for IoT-based water monitoring.

Furthermore, Toyosada, Otani, and Shimizu (2016) address the pressing issue of water conservation in rapidly developing countries like Vietnam, where water resource management challenges are becoming increasingly prominent. Focusing specifically on the water-use patterns within a hotel in Ho Chi Minh City, their study seeks to quantify the effectiveness of water-saving devices by modelling toilet and shower usage. Through detailed analysis, the researchers aimed to understand the dynamics of water consumption in these facilities, including the frequency of toilet flushes, shower usage, and associated water temperatures. The findings reveal valuable insights into average water consumption patterns, with full toilet flushes occurring 3.3 times per day, half flushes 3.0 times per day, and shower usage resulting in daily consumption of 48.1 L (Toyosada et al. 2016). Further, the study identifies fluctuations in hot and cold-water usage throughout the day, highlighting peak periods of activity in the morning and evening. By shedding light on these usage patterns, the research contributes to developing targeted water conservation and management strategies in hotel settings to mitigate water scarcity challenges in Vietnam.

3   IoT and Water Monitoring Management

Focusing on the technology side, Jamaluddin et al. (2016) present a study focusing on developing a wireless data acquisition system for wireless water flow monitoring in a closed-channel pipeline system utilising an Android smartphone. The system incorporates an ATMEGA 328 single-chip microcontroller equipped with a Bluetooth module and Near Field Communication (NFC) tag. The microcontroller receives a pulse train frequency input from the hall-effect water flow sensor and transfers the data to the Android smartphone via Bluetooth (Jamaluddin et al. 2016). An NFC tag is also utilised to identify and locate water flow sensors. In detail, the Android application enables the display of sensor ID based on the smartphone's NFC tag and real-time water flow rate. In this paper, experimental tests were conducted to demonstrate the successful operation of the flow monitoring system, exhibiting real-time monitoring capabilities and achieving a high accuracy rate of 99.98% in wireless water flow measurement (Jamaluddin et al. 2016). Importantly, this study represents a significant advancement in wireless water flow monitoring systems by leveraging the capabilities of Android smartphones, integrating Bluetooth wireless serial communication and NFC technology, and achieving a high accuracy rate in water flow measurement. Thus, findings obtained from this study are instrumental in developing a robust water monitoring framework that effectively tracks hot and cold temperatures.

Exploring other applications, Rajurkar, Prabaharan, and Muthulakshmi (2017) delve into water management using IoT technology, focusing on monitoring water usage within a residential block. With water being a fundamental resource for various human activities, including domestic, agricultural, and industrial purposes, efficient management becomes imperative. The study highlights the stark reality of water scarcity, with only a minute fraction of freshwater available for direct human use (Rajurkar et al. 2017). In response to the pressing need for effective water management, the project proposes a sophisticated system equipped with sensors to monitor water flow within pipelines, providing real-time data on usage. In detail, this data is transmitted to the cloud through IoT infrastructure, facilitating remote access and monitoring. Residents then receive alerts via a mobile application, enabling them to track their water usage and adhere to predefined limits set by municipal authorities (Rajurkar et al. 2017). By integrating IoT technology with water management practices, the system aims to minimise water wastage and promote responsible consumption, thus addressing the challenges posed by water scarcity and sustainability. Hence, the insights drawn from this study are also instrumental in understanding the role of IoT sensors in effective water management, especially when considering hot and cold temperature controls.

From a wider perspective, Zulkifli et al. (2022) offer a comprehensive systematic review focusing on IoT-based water monitoring systems, emphasising the pivotal role of water quality monitoring in advancing smart agriculture and meeting human needs. With a transition toward automated monitoring, reliable models and accurate datasets become imperative for effective water quality management. By extensively reviewing water quality literature from 2018 to 2022, their study identifies critical concerns, challenges, and research gaps in the field (Zulkifli et al. 2022). Their taxonomy categorises the literature into three groups, facilitating a structured analysis of IoT-based water monitoring methodologies. Through synthesis, the review identifies shortcomings in model accuracy, data-gathering systems' development, and types of data used, offering recommendations to expedite progress in this field. Thus, this systematic review is a vital resource for researchers and practitioners, aiding in understanding IoT-enabled water monitoring systems and addressing critical challenges in water quality management.

3.1 IoT and Water Quality Monitoring

The capability of IoT for water usage monitoring extends far beyond simple data collection. It offers a powerful tool for improving water resource management, enhancing efficiency, promoting sustainability, and ensuring access to clean and safe water for all. For example, Ramadhan (2020) addresses the pressing issue of deteriorating drinking water quality in Najaf, Iraq, by proposing an electronic water-monitoring system capable of proactive response to unsafe water supply. Based on wireless sensor networks and Internet of Things technologies, the system aims to replace conventional methods with real-time monitoring capabilities. Statistical analysis conducted on data from five Najaf water stations underscores the urgency, revealing poor water quality that fails to meet World Health Organization standards (Ramadhan 2020). The proposed system offers remote and intelligent monitoring, measuring various parameters, including pH level, temperature, nitrate, chloride, dissolved oxygen concentration, turbidity, oxidation-reduction potential (ORP), conductivity or total dissolved solids (TDS), and sodium content. The study's findings demonstrate the efficiency of IoT in monitoring water temperature, providing a robust framework for tracking hot and cold water.

More so, Prasad et al. (2015) delve into IoT-enabled water quality monitoring systems in their paper on the "Smart Water Quality Monitoring System". With a global concern over deteriorating water quality due to industrial activities and chemical usage, robust monitoring mechanisms are needed. The authors emphasise the significance of ensuring the availability of high-quality water for human consumption and environmental health. By leveraging IoT and remote sensing technologies, their proposed system aims to enhance the existing measurement infrastructure for water quality monitoring, particularly in regions like the Fiji Islands in the Pacific Ocean (Prasad et al. 2015). While the authors' focus is primarily on water quality, their utilisation of IoT technology aligns closely with the broader research topic of IoT-enabled water flow monitoring systems for hot and cold water usage. Their work underscores the potential of IoT-based solutions in monitoring and improving water quality, thereby contributing to more effective water resource management strategies.

Also, Muangprathub et al. (2019) present a pioneering approach to smart agriculture through their paper on IoT and agriculture data analysis for intelligent farms. Although they focus on optimising watering systems for crops, their utilisation of wireless sensor networks (WSNs) aligns closely with the broader research topic of IoT-enabled water flow monitoring systems for hot and cold water usage. The authors propose a comprehensive system comprising hardware, a web application, and a mobile application to monitor and manage crop fields effectively (Muangprathub et al. 2019). By incorporating soil moisture sensors and employing data mining techniques, the system analyses crop data to predict suitable environmental conditions for optimal crop growth. Additionally, the mobile application allows for both automatic and manual control of crop watering, enhancing flexibility and efficiency in agricultural management. Furthermore, Mukta et al. (2019) contribute to IoT-enabled water flow monitoring systems for hot and cold-water usage by studying an IoT-based Smart Water Quality Monitoring System. While their primary focus is assessing water quality parameters such as temperature, pH, electric conductivity, and turbidity, their innovative use of IoT technology aligns closely with the broader research topic. By continuously measuring water conditions and transmitting data to a desktop application, their system demonstrates the potential for real-time water quality monitoring. Although their emphasis is on water quality analysis and classification, their underlying IoT framework presents opportunities for expanding into monitoring water flow rates for both hot and cold water usage. Further, the paper's incorporation of IoT technology lays a foundation for further exploration into comprehensive water flow monitoring systems.

Finally, Rantanen et al. (2021) contribute to advancing IoT-enabled water flow monitoring systems for hot and cold-water usage with their research on "Utilizing Cost-effective NB-IoT-based Sensors for Detecting Water Temperature and Flow". While their primary focus is on detecting ambient room temperature and water pipe temperatures, their innovative prototype showcases the potential for monitoring water flow within pipes. By leveraging wireless temperature measurement technology, their system offers a cost-effective solution for detecting anomalies in water applications and devices (Rantanen et al. 2021). This approach aligns closely with the broader research topic of IoT-enabled water flow monitoring systems, as their prototype demonstrates the feasibility of using NB-IoT-based sensors for monitoring water flow in both hot and cold-water systems. For instance, it provides a design framework for designing and implementing temperature detection IoT modules for water monitoring hot and cold temperature categories. Such information is valuable in handling the current research needs. Thus, this paper presents an exciting advancement in the field, offering a glimpse into the future of efficient and cost-effective water flow monitoring technologies.

In conclusion, while these studies extensively examine IoT-enabled water flow monitoring systems, they demonstrate substantial gaps that the current research strives to address. Firstly, while studies like Cazcarro et al. (2014) and Anjana et al. (2015) provide insights into water consumption patterns in specific contexts such as tourism and urban environments, there's a lack of comprehensive research integrating invasive and non-invasive IoT sensors for accurate data collection across various settings. Secondly, while research by Toyosada et al. (2016) and Jamaluddin et al. (2016) explores water consumption patterns and wireless monitoring systems, there's limited focus on how these systems can effectively track real-time hot and cold water temperatures. Thirdly, while studies like Rajurkar et al. (2017) and Zulkifli et al. (2022) delve into IoT technology's role in water management and monitoring, there's a gap in understanding how insights derived from IoT-enabled systems can inform decision-making and resource management practices regarding hot and cold-water usage. Additionally, while studies like Ramadhan (2020) and Prasad et al. (2015) highlight the importance of real-time water quality monitoring using IoT, this research must be expanded to include temperature monitoring for hot and cold water systems. While research by Muangprathub et al. (2019) and Mukta et al. (2019) demonstrates the potential of IoT technology in agricultural and water quality monitoring, respectively, there's a limited exploration into comprehensive IoT-based water flow monitoring systems that encompass both hot and cold water usage. Hence, the current study aims to bridge these gaps by developing an integrated IoT-enabled water flow monitoring system that effectively tracks hot and cold-water temperatures, collects accurate data using invasive and non-invasive sensors, and provides valuable insights for sustainable water management practices.

4   Study of Integrated Invasive and Non-invasive Sensor Water Monitoring Prototype

The scope of the research encompasses the comprehensive development and implementation of an IoT-enabled water flow monitoring system tailored to analyse hot and cold-water usage patterns. The system's primary objective is integrating invasive and non-invasive IoT sensors to gather water flow rates and temperature data accurately. A microcontroller then processes this data to facilitate real-time transmission to a cloud server for further analysis and remote monitoring. The research also addresses critical questions about water consumption patterns, including the frequency and volume of water flow through hot and cold water pipes and temperature variations. The scope includes investigating the potential applications and benefits of deploying such a system for sustainable water resource management and conservation efforts in residential and commercial settings.

4.1   Design Process

The proposed system for water flow monitoring integrates several vital components to facilitate efficient data collection and analysis. At the core of this system is the FireBeetle ESP32 microcontroller. This module provides dual Wi-Fi and Bluetooth capabilities, ensuring reliable connectivity for real-time data transmission (He & Iqbal 2023). This microcontroller is the central hub for collecting and transmitting data from various sensors to the designated endpoints. The system incorporates 2 x DIGITEN G1/2″ Invasive Water Flow Sensors into its setup to enable accurate measurement of water usage. These sensors are strategically positioned within the water pipes to capture precise flow rates of both hot and cold water streams. The proposed system implements two ultrasonic clamp-on non-invasive water flow sensors, leveraging their inbuilt temperature sensors and time-of-flight principle. The sensors utilise LoRaWAN technology for data transmission, ensuring reliable and efficient communication (Piechota et al. 2023). The system features a LoRaWAN gateway and Wi-Fi router to facilitate communication between the microcontroller and the cloud dashboard. This proposed system and its subcomponents are illustrated in the diagrams below (Figs. 1, 2, 3 and 4).

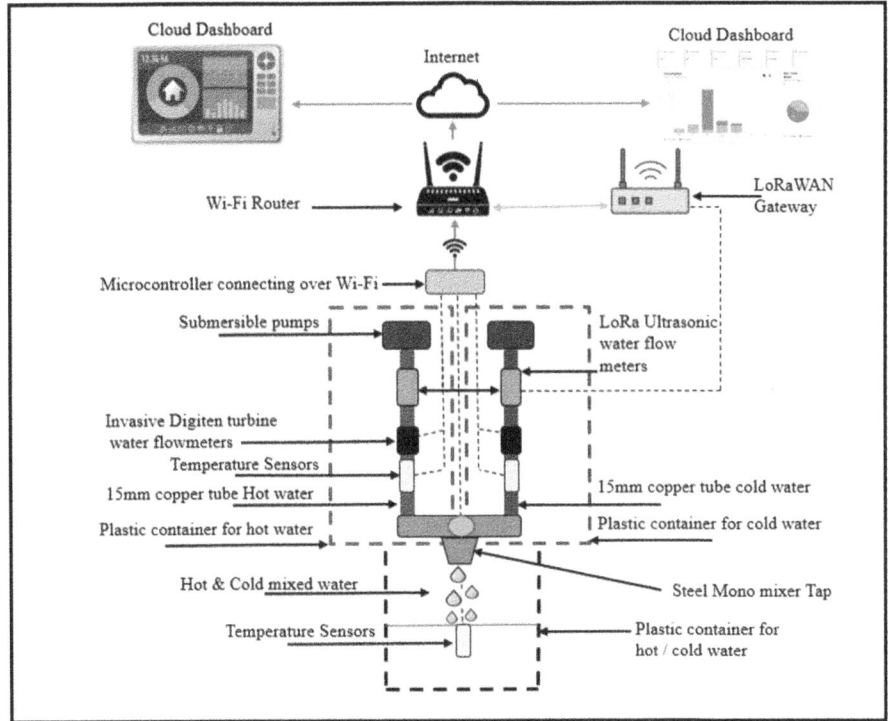

Fig. 1.   Overview of IoT water sensor usage

The sensors implemented in the proposed system are highly effective in capturing precise water usage readings and measuring the volume of water flowing through both

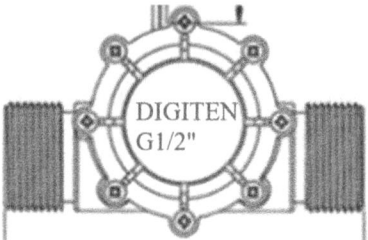

Fig. 2. Turbine water flow sensor

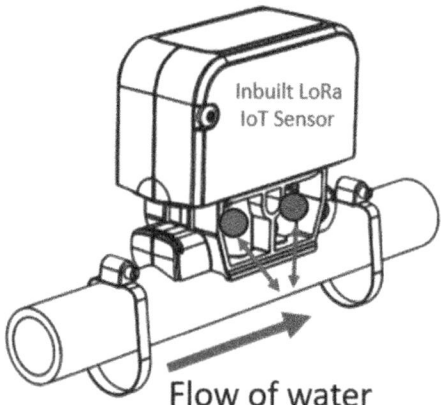

Fig. 3. LoRa Ultrasonic water flow sensor

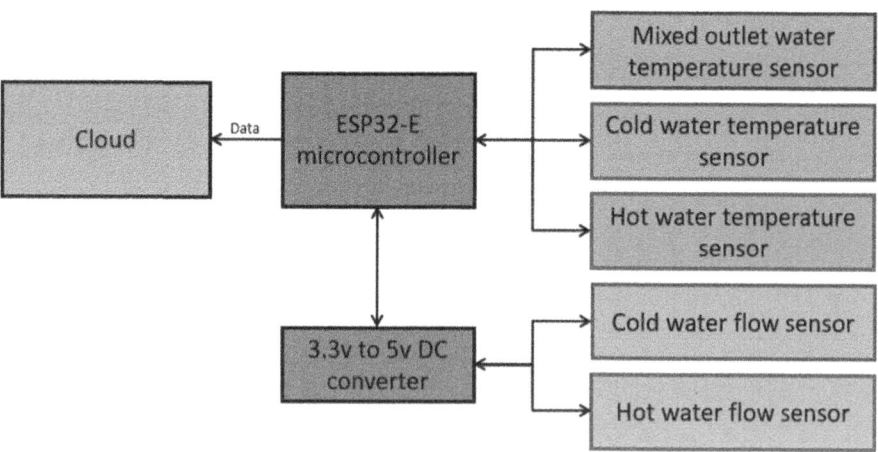

Fig. 4. Conceptual framework for the proposed system

hot and cold water pipes. Designed explicitly for invasive measurement capabilities, the DIGITEN Invasive Water Flow Sensors are strategically placed within the water flow path to deliver precise temperature and flow rate data. The G1 ½ water flow sensor utilises the Hall effect method to detect the flow rate. It offers a cost-effective and durable alternative to the Doppler-effect method due to its non-contact detection. The rotational speed for the sensors changes with water flow, generating pulses detected by the Hall-effect sensor. This electrical signal's frequency, proportional to the water flow velocity, is processed on a microcontroller. These sensors were carefully onboarded using a carefully designed circuit board that ensured optimal performance and efficiency. This approach effectively integrated all the sensors into the circuit board by soldering different pins and component segments. This circuit architecture for the microcontroller and onboarding components is illustrated in the diagram below (Fig. 5 and Table 1).

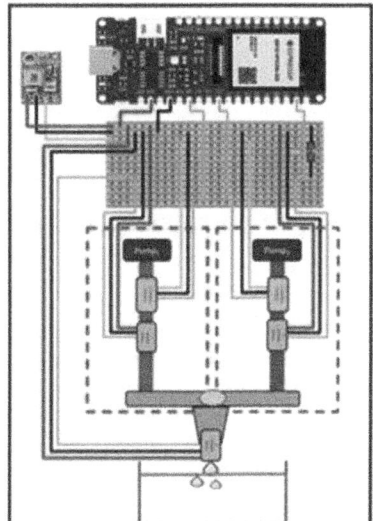

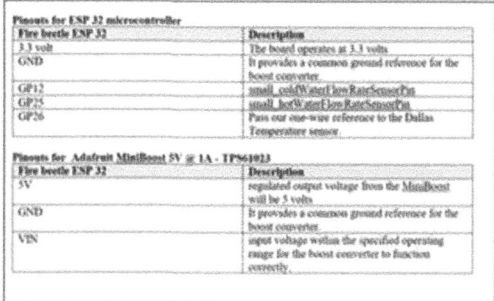

Pinouts for ESP 32 microcontroller

Fire beetle ESP 32	Description
3.3 volt	The board operates at 3.3 volts
GND	It provides a common ground reference for the boost converter
GP12	small_coldWaterFlowRateSensorPin
GP25	small_hotWaterFlowRateSensorPin
GP26	Pass our one-wire reference to the Dallas Temperature sensor.

Pinouts for Adafruit MiniBoost 5V @ 1A - TPS61023

Fire beetle ESP 32	Description
5V	regulated output voltage from the MiniBoost will be 5 volts
GND	It provides a common ground reference for the boost converter
VIN	input voltage within the specified operating range for the boost converter to function correctly.

Fig. 5. Hardware Circuit Design

4.2 Experimental Setup

The system's software uses the Arduino IDE for microcontroller programming and the App Inventor for Android (AIA). Arduino IDE is an open-source software for embedded systems that facilitates code writing and microcontroller uploading. On the other hand, AIA, developed by MIT Media Lab, offers a visual programming platform for Android app development, allowing easy creation through a block-based interface. These programming environments provide software functionalities that precisely measure and record flow rates and temperatures using diverse sensors. Various libraries are integrated into the development process, providing an efficient design strategy. For instance, they facilitate robust sensor interrupt handling, ensure safe data processing and transmission, facilitate serial communication among the modules, and enhance remote data transmission to the cloud platform and dashboard. Hypertext transfer protocol (HTTP) and

Table 1. Hardware components used for onboarding

Hardware	Description
DIGITEN G1/2″ water flow sensor	G1 ½ (Diameter ½ in.), Produced by SEA, Model YF-S201, Rate 1–30 L/min, Water pressure ≤1.754 Mpa
Ultrasonic water flow sensor	Manufactured by Quandify – CubicMeter, the sensor complies with MID/OIML R-
LoRaWAN Gateway	RAK7268 Wis Gate LoRaWAN Edge Lite 2
Wi-Fi internet router	BT Hub3 Wi-Fi router
Submersible pond pumps	capacity of 1500 L/H and a power of 40 W
Copper, plastic tubing	15 mm copper tubing, 10 mm plastic tubing using reducers
Sink mono tap mixer	The mono mixer tap used in our test was purchased from IKEA
15-litre plastic containers	Transparent materials in constructing these boxes offer an efficient and effortless means of identifying the contents within
DS18B20 temperature sensors	DS18B20 sensors have a waterproof design and a built-in resistor

application programming interfaces (APIs) are extensively leveraged in the design, providing an efficient communication and software-based interaction framework for the components.

The application logic for the proposed system begins with initialising key components, including setting up the interrupt pin, configuring the one-wire interface, and establishing a Wi-Fi connection. The system enables interrupt detection and operates within a time interval of less than 1 s. During this time, the interrupt detection is temporarily disabled. The system calculates water flow rates (wf1 and wf2) within this time frame using the interrupt-generated counter. Simultaneously, it reads temperature values (t1 and t2) from the sensors. The collected data, including water flow rates and temperatures, is then posted to a cloud endpoint. Following data transmission, the system increments the interrupt-generated counter, indicating that an interrupt event has occurred. This summary outlines the sequential steps involved in the application logic, highlighting the initialisation, data acquisition, transmission to the cloud, and management of interrupt events. This logic is illustrated in the diagram below (Fig. 6).

4.3   Results and Analysis

4.3.1   Turbine Flow Data

Data from the turbine flow sensor was captured for seven days, with the results being presented in the table below (Table 2).

These results are visualised in the graph below (Fig. 7).

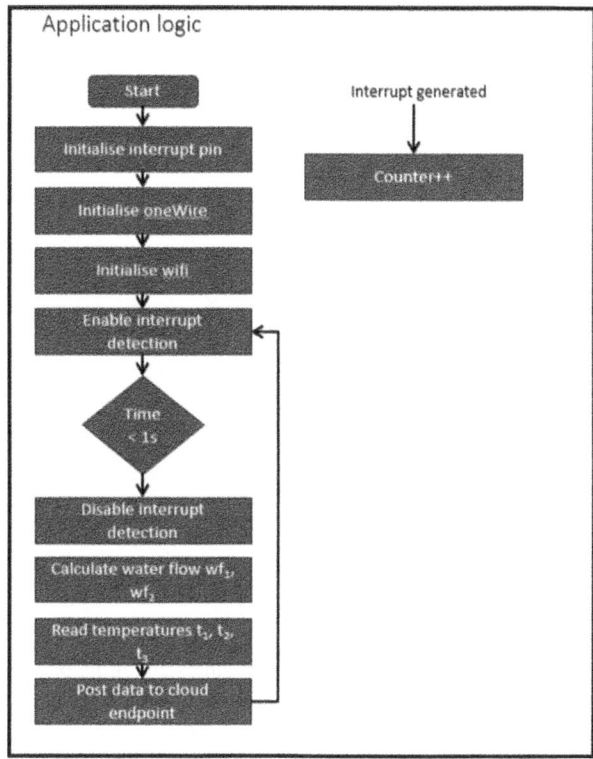

Fig. 6. Application logic

Table 2. Data obtained from the Turbine Water flow sensor

Date	Time	Turb Cold water L	Turb Cold Temp c	Turb Hot water L	Turb HotTemp c	Mix Turb HotTemp c
17/11/2023	09:00	9.1	14.9	12.1	17	31.9
18/11/2023	10:00	12.1	12.3	18.3	19.3	31.6
19/11/2023	12:00	6.1	10.1	2.4	21.2	31.3
20/11/2023	15:00	7.3	11.3	10	19.2	30.5
23/11/2023	00:00	15.2	4.4	17.1	16.5	20.9
24/11/2023	16:00	13.3	14.2	10.4	22.2	36.4
25/11/2023	17:00	16.4	14.2	4.2	23.2	37.4

Data analysis reveals varying trends in the parameters measured, reflecting dynamic changes in water consumption patterns. The flow rate of cold water, as indicated by the Turb Cold water L parameter, exhibits variability throughout the recorded period, ranging from 6.1 to 16.4 L per minute. Concurrently, the temperature of cold water, denoted by

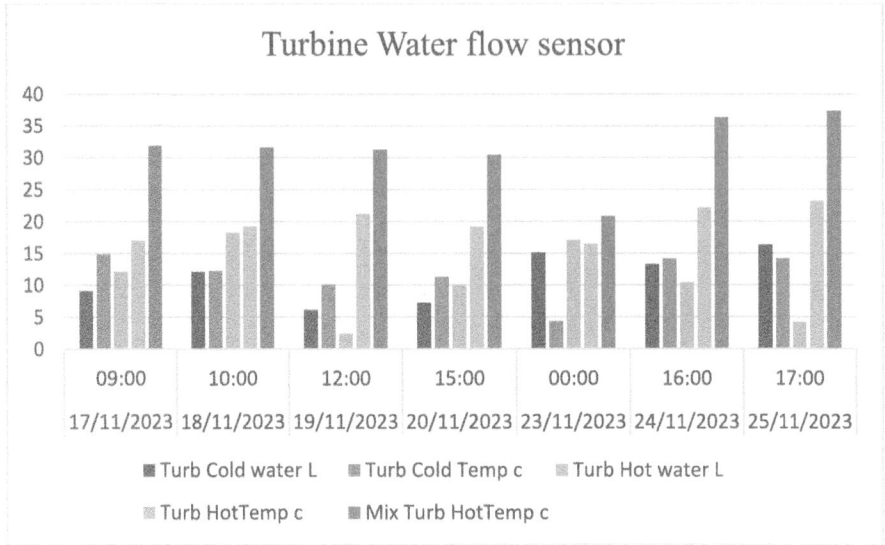

Fig. 7. Visualisation of data obtained from the Turbine Water flow sensor

Turb Cold Temp c, demonstrates fluctuations with values changing between 4.4 and 14.9 °C. Similarly, the flow rate and temperature of hot water display dynamic patterns. Turb Hot water L, representing the flow rate of hot water, fluctuates between 2.4 and 18.3 L per minute, indicating changes in demand or usage patterns over time. Correspondingly, Turb HotTemp c, depicting the temperature of hot water, varies from 16.5 to 23.2 °C, reflecting adjustments in the heating system or shifts in demand throughout the recorded period. The mixed turbine hot water temperature, represented by Mix Turb HotTemp c, illustrates adjustments made to the water supply temperature based on the flow rates and temperatures of both cold and hot water.

The detailed review of specific time points within the dataset provides valuable insights into the system's dynamics of water flow and temperature regulation. For instance, the examination of the data at 12:00 PM on November 19, 2023, reveals a notable balance between cold and hot water parameters. While the cold water flow rate remains moderate with a relatively low temperature, the hot water flow rate is lower but accompanied by a higher temperature. Subsequently, the mixed hot temperature experiences a significant increase, indicating an effective mixing process that adjusts water temperature to meet desired levels. This observation highlights the system's ability to dynamically regulate water temperatures based on variations in flow rates and temperatures of cold and hot water sources, emphasising the importance of understanding the interplay between these variables for efficient water management.

Further, the data analysis at 4:00 PM on November 24, 2023, underscores the system's dynamic nature of water flow and temperature fluctuations. The substantial increase in cold and hot water flow rates compared to previous readings suggests a surge in water demand or usage during this time. The significant rise in hot water temperature contributes to a considerable increase in the mixed hot temperature, indicating efficient heat transfer mechanisms within the system. These fluctuations illustrate the system's responsiveness to changes in demand and its capability to maintain desired water temperatures even under varying conditions, highlighting its effectiveness in meeting user requirements. The examination of data at 5:00 PM on November 25, 2023, emphasises the system's adaptability and performance in regulating water temperatures. Despite a decrease in hot water flow rate compared to previous readings, the substantial flow of cold water and the relatively high temperature of hot water result in the highest recorded mixed hot temperature. This observation suggests efficient mixing processes and temperature adjustments within the system to ensure consistent water delivery at desired temperatures.

The experiment shows a relationship between water flow rates and temperatures observed in cold and hot water streams. Understanding this relationship is pivotal for thoroughly assessing the Turbine Water Flow Sensor's efficacy and susceptibility to environmental fluctuations. The sensor's ability to accurately capture flow rate changes and temperature changes emphasises its reliability in monitoring water systems. This phenomenon implies that alterations in flow rates directly influence water temperatures and vice versa. This situation illustrates the sensor's sensitivity to dynamic changes within the water system. Such insights are crucial for optimising the sensor's performance and ensuring its applicability across various environmental conditions. Further, comprehending this connection enables researchers and practitioners to interpret sensor data accurately, leading to informed decision-making in water management and conservation efforts. Hence, acknowledging and analysing the relationship between water flow rates and temperatures is fundamental for harnessing the full potential of the Turbine Water Flow Sensor in real-world applications.

The data analysis also highlights significant temporal variability within the monitored water system. The fluctuations observed across different dates and times underscore the dynamic nature of the system, emphasising the importance of accounting for time-dependent factors in water system analysis. These temporal changes may result from various factors such as daily patterns, seasonal variations, or external influences. Understanding and quantifying these temporal variations are essential for developing robust monitoring and control strategies that adapt to changing environmental conditions. Additionally, recognising the temporal variability within the water system enables researchers to identify trends, patterns, and anomalies over time, facilitating proactive interventions to mitigate potential risks or optimise system performance. Thus, incorporating temporal variability analysis into water system monitoring practices enhances the effectiveness and reliability of Turbine Water Flow Sensors, ultimately contributing to more efficient and sustainable water management practices.

The Turbine Water Flow Sensor experiment offers a thorough understanding of the dynamics governing water flow and temperature variations over time. The sensor effectively captures the system's fluctuations through random time selections during the experiment. This feature showcases the sensor's capability to provide detailed insights into the water system's behaviour. The significance of incorporating both cold and hot water parameters and the mixed hot temperature is evident as these factors collectively contribute to the system's overall performance. These variables offer a holistic perspective on how the water system functions under different conditions, enabling informed decision-making in water management practices. Further, the analysis and exploration of the observed trends may unveil additional insights into the efficiency and effectiveness of water flow systems across diverse applications. The comprehensive understanding derived from the experiment underscores the importance of continuous research and exploration to enhance the performance and applicability of water flow sensors in various real-world scenarios. Thus, the findings obtained from the turbine water flow sensor offer valuable insights on the relationship between flow rates and temperatures for mixing cold and warm water, leading to effective water management and energy utilisation.

4.3.2 Ultrasonic Water Flow Analysis

Data from the ultrasonic water flow sensor was captured for seven days, with the results being presented in the table below (Table 3 and Fig. 8).

Table 3. Data obtained from the ultrasonic water flow sensor

Date	Time	U/S Cold water L	U/S Cold Temp c	U/S Hotwater L	U/S Hot Temp c
17/11/2023	09:00	9.2	15.1	12.2	16.9
18/11/2023	10:00	12.3	12.2	18.1	19
19/11/2023	12:00	6.1	10.2	2.1	21.1
20/11/2023	15:00	7.2	11.1	10.3	19.2
23/11/2023	00:00	15.4	4.1	17.2	16.3
24/11/2023	16:00	13.1	14.2	10.3	22.1
25/11/2023	17:00	16.5	14.1	4.1	23.4

This data is visualised in the graph presented below.

The data from the ultrasonic water flow sensors offers valuable insights into the dynamic behaviour of water flow and temperature variations over time. One prominent observation is the apparent relationship between flow rates and cold and hot water temperatures. When the flow rate increases, there tends to be a corresponding temperature change, indicating a relationship between these two variables. This relationship underscores the importance of comprehensively considering flow rates and temperatures to understand the water system's behaviour. Also, the data showcases significant temporal variability, suggesting that time-dependent factors play a crucial role in influencing water flow and temperature dynamics. The fluctuations observed across different dates and times highlight the system's dynamic nature, which usage patterns or external environmental conditions may influence. This temporal variability emphasises the need for continuous monitoring and analysis to accurately capture the system's fluctuations.

A close review of the ultrasonic water flow sensor data provides valuable insights into specific instances, highlighting the sensor's performance under varying conditions. For example, on December 7, 2023, at 14:00, the data indicates a balanced and stable water flow condition, with moderate cold water flow rates accompanied by typical temperatures. This observation suggests a consistent and reliable sensor operation, accurately

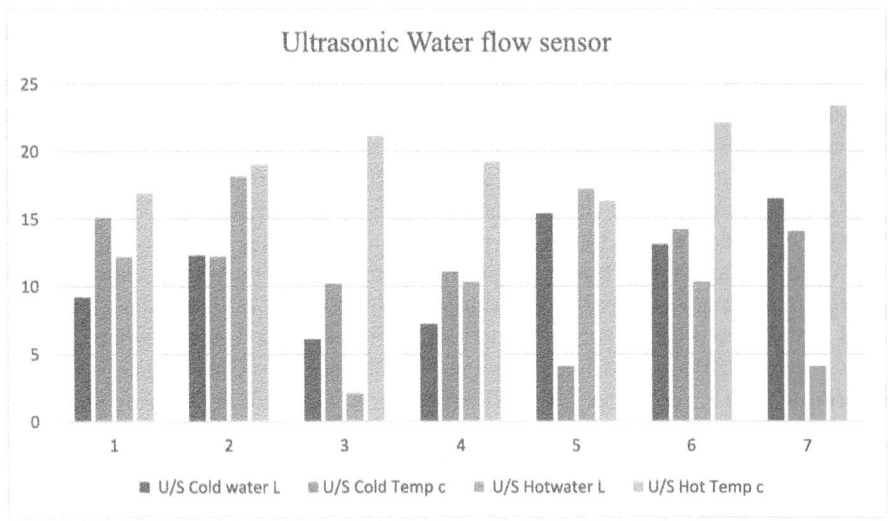

Fig. 8. Visualisation of data obtained from the ultrasonic water flow sensor

capturing data during normal flow conditions. Similarly, the data recorded on January 15, 2024, at 9:30 AM, showcases an increase in both cold and hot water flow rates compared to previous readings, demonstrating the sensor's ability to accurately detect dynamic changes in water flow. Despite the higher flow rates, the temperatures remain relatively constant, indicating the sensor's capability to maintain accuracy across varying flow conditions. Furthermore, the randomly selected instance from March 23, 2023, at 17:45, reveals a disparity between cold and hot water flow rates, with cold water flow reaching its peak while hot water flow significantly decreases. This intricate dynamic, coupled with the elevated hot water temperature, underscores the advanced capabilities of the Ultrasonic Water Flow Sensor in capturing and monitoring fluctuations in water flow rates and temperatures.

There is a discernible trend of consistency and stability across consecutive readings. This stability indicates a commendable level of reliability in the ultrasonic water flow sensors' performance, as they consistently capture flow rates and temperatures with a notable degree of accuracy. Such consistency strengthens the sensors' effectiveness in monitoring water flow systems, providing stakeholders with dependable data for decision-making processes. Further, the data enhances understanding of the water flow system's efficiency and functionality, offering insights that can inform future strategies in water management and resource allocation. By analysing the data's trends and variations, stakeholders leverage informed decision-making frameworks to optimise water usage, enhancing system efficiency. Further, this consideration helps rectify any anomalies in the water flow system. This proactive approach ensures the sustainable utilisation of water resources while maintaining the integrity and reliability of the monitoring system.

4.4 Discussion

By comparing the turbine and ultrasonic water flow sensors data, we can demonstrate consistent performance despite minor variations between the two sensor types. Across the random instances examined, the ultrasonic sensor consistently registers slightly higher values than the turbine sensor, though within negligible margins. This systematic pattern in the discrepancies between the two sensors suggests inherent differences in their measurement methodologies. While the variances are minimal, they underscore the importance of conducting thorough assessments of each sensor type's accuracy and reliability. Such evaluations are essential to guarantee the effectiveness of these sensors in real-world applications where precise measurements are paramount. Understanding the variations in sensor technologies and measurement principles is crucial for optimising their performance and ensuring their suitability for diverse water flow monitoring scenarios. Hence, there is a need for continued research and assessment to enhance the accuracy and reliability of turbine and ultrasonic water flow sensors in practical applications.

At the core of this comparative analysis is an understanding of the fundamental principles guiding the measurement methodologies of each sensor type. This understanding enhances the practical interpretation of data in water flow monitoring applications. Turbine sensors operate by employing mechanical rotation to measure water flow, a process that may result in mechanical degradation over extended periods of use. On the other hand, ultrasonic sensors utilise sound waves for measurement, with their accuracy potentially affected by factors such as the speed of sound in the medium, which can fluctuate with changes in temperature and pressure. These disparities in measurement principles underscore the significance of acknowledging each sensor type's distinct characteristics and potential constraints. Considering these variations, stakeholders make informed decisions when analysing data and implementing strategies to optimise water flow monitoring systems. Thus, this holistic understanding enables business users to address challenges effectively, maximise sensor performance, and enhance the reliability of data-driven insights for efficient water management practices.

Similarly, it is crucial to ensure the accuracy of the sensors through rigorous calibration procedures as they align the data collected with the standardised measures. The turbine and ultrasonic sensors require careful calibration to mitigate potential inaccuracies from sensor drift and environmental influences. As a result, regular calibration checks are essential for preserving accuracy over time and promptly addressing any deviations from anticipated values. Also, environmental conditions such as ambient temperature and pressure significantly impact sensor performance. Turbine sensors are sensitive to alterations in water density caused by temperature shifts, while ultrasonic sensors could face interference from air bubbles or impurities within the water. Recognising and accommodating these environmental factors is essential for optimising sensor accuracy and reliability in real-world applications. The meticulous calibration and accounting for environmental variables through this configuration of the sensors enhances precision and effectiveness during the monitoring process. This consideration ensures the integrity of the collected data, effectively supporting the informed decision-making frameworks in the organisation.

4.5   Limitations

While the experimental approach chosen for this study offers several advantages, there are certain limitations. Firstly, experimental designs may not always fully capture the complexity of real-world scenarios, as they often involve controlled laboratory settings that may not fully represent the dynamics of natural environments. Additionally, manipulating variables in an experimental setting may not always accurately reflect how these variables interact in everyday situations, potentially limiting the generalizability of the findings. Furthermore, there may be additional factors and variables that may influence water consumption patterns that are not fully accounted for in the study design. Hence, this research approach has limitations despite its effectiveness in addressing the specific study aim.

5   Conclusion

The comparative analysis of Turbine and Ultrasonic Water Flow Sensors emphasises the critical role of reliability and precision in sensor measurements for fluid system monitoring applications. Despite the minimal variances observed in litres measured between the two sensor types, the mere existence of discrepancies underscores the necessity for meticulous calibration and environmental considerations. Achieving and maintaining accuracy demands a proactive approach encompassing ongoing monitoring, regular calibration checks, and strict adherence to best practices in sensor installation and maintenance. By implementing these measures, stakeholders can effectively address potential inaccuracies and optimise the performance of water flow monitoring systems, thereby enhancing their overall reliability and efficacy across diverse applications. Organisations can make informed decisions, improve operational efficiency, and ensure the integrity of fluid management practices in various industries by prioritising accuracy and reliability in sensor measurements.

In addition to calibration and maintenance, the choice between Turbine and Ultrasonic sensors should be informed by specific application requirements to maximise effectiveness. Each sensor type presents unique advantages and limitations that necessitate careful evaluation. Parameters such as measurement precision, response time, and compatibility with existing infrastructure are critical factors in determining the most suitable sensor for a given application. Additionally, the selection of communication technology, whether Wi-Fi for Turbine sensors or LoRaWAN for Ultrasonic sensors, introduces further considerations such as range, power consumption, and network compatibility. Making well-informed sensor type and communication technology decisions is paramount for optimising system performance and ensuring seamless integration into existing monitoring frameworks. By carefully weighing these factors, stakeholders can ensure that the chosen sensors meet the precise needs of their application, thereby enhancing operational efficiency and facilitating reliable data collection for informed decision-making in various industries.

Finally, turbine and ultrasonic water flow sensors play pivotal roles in fluid system monitoring across diverse industries, from HVAC to environmental science. Their capacity to deliver reliable data empowers stakeholders with valuable insights into fluid dynamics and system behaviour, facilitating informed decision-making and resource optimisation. By acknowledging the variances between sensor types and placing a premium on measurement accuracy, organisations can unlock the full potential of water flow monitoring systems, thereby fostering efficiency and sustainability in fluid management practices. As technological advancements continue to enhance sensor capabilities and data processing techniques, the future holds promising prospects for further refining fluid monitoring methodologies and driving continuous improvement in fluid system management across various sectors.

References

Ahmed, G.: Improving IoT privacy, data protection and security concerns. Int. J. Technol. Innov. Manag. **1**(1), 19–34 (2021)

Anjana, S., Sahana, M.N., Ankith, S., Natarajan, K., Shobha, K.R., Paventhan, A.: An IoT-based 6LoWPAN-enabled experiment for water management. In: 2015 IEEE International Conference on Advanced Networks and Telecommunications Systems, pp. 1–6. IEEE, December 2015

Cazcarro, I., Hoekstra, A.Y., Chóliz, J.S.: The water footprint of tourism in Spain. Tour. Manag. **40**, 90–101 (2014)

da Silva, T.B., dos Santos Chaib, R.P., Arismar, C.S., da Rosa Righi, R., Alberti, A.M.: Toward future Internet of Things experimentation and evaluation. IEEE Internet Things J. **9**(11), 8469–8484 (2021)

He, W., Iqbal, M.T.: Power consumption minimisation of a low-cost IoT data logger for photovoltaic system. J. Electron. Electr. Eng. **2**(2), 241–261 (2023)

Jamaluddin, A., Harjunowibowo, D., Rahardjo, D.T., Adhitama, E., Hadi, S.: Wireless water flow monitoring based on Android smartphone. In: 2016 2nd International Conference of Industrial, Mechanical, Electrical, and Chemical Engineering, pp. 243–247. IEEE, October 2016

Muangprathub, J., Boonnam, N., Kajornkasirat, S., Lekbangpong, N., Wanichsombat, A., Nillaor, P.: IoT and agriculture data analysis for the smart farm. Comput. Electron. Agric. **156**, 467–474 (2019)

Mukta, M., Islam, S., Barman, S.D., Reza, A.W., Khan, M.S.H.: IoT-based smart water quality monitoring system. In: 2019 IEEE 4th International Conference on Computer and Communication Systems, pp. 669–673. IEEE, February 2019

Piechota, P., Synowiec, P., Andruszkiewicz, A., Wędrychowicz, W., Wróblewska, E., Mrowiec, A.: Experimental determination influence of flow disturbances behind the knife gate valve on the indications of the ultrasonic flow meter with clamp-on sensors on pipelines. Sensors **23**(10), 1–28 (2023)

Prasad, A.N., Mamun, K.A., Islam, F.R., Haqva, H.: Smart water quality monitoring system. In: 2015 2nd Asia-Pacific World Congress on Computer Science and Engineering, pp. 1–6. IEEE, December 2015

Rajurkar, C., Prabaharan, S.R.S., Muthulakshmi, S.: IoT-based water management. In: 2017 International Conference on Nextgen Electronic Technologies: Silicon to Software, pp. 255–259. IEEE, March 2017

Ramadhan, A.J.: Smart water-quality monitoring system based on enabled real-time Internet of Things. J. Eng. Sci. Technol. **15**(6), 3514–3527 (2020)

Ramli, I.A., Hanafi, D.: Water flow, temperature, and PH monitoring based on the Internet of Things (IoT). Evol. Electr. Electron. Eng. **3**(2), 793–800 (2022)

Rantanen, P., Mäkivaara, J., Saari, M., Sillberg, P., Jaakkola, H.: Utilising cost-effective NB-IoT-based sensors for detecting water temperature and flow. In: 2021 IEEE 25th International Conference on Intelligent Engineering Systems (INES), pp. 000165–000170. IEEE, July 2021

Toyosada, K., Otani, T., Shimizu, Y.: Water use patterns in Vietnamese hotels: modeling toilet and shower usage. Water **8**(3), 1–12 (2016)

Zulkifli, C.Z., et al.: IoT-based water monitoring systems: a systematic review. Water **14**(22), 1–29 (2022)

Atmospheric River Detection - A Survey on Deep Learning and Quantum Neural Networks

S. Sivachitralakshmi[1]($\boxtimes$) (iD), P. Chitra[1], and S. Bharathi[2] (iD)

[1] Department of Computer Science and Engineering (Emerging Technologies), SRM Institute of Science and Technology, Chennai, India
shivachitra2k8@gmail.com, Chitrap1@srmist.edu.in
[2] Department of Computer Science and Engineering, SRM Institute of Science and Technology, Chennai, India

Abstract. Atmospheric rivers (ARs) represent narrow corridors that facilitate the predominant poleward transportation of water vapor in the midlatitudes. These corridors exhibit notable features such as elevated water vapor levels and robust lower-level winds, playing a role in the expansive warm conveyor belt associated with extratropical cyclones. The meridional movement of water vapor within ARs holds significant importance for water reserves, yet their interaction with mountainous regions can lead to severe flooding events. Quantum neural networks are an emerging field combining quantum computing with artificial neural networks. The idea is that the computational advantage of quantum computing could potentially improve the performance of neural networks including those used for the complex task of atmospheric river detection.

Keywords: Atmospheric Rivers · Flood · Mid latitudes · Artificial intelligence · Convolutional Neural Networks · Quantum Neural Networks

1 Introduction

Atmospheric researchers must analyze and investigate both climatological and meteorological components of moisture movement within the atmosphere (Gimeno et al. 2012; Gimeno 2013) [1, 2]. It is imperative to scrutinize the conceptual frameworks of moisture transportation to facilitate investigations into the source of continental precipitation. The concept of atmospheric rivers (AR) plays a pivotal role in the examination of water vapor conveyance in extratropical zones. Within this concise examination, we outline the fundamental attributes of ARs, which play a significant role in conveying substantial quantities of water through relatively narrow "corridors" across mid-latitudes towards higher latitudes.

1.1 Understanding Quantum Neural Networks

1. **Quantum Bits (Qubits):** Unlike classical bits, qubits can exist in multiple states simultaneously (superposition) and be entangled with other qubits, providing a richer and more dynamic form of computation.

P. D. Sivakumar et al. (Eds.): IRCCTSD 2024, CCIS 2362, pp. 100–109, 2025.
https://doi.org/10.1007/978-3-031-82386-2_8

2. **Quantum Gates:** Analogous to neural network activation functions, quantum gates manipulate the probabilities of qubit states, allowing for complex transformations.
3. **Quantum Entanglement:** This property can be used to represent complex correlations between data points, potentially capturing the multifaceted dynamics of atmospheric data.
4. **Hybrid Models:** Most practical implementations of QNNs currently are hybrid, combining quantum processing with classical neural networks to leverage the strengths of both technologies.

1.2 Observation of Atmospheric Rivers

A standard AR is situated within the warm conveyor belt located in the pre-cold-frontal sector of an extratropical cyclone. Its characteristics consist of a concentrated strip of heightened low-level specific humidity, where frontal convergence induces the air to rise, leading to the vertical extension of this band of intensified humidity. Additionally, there exists a pre-cold-frontal low-level jet attributed to the temperature contrast along the cold front, a result of the thermal wind relationship. Moreover, the vertical profile of the equivalent potential temperature displays a moist-neutral stratification within the AR area, low-level potential instability on the cold side of the front, and a region of subsidence at lower levels preceding the AR, linked to the dry cap atop the trade wind inversion. The moisture content in an AR emanates from two primary sources:

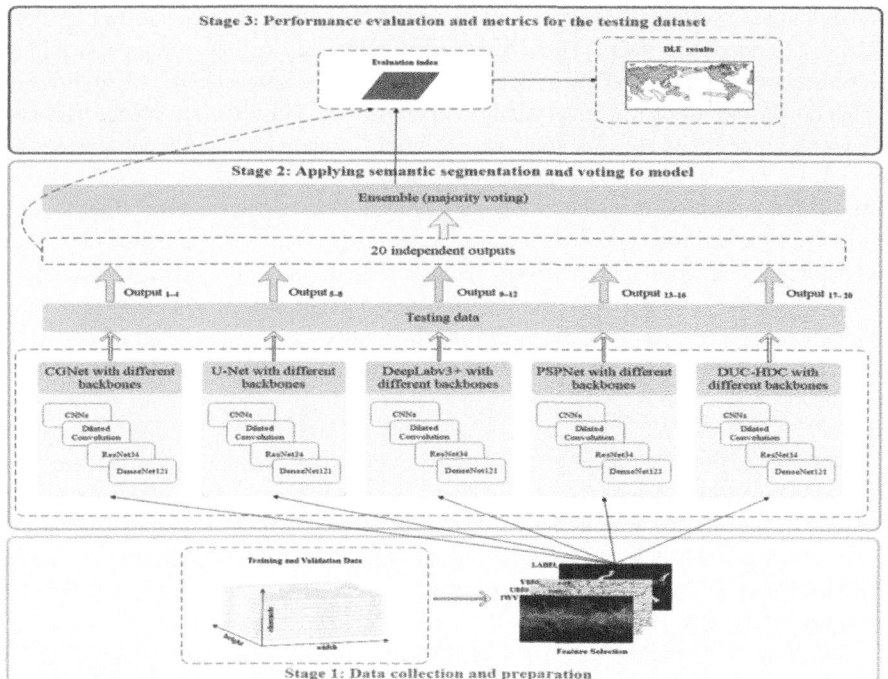

Fig. 1. Three stages of Atmospheric river detection using DLE

local moisture convergence along the cold front of the extratropical cyclone, and direct transport of tropical moisture towards higher latitudes. The diverse precipitation patterns witnessed in an AR occurrence encompass "cold" precipitation (chiefly originating from ice particles above the freezing level in more northern latitudes), "warm" precipitation (characterized by minimal ice presence in precipitation at temperatures surpassing the freezing level), and mixed precipitation zones exhibiting a combination of warm and cold rainfall. These three precipitation regimes manifest with comparable frequency on average. The schematic representation in Fig. 1 delineates the typical configuration and dimensions of an Atmospheric River.

1.3 Advantages in Atmospheric River Detection

- Handling Complexity: QNNs can theoretically model complex phenomena involving numerous variables and nonlinear interactions, common in atmospheric sciences.
- Speed and Scalability: Quantum computing promises exponential speed-ups in certain computations, potentially accelerating the processing of large climatic datasets.
- Enhanced Pattern Recognition: The quantum-enhanced feature space exploration could reveal new patterns or correlations obscured in classical-only approaches.

2 Literature Survey

Most of the water vapor transported meridionally across the mid-latitudes (>90% of the total midlatitude vertically integrated water vapor flux) occurs through narrow corridors known as atmospheric rivers. These temporary filamentary regions are present within the warm conveyor belt (WCB) of extratropical cyclones, characterized by high water vapor content and strong low-level winds (a low-level jet). While their existence has been known for some time [1], we commence our brief review with the influential study by Newell et al. (1992), who described these elongated (about 2000 km), narrow (about 300–500 km wide) bands of intensified water vapor flux as "tropospheric rivers" (Fig. 1). The term "rivers" was chosen due to their capacity to transport water at volumetric flow rates comparable to the world's major rivers. However, the term "tropospheric" is no longer widely used, and "atmospheric rivers" (ARs, Zhu and Newell 1998) is now preferred, although there is ongoing debate regarding the suitability of this terminology, with alternate suggestions such as "tropical moisture exports" or "moisture conveyor belts" [2]. The latter term draws on the analogy with the conveyor belt concept of extratropical cyclones, where the WCB is a broader element of extratropical cyclones crucial for transporting sensible and latent heat polewards, counterbalancing other components of the cyclone that move relatively cool and dry air equatorwards. ARs are also known by informal names like "Hawaiian fire hose" or "Pineapple Express" [3] terms commonly used by forecasters to describe ARs linking tropical moisture near the Hawaiian Islands to the west coast of North America; in the central United States, ARs have been dubbed the "Maya Express".

At any given moment, there are typically three-to-five significant conduits present in each Hemisphere, each responsible for the transportation of substantial quantities of water vapor through narrow streams across the midlatitudes [4].

Very little is known about the relationship between the presence and activity of ARs and large-scale ocean-atmosphere interactions. [5] used a single case study—the high-impact AR landfall in the Pacific Northwest in March 2005—to discover that a series of distinct planetary-scale phenomena—(1) a convective signal initially forming over the tropical Indian Ocean; (2) eastward-propagating Kelvin waves in the tropics; (3) a wave packet (EWP) propagating from western Asia to the Pacific in the extratropics, amplifying ridges and troughs in the eastern Pacific; and (4) deep penetration into the tropics by the EWP, which, in conjunction with the Kelvin waves, favored the uptake of tropical moisture by the AR.

[6, 7] found that during the negative phase of Arctic Oscillation (AO) and Pacific/North American (PNA) more winter ARs occurred in California for the period 1998–2011. The estimated increase in ARs during the negative and positive phases (as opposed to their opposite phases) was 90% for the AO and 50% for the PNA. Studies of ARs impacting Britain found a significant negative relationship between winter ARs and the Scandinavian Pattern [8]. In subsequent work covering Western Europe, [9] found that the North Atlantic Oscillation (NAO) affected AR activity in different parts of Europe; in southern Europe ARs are concurrent with negative NAO phases, whereas in northern Europe a more positive NAO-type pattern is associated with AR occurrence.

Despite the benefits that DL methodologies offer, the challenge remains in rectifying errors in imbalanced datasets, referred to as the asymmetry challenge. This issue arises in scenarios where one or more classes in a binary or multi-classification problem have significantly fewer samples than other classes. In order to mitigate these errors and uncertainties, various ensemble techniques have been suggested in the literature [10]. These ensemble approaches involve the integration of multiple DL algorithms using specific strategies [11]. Their effectiveness has been demonstrated in fields like geophysics, such as ozone forecasting [12, 13] and disaster prediction [14]. Nonetheless, the application of ensemble techniques incorporating DL algorithms in the realm of extreme weather and climate studies, particularly in the context of tropical cyclones, extratropical cyclones, and ARs, remains limited. Building upon prior research, the present investigation introduces a DL-based ensemble methodology for the identification of ARs.

3 Methodology

Atmospheric rivers (ARs) play a crucial role in extreme precipitation events globally, including India. Utilizing machine learning for AR tracking has shown promise in enhancing prediction accuracy and computational efficiency. Recent research highlights the application of deep learning methods, such as DeepLabv3+, in identifying ARs with high performance and reduced computational costs. These methods offer a novel approach to tracking ARs, especially in regions like India where precipitation prediction is challenging. By leveraging machine learning techniques, researchers can overcome computational limitations associated with traditional AR identification methods, providing valuable insights into the behavior and impact of ARs in India's atmospheric dynamics.

3.1 Numerical Methodology

Various methodologies have been developed to detect Atmospheric Rivers (ARs) from either the integrated water vapor (IWV) or integrated water vapor transport (IVT) field, given their significance to the global water cycle and regional precipitation. These methods exhibit different approaches and utilize varying thresholds, often determined subjectively, leading to diverse outcomes even within the same geographical area. For example, Prabhat et al. (2012) introduced the Toolkit for Extreme Climate Analysis, which identifies global ARs by highlighting mesh points where IWV exceeds 2 cm. Guan and Waliser (2015) proposed a global AR detection approach based on an IVT threshold set at the 85th percentile or 100 kg m^{-1} s^{-1}, choosing the greater value between the two. The research spanned from 1997 to 2014, with data collected at 6-h intervals. Pan and Lu (2019) characterized summer ARs in East Asia by employing dual thresholds: the 80th percentile of local IVT and the 85th percentile of spatially-smoothed IVT. Mundhenk et al. (2016) identified ARs in the North Pacific region using a minimum anomalous IVT threshold of 250 kg m^{-1} s^{-1}. These methodologies heavily rely on the subjective determination of thresholds, as highlighted by Muszynski et al. (2019) and Rutz et al. (2019).

3.2 Deep Learning Method

An alternative approach involves the utilization of a machine learning model, particularly a deep learning (DL) technique. In contrast to the previously mentioned approaches, DL techniques have the capacity to discern a broad array of patterns from various variables, thus eliminating subjective thresholds on the geophysical parameters (Muszynski et al. 2019). Several DL techniques have been employed in the identification of Atmospheric Rivers (ARs). For example, convolutional neural networks and support vector machines have been utilized for the classification of severe weather events (Liu et al. 2016; Muszynski et al. 2019). Racah et al. (2016) introduced a semi-supervised approach that integrates temporal and unlabeled data to detect extreme weather patterns in situations where data labeling is incomplete. Prabhat et al. (2021) successfully detected ARs through the implementation of a semantic segmentation algorithm.

3.3 Atmospheric River Pattern Recognition Method

Step 1: In this step, topology techniques are applied and TDA methods to automatically extract pertinent topological aspects from complicated climate data, such as reanalysis products and the output of climate models. The algorithm TDA draws topological feature descriptors of weather patterns, i.e., characteristics of atmospheric rivers (ARs) and non-atmospheric rivers (non-ARs), in a threshold-free manner, from the Union-Find data structure (Hopcroft and Ullman 1973; Tarjan 1975). These topological feature descriptors, which are derived from global picture snapshots on a latitude-longitude grid, are referred to as linked areas (Edelsbrunner and Harer 2010). The ML classifier in Stage 2 receives 10 topological feature descriptors as input.

Step 2: In this step, topology t· Step 2: Using the Support Vector Machine (SVM) ML classifier, a binary classification job is carried out in this stage (Cortes and Vapnik,

1995; Chang and Lin, 2011). There are two phases in the categorization task: i) Using the created SVM model to test the model's ability to split events into two categories (i.e., ARs and non-ARs) based on unlabeled descriptors; ii) training the classifier to identify ARs from other weather events in the pictures. Topological feature descriptors (from Stage 1) supplied by TECA (Prabhat et al. 2015) are used in the training procedure. The performance of the classifier is assessed in terms of sensitivity, accuracy, and precision.

4 Quantum Neural Network in AR

Quantum neural networks are neural network models that use the concepts of quantum mechanics for computing. In 1995, Subhash Kak and Ron Chrisley separately presented the first concepts of quantum brain computing. These concepts were related to the notion of quantum mind, which suggests that cognitive function involves the influence of quantum phenomena. Conventional research in quantum neural networks often entails merging classical artificial neural network models, commonly used in machine learning for pattern recognition, with the benefits of quantum information to create more effective algorithms.

4.1 Quantum Classification

Development of quantum neural networks is warranted for several reasons. To begin with, there is a chance that quantum computers may surpass traditional computers in a number of areas: When compared to the most well-known classical methods, some quantum resources, like contextuality and quantum nonlocality, have been shown to provide unconditional quantum advantages in the resolution of specific computational issues [22, 23]. Additionally, some quantum Fourier transform based algorithms, like Shor's factoring algorithm [21], provide exponential speedups. In the era of big data, these intriguing findings encourage more research into the possible benefits of QNN models.

Using the provided dataset, a quantum classification model is expected to automatically and very probabilistically assign the proper labels to the characteristics. Additionally, the classifier should be able to generalize effectively to some unknown data after training.

4.2 Variational Quantum Circuits

A common way to portray quantum neural networks is as parameterized quantum circuits, in which certain quantum gates' rotation angles may be used to represent variational parameters. Figure 1 shows the fundamental framework, which is mostly composed of the quantum circuit ansatz, the traditional optimization approach and the cost function ansatz. Commonly utilized options for the fundamental building blocks include Controlled-NOT gates, Controlled-Z gates, and parameterized single-qubit rotation gates (Rx (θ), Ry (θ), and Rz (θ)), as shown below (Fig. 2).

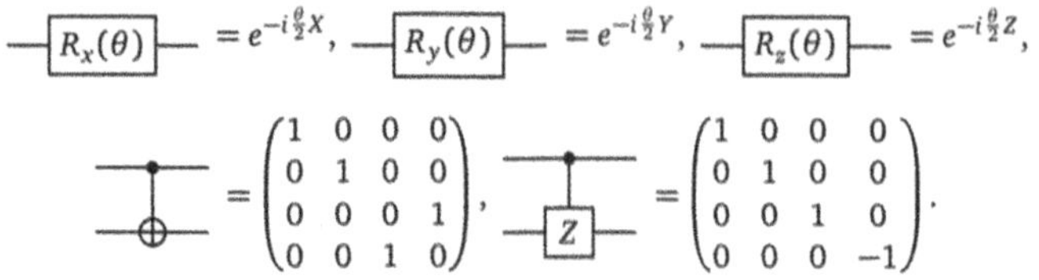

$$\boxed{R_x(\theta)} = e^{-i\frac{\theta}{2}X}, \quad \boxed{R_y(\theta)} = e^{-i\frac{\theta}{2}Y}, \quad \boxed{R_z(\theta)} = e^{-i\frac{\theta}{2}Z},$$

$$= \begin{pmatrix} 1 & 0 & 0 & 0 \\ 0 & 1 & 0 & 0 \\ 0 & 0 & 0 & 1 \\ 0 & 0 & 1 & 0 \end{pmatrix}, \qquad \boxed{Z} = \begin{pmatrix} 1 & 0 & 0 & 0 \\ 0 & 1 & 0 & 0 \\ 0 & 0 & 1 & 0 \\ 0 & 0 & 0 & -1 \end{pmatrix}.$$

Fig. 2. Single qubit rotation

4.3 Optimization Using QNN

Simply comparing the likelihood of various measurement results allows us to assign the input the label with the highest probability during the prediction phase. We must optimize the trainable parameters in order to provide desired predictions in order to accomplish this aim. Generally speaking, the first step in determining the distance between the present output and the desired output is to create a cost function. The cross entropy (CE) and mean square error (MSE) are common cost functions:

QNN can provide better classification for large datasets, optimisation technique can provide with better accuracy in prediction by using multiple features as parameters. Applying Quantum Neural Networks on detecting atmospheric river potentially increases the rate of detection than the existing neural network models.

5 Atmospheric Rivers and Flooding in India

According to data, there is a significant amount of geographical heterogeneity in the moisture transport associated with AR over India. Specifically, the peninsular India and the eastern portion of the Indo-Gangetic plain have more IVT (>500 kg m^{-1} s^{-1}) than the other locations. The lower Indo-Gangetic plain and peninsular India has more AR occurrences during the summer monsoon season, which is consistent with comparatively greater IVT [15–20]. Peninsular India is affected by ARs mostly in July and August, whereas northern India is impacted by ARs throughout the summer monsoon season.

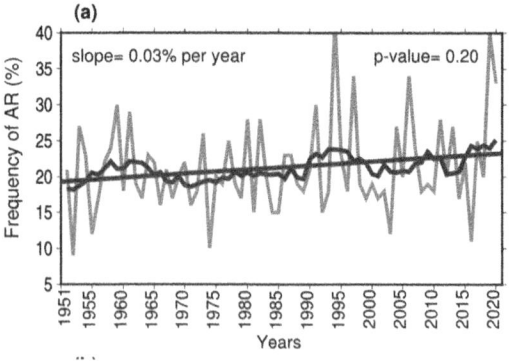

Fig. 3. Trend in frequency of occurance of ARduring 1951–2020

The mean IVT anomaly (40-120E, 10S-25N) during the summer monsoon season has significantly (p-value = 0.0001) increased (Fig. 1b). The increase in IVT during the summer monsoon season can be partially associated with the warming climate that accelerates anomalous moisture uptake10, leading to the rising trend in the frequency of ARs. Results show that the Arabian Sea, northern India, and peninsular India experienced a significant increase in the AR frequency during the summer monsoon season (Fig. 4) (Fig. 3 and Fig. 5).

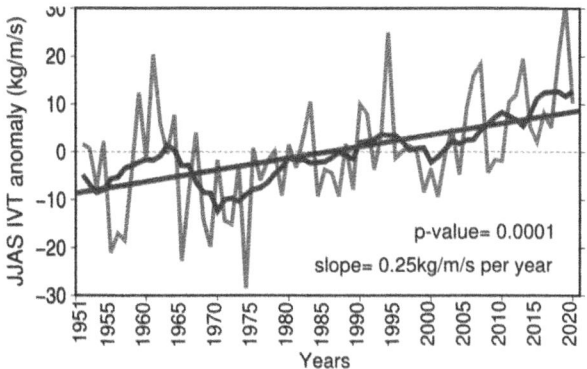

Fig. 4. Trend in summer monsoon season IVT anomaly

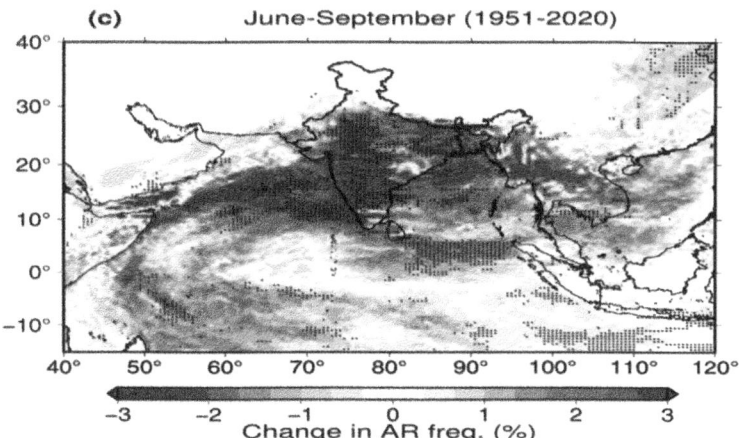

Fig. 5. Long term (1951–2020) change in frequency of AR's in India

6 Conclusion

Our study focused on landfalling ARs during the summer monsoon, which has significant effects on India's water supplies and flood threats because to their ability to produce massive volumes of precipitation in a matter of hours 9, 13, 31. The ability of ARs to store more moisture is rising due to climate change10, 28, 61, raising fears about more catastrophic floods in the future. The summer monsoon season in India, which spans the Indian subcontinent from west to east, is when the ARs mostly occur. Between 1951 and 2020, there were 574 ARs in the summer monsoon season; the frequency of ARs increased with time. Furthermore, floods have been induced in India in roughly 80% of the most severe ARs (top 1/3rd AR occurrences, Supplemental Data 1) over the past 20 years. With the use of Machine Learning, Deep Learning and Quantum computing the prediction of ARs will not be humongous Task. The modern day heavily depend upon HPCs for weather forecasting, with the invent of quantum computing and Artificial intelligence AR forecasting can be done at the ease, reducing damage due to Floods and Droughts in india.

7 Future Work

- Noise and Error Rates: Current quantum computers are noisy and prone to errors, which can limit the reliability of QNN outputs.
- Hardware Limitations: The number of qubits and the quality of quantum gates currently available may not be sufficient for large-scale applications like comprehensive climate modeling.
- Algorithmic Development: Developing robust quantum algorithms for complex tasks like atmospheric river detection is still an active area of research.

References

1. Gimeno, L.: Grand challenges in atmospheric science. Front. Earth Sci. **1**, 1 (2013). https://doi.org/10.3389/feart.2013.00001
2. Gimeno, L., et al.: Oceanic and terrestrial sources of continental precipitation. Rev. Geophys. **50**, RG4003 (2012). https://doi.org/10.1029/2012RG00038
3. Namias, J.: The use of isentropic analysis in short term forecasting. J. Aeronaut. Sci. **6**, 295–298 (1939). https://doi.org/10.2514/8.86
4. Neiman, P.J., Ralph, F.M., Wick, G.A., Lundquist, J.D., Dettinger, M.D.: Meteorological characteristics and overland precipitation impacts of atmospheric rivers affecting the West Coast of North America based on eight years of SSM/I satellite observations. J. Hydrometeorol. **9**, 22–47 (2008). https://doi.org/10.1175/2007JHM855.1
5. Neiman, P.J., Schick, L.J., Ralph, F.M., Hughes, M., Wick, G.A.: Flooding in Western Washington: the connection to atmosphericrivers. J. Hydrometeorol. **12**, 1337–1358 (2011). https://doi.org/10.1175/2011JHM1358.1
6. Ralph, F.M., Coleman, T., Neiman, P.J., Zamora, R.J., Dettinger, M.D.: Observed impacts of duration and seasonality of atmospheric-river landfalls on soil moisture and runoff in coastal northern California. J. Hydrometeorol. **14**, 443–459 (2013). https://doi.org/10.1175/JHM-D-12-076.1
7. Ralph, F.M., Intrieri, J., Andra, D., Jr., Atlas, R., Boukabara, S., Bright, D., et al.: The emergence of weather-focused testbeds linking research and fore-casting operations. Bull. Amer. Meteor. Soc. **94**, 1187–1211 (2013). https://doi.org/10.1175/BAMS-D-12-00080.1
8. Ralph, F.M., Dettinger, M.D.: Storms, floods, and the science of atmospheric rivers. Eos **92**, 265–266 (2011). https://doi.org/10.1029/2011EO320001

9. Ralph, F.M., Neiman, P.J., Kiladis, G.N., Weickman, K., Reynolds, D.W.: A multi-scale observational case study of a Pacific atmospheric river exhibiting tropical–extratropical connections and a mesoscale frontal wave. Mon. Weather Rev. **139**, 1169–1189 (2011). https://doi.org/10.1175/2010MWR3596

10. Ralph, F.M., Neiman, P.J., Wick, G.A., Gutman, S.I., Dettinger, M.D., Cayan, D.R., et al.: Flooding on California's Russian river: role of atmospheric rivers (2006)

11. Long, J., Shelhamer, E., Darrell, T.: Convolutional networks for images, speech, and time-series. In: Arbib, M.A. (ed.) The Handbook of Brain Theory and Neural Networks, pp. 255–258. MIT Press, Cambridge (2015)

12. An introduction to convolutional neural networks. arXiv preprint arXiv:1511.08458 (2015)

13. Wang, P., Chen, P., Yuan, Y., Liu, D., Huang, Z., Hou, X., Cottrell, G.: Understanding convolution for semantic segmentation. In: 2018 IEEE Winter Conference on Applications of Computer Vision (WACV), pp. 1451–1460. IEEE (2018). https://doi.org/10.1109/WACV.2018.00163

14. Wu, T., Tang, S., Zhang, R., Cao, J., Zhang, Y.: CGNet: a light-weight context guided network for semantic segmentation. IEEE Trans. Image Process. **30**, 1169–1179 (2020). https://doi.org/10.1109/TIP.2020.3042065

15. Pattanaik, D.R., Rajeevan, M.: Variability of extreme rainfall events over India during southwest monsoon season. Meteorol. Appl. **17**, 88–104 (2010)

16. Martha, T.R., et al.: Landslides triggered by the June 2013 extreme rainfall event in parts of Uttarakhand state, India. Landslides **12**, 135–146 (2015)

17. Srivastava, A.K., Guhathakurta, P.: Climate Diagnostics Bulletin of India (2013). http://www.indiaenvironmentportal.org.in/content/446964/climatediagnostics-bulletin-of-india/

18. Ray, P.K., et al.: Kedarnath disaster 2013: causes and consequences using remote sensing inputs. Nat. Hazards 81, 227–243 (2016)

19. Hunt, K.M.R., Menon, A.: The 2018 Kerala floods: a climate change perspective. Clim. Dyn. **54**, 2433–2446 (2020)

20. Lyngwa, R.V., Nayak, M.A.: Atmospheric river linked to extreme rainfall events over Kerala in August 2018. Atmos. Res. **253**, 105488 (2021). ISSN 0169-8095. https://doi.org/10.1016/j.atmosres.2021.105488

21. Shor, P.W.: Polynomial-time algorithms for prime factorization and discrete logarithms on a quantum computer. SIAM Rev. **41**(2), 303–332 (1999). https://doi.org/10.1137/S0036144598347011

22. Aïmeur, E., Brassard, G., Gambs, S.: Machine learning in a quantum world. In: Proceedings of the 19th Canadian Conference on Artificial Intelligence (Canadian AI'06), pp. 433–444 (2006)

23. Bernstein, E., Vazirani, U.: Quantum complexity theory. In: Proceedings of the 25th Annual ACM Symposium on Theory of Computing (STOC'93), pp. 11–20 (1993)

Inclusive Communication Techniques

Emotion Recognition Using Text in a Chatbot Using Logistic Regression

Hemanth Arora, B. Darshan, Bhavya Jain, and S. K. B. Sangeetha[✉]

Department of Computer Science and Engineering (Emerging Technologies), SRM Institute of Science and Technology, Vadapalani Campus, Chennai, India
{ha3968,bb4499,bj9082,sangeets8}@srmist.edu.in

Abstract. Emotion recognition in text plays a crucial role in enhancing the effectiveness of chatbots by enabling them to better understand and respond to users' feelings. This work focuses on using a popular machine learning algorithm called logistic regression to categorize emotions from textual inputs. Through the use of labeled data with examples of different emotional expressions, the chatbot is able to predict the user's emotional state during interactions with accuracy. The logistic regression model evaluates the text input to assign a probability distribution over various emotions by utilizing features like word frequency, sentiment analysis, and linguistic patterns. This makes it possible for the chatbot to adjust its responses appropriately, resulting in more sympathetic and customized exchanges. Through this approach, the chatbot can exhibit a higher level of emotional intelligence, thereby improving user satisfaction and overall user experience.

Keywords: Emotion recognition · text analysis · chatbot · logistic regression · machine learning · user interaction

1 Introduction

A state-of-the-art use of machine learning and natural language processing (NLP) methods in chatbots is emotion recognition through text. Its goal is to recognize and extract emotional content from textual inputs so that chatbots can respond with greater empathy and personalization. Logistic regression's simplicity, efficiency, and interpretability make it a fundamental algorithm for developing strong emotion recognition systems. Logistic regression, a popular statistical technique for binary classification tasks, is well-suited for emotion recognition. It uses input features that are taken from text data to model the likelihood of a particular emotion class. Linguistic cues, sentiment scores, word frequencies, contextual data, and other patterns suggestive of various emotional states are a few examples of these features [11].

Logistic regression gains an understanding of the association between these features and emotions through training. This makes it possible for the chatbot to categorize text inputs into pre-established emotion groups like joyful, depressed, angry, or excited. Additionally, logistic regression offers a distinct decision boundary that facilitates precise forecasting and sheds light on the significance of particular characteristics in identifying

P. D. Sivakumar et al. (Eds.): IRCCTSD 2024, CCIS 2362, pp. 113–123, 2025.
https://doi.org/10.1007/978-3-031-82386-2_9

emotions, improving interpretability and transparency. There are various advantages when emotion recognition-enabled chatbots incorporate logistic regression. It enhances conversational experiences through empathetic interactions, improves user engagement by customizing responses to users' emotional states, makes adaptive response generation easier, and allows targeted interventions based on detected emotions [12].

In real-world settings, logistic regression-powered chatbots can be extremely useful for social companionship, education, mental health support, and customer service. These chatbots create stronger bonds with users, develop emotional intelligence, and raise user satisfaction levels by correctly recognizing and reacting to user emotions. Overall, the use of text and logistic regression for emotion recognition marks a significant breakthrough in conversational agent capabilities, allowing them to negotiate challenging emotional environments and provide more meaningful user interactions across a range of domains [13].

There are a number of drawbacks to the current system for text-based emotion recognition in chatbots that uses logistic regression. First and foremost, the linear nature of the logistic regression model limits its ability to represent more intricate and non-linear relationships between textual data and emotions. This may lead to decreased efficacy and accuracy in picking up on minute details and variances in emotions conveyed through text. Furthermore, in the context of natural language processing and emotion recognition, it may not always be the case that the relationship between input features (text data) and output labels (emotions) is independent of one another. This is the assumption made by logistic regression [14].

Additionally, noise and outliers in the training set can cause overfitting or under-fitting problems with Logistic Regression, which further reduces the model's ability to predict emotions accurately. Another drawback of logistic regression is that its results are difficult to interpret and explain, which makes it difficult for users and developers to comprehend how the model arrived at its predictions for various emotions. Additionally, imbalanced datasets—where some emotions are over- or underrepresented—may be difficult for logistic regression to handle, which could result in biased predictions and poor performance overall.

Finally, a challenge with the scalability of Logistic Regression for emotion recognition in large datasets is that a large volume of text data may be difficult for the model to process and analyze effectively. All things considered, the shortcomings of Logistic Regression when it comes to text-based emotion recognition in chatbots underscore the necessity for more sophisticated, non-linear models and methodologies to enhance precision, resilience, and efficiency in identifying intricate emotional cues conveyed via text.

The main contributions are

1 To establish more meaningful and interesting interactions by recognizing and reacting to user emotions in real time
2 To respond to users in a sympathetic manner by understanding their emotions and experiences.

Chatbots can improve the quality of conversations by fostering a supportive and understanding environment by accurately recognizing and responding to user emotions. This goal is crucial to achieving the intended results of the interaction—whether that be helping, giving advice, or just having a casual conversation—as well as building rapport and trust between users and chatbots. In the end, enhancing conversational experiences helps to foster closer bonds and increase sustained user satisfaction. With this detailed introduction, Sect. 2 explains the related work, Sect. 3 describes the system methodology, Sect. 4 depicts the experimental results followed by conclusion in Sect. 5.

2 Related Study

Emotional intelligence has been incorporated into chatbots in a number of studies to improve mental health support and intervention. In an effort to provide users with responses that are specifically tailored to their emotional needs, researchers have used machine learning and logistic regression techniques for emotion recognition in both Polish and English. These initiatives highlight how crucial cross-linguistic emotion recognition is for a variety of user bases. Through emotion simulation techniques, chatbots have been specifically designed to detect depression, support students' mental health, and mimic empathy. Users who are experiencing emotional distress can receive comprehensive support when emotion recognition and solution recommendation systems are integrated.

A conversational interface for mental health assistance and diagnosis that uses a logistic regression model for sentiment and emotion analysis [1]. Their work concentrated on helping people with mental health issues through conversational interactions, in which the ability to recognize emotions was essential to comprehending users' emotional states and offering suitable support. In order to enable the chatbot to provide users with personalized responses and interventions based on their emotional needs, they sought to accurately classify emotions expressed in text inputs by utilizing logistic regression.

For a therapeutic chatbot, text-based emotion recognition in Polish and English is presented [2]. Their research aided in the creation of chatbots that can effectively identify users' emotions from textual inputs and provide therapeutic interventions. Through their research, they were able to demonstrate how crucial cross-linguistic emotion recognition is to serving a variety of user demographics and enhancing the availability of mental health support services.

A chatbot created to use emotion recognition to support students' mental health [3]. A review is done on machine learning and emotional artificial intelligence's use in depression detection [5]. Their research shed light on the possibility of using machine learning and emotional AI to detect and treat depression-related problems early on. Through a comprehensive analysis of current literature, they brought attention to how emotion recognition technologies can complement conventional diagnostic methods and enable prompt interventions for individuals who may be at risk of depression.

Their chatbot attempted to provide students experiencing emotional difficulties with individualized support by utilizing machine learning techniques for emotion recognition. Their work contributed to the development of proactive approaches to student well-being within educational settings by addressing the growing need for mental health interventions that are specific to the emotional experiences of individual students.

Machine learning techniques for human conversational emotion recognition are investigated [4]. Their work clarified how machine learning can be applied to comprehend emotional cues present in human interactions, which is crucial for creating chatbots that are sympathetic. They laid the foundation for improving the emotional intelligence of conversational agents by analyzing conversational data and finding patterns suggestive of various emotional states.

LENNA, an empathetic chatbot built to mimic emotions and analyze conversations [6]. Their research helped to clarify how emotion simulation can be used by chatbots to simulate empathy and improve user experience during dialogue. They showed that it is possible to use emotion simulation techniques in chatbot design to make conversational agents that are more engaging and sensitive to emotions through LENNA.

A novel chatbot for emotion recognition and solution suggestion is introduced [7]. Their research was centered on creating a chatbot that could identify the emotions of its users and offer suggestions or suitable solutions to effectively address their emotional needs. They sought to provide comprehensive support to users going through emotional distress by fusing emotion recognition with recommendation systems for solutions, thus increasing the overall efficacy of chatbot interventions.

There is discussion about the acceptability of a chatbot with emotional intelligence that uses machine learning algorithms [8]. Their research shed light on users' acceptance of emotionally intelligent chatbots, which is important for creating conversational systems that are both intuitive to use and efficient. Through an assessment of user attitudes and perceptions regarding emotionally intelligent chatbots, they were able to pinpoint acceptance factors and suggest ways to improve user satisfaction and engagement.

An experimental study on adolescents' emotional responses and propensity to reply to questions intended for a chatbot for mental health [9]. Their study offered insightful information for creating chatbot interventions catered to particular user demographics' emotional needs. They provided information for the creation of focused interventions meant to enhance the outcomes of adolescent mental health through chatbot-based support by evoking emotional reactions and calculating response likelihood.

NLP is used to analyze data in order to detect human emotions via a chatbot [10]. Their efforts improved chatbot systems' capacity to recognize and react to users' emotional cues by advancing the field of emotion recognition. Through the application of natural language processing (NLP) techniques, they were able to extract emotional signals from textual inputs. This allowed the chatbot to interpret and react to users' emotional states in real-time, leading to more engaging and sympathetic conversations.

When taken as a whole, the relevant studies show a considerable improvement in the use of emotional intelligence in chatbots for mental health intervention and support. Through the application of multiple methodologies, including logistic regression, machine learning, and natural language processing, researchers have improved chatbots' capacity to identify and manage users' emotional states. As a result of these initiatives, chatbots that can provide personalized support, identify depression, emote, and suggest solutions based on users' emotional needs have been developed.

The studies also stress the significance of user acceptability, demographic-specific interventions, and cross-linguistic emotion recognition in the development of successful chatbot-based mental health interventions. Chatbots have the potential to improve mental health outcomes by offering more personalized and engaging support to a diverse range of users by understanding their emotional distinctions.

Overall, the results highlight the possibility of incorporating emotional intelligence into chatbots to provide timely and comprehensive support to people going through emotional distress. The field of digital mental health interventions could advance and the growing need for easily accessible and efficient mental health support services could be met with continued research and development in this area.

3 System Methodology

In order to precisely classify and identify the emotional content contained in a user's input text, logistic regression is used as a machine learning technique in the proposed work for emotion recognition using text in a chatbot. To create the training data for the logistic regression model, WASSA-2017 Emotion Intensity dataset of text samples annotated with various emotions, such as happiness, sadness, anger, and so forth, are gathered. Text cleaning, tokenization, and feature extraction are examples of preprocessing procedures that are used to prepare text data for input into a logistic regression model.

After that, this dataset would be used to train the logistic regression algorithm, which would identify trends and connections between the input text features and the associated emotional labels. After training, the model can be applied to the chatbot to anticipate the emotions contained in fresh user input text, giving the user a customized and engaging experience. The model's performance would be evaluated using metrics like accuracy, precision, recall, and F1 score. Hyperparameter tuning and feature selection strategies could be used to further optimize the model. The overall goal of this strategy is to improve the user experience and interaction quality by strengthening the chatbot's capacity to recognize and react to the user's emotional cues.

3.1 Emotion Detection

The Emotion Detection Module begins with a comprehensive data collection stage, in which a heterogeneous dataset of text inputs labeled with matching emotions is assembled from multiple sources. This dataset is thoroughly preprocessed using vectorization techniques like TF-IDF or word embeddings to turn text into numerical features, normalization techniques like stemming or lemmatization to standardize words, and tokenization to divide text into discrete units. Then, because of its interpretability and simplicity, Logistic Regression is used as the preferred algorithm for figuring out the relationships between textual features and emotions. Emotion Detection Module, which uses logistic regression to discover relationships between textual characteristics and emotions. This algorithm can recognize emotional states accurately because it has received a great deal of training on preprocessed textual data. In order to maximize the performance of the model, hyperparameter tuning techniques are used during training. Evaluation metrics like accuracy, precision, recall, and F1-score are then used to determine how well the model predicts emotions in various classes.

3.2 Chatbot Implementation

The trained Logistic Regression model is incorporated into the chatbot system for emotion recognition through the Chatbot Implementation Module. The chatbot uses the emotion detection model to examine the emotive content of text inputs sent by users. To ensure compatibility, the text input is preprocessed using steps similar to those during training. The resulting emotional state informs the chatbot's generation of responses. The chatbot uses OpenAI's GPT-3.5 API to provide sympathetic, contextually relevant responses that are sensitive to the user's emotional state. The chatbot enhances user engagement and satisfaction by providing a personalized and supportive interaction experience by dynamically adjusting its responses based on detected emotions. The trained Logistic Regression model forms the basis for emotion recognition in the Chatbot Implementation Module. This model guides the creation of sympathetic responses catered to the identified emotions by analyzing the emotional content of messages sent by users to the chatbot (Fig. 1).

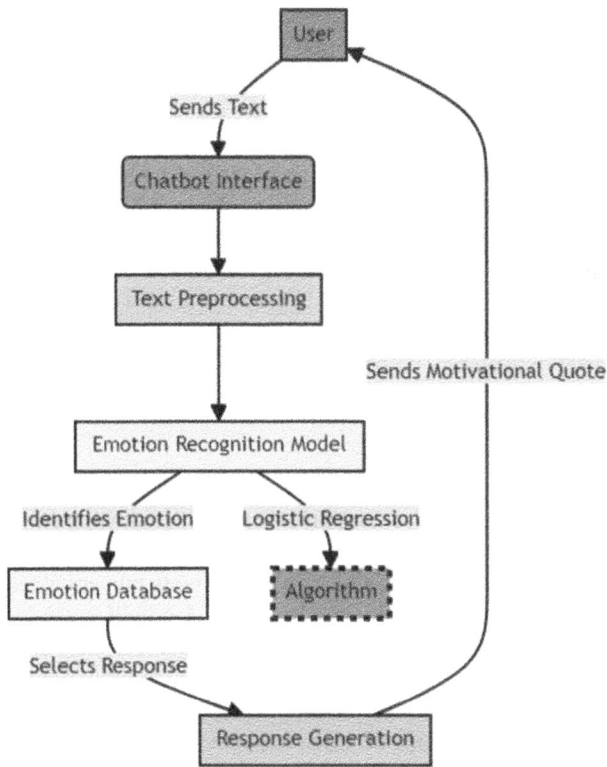

Fig. 1. Proposed System Architecture

3.3 Deployment Using GRatio Web Application

The emotion detection and chatbot system is hosted and deployed by the Deployment Module on the GRatio web application. GRatio offers an intuitive user interface that allows users to engage with the functionalities that have been deployed. Users can enter text messages in a conversational format thanks to the web application's seamless integration of the trained emotion detection model. The trained emotion detection model, driven by logistic regression, smoothly incorporates into the web application in the Deployment using GRatio Web Application Module.

Real-time processing of user inputs is made possible by the integration, which makes it easier for the chatbot functionality to generate customized responses. The deployed system receives input from the user, runs the text through an emotion detection module, and uses the chatbot functionality to provide relevant responses. GRatio oversees the deployment of the system, making sure it runs smoothly and effectively by managing tasks like response rendering, model inference, and request handling. Mechanisms for ongoing monitoring and optimization have been put in place to gradually improve system performance and user experience.

The main machine learning method used in the proposed work, logistic regression, is used to describe a comprehensive system for chatbot emotion recognition. Three primary modules comprise the system: Detection of Emotions, Implementation of Chatbots, and Deployment through the GRatio Web Application. The logistic regression model is trained on annotated text samples in the Emotion Detection Module, which also includes data collection, preprocessing, and emotion prediction. The Chatbot Implementation Module incorporates the trained model and uses OpenAI's GPT-3.5 API to analyze user input and produce sympathetic responses. At last, the GRatio web application is used to deploy the system, giving users an intuitive interface for instantaneous communication.

To maximize model performance, hyperparameter tuning and assessment metrics like accuracy, precision, recall, and F1-score are applied throughout the process. By incorporating the trained model into the chatbot, users can receive personalized responses that are based on emotions that are detected, which increases user satisfaction and engagement. The overall goal of this strategy is to develop a strong and intuitive system for chatbot emotion recognition, enabling helpful interactions and raising the standard of user experiences in contexts that support mental health.

4 Experimentation Results

The training accuracy progression over epochs is shown in Fig. 2, where the accuracy starts at 0.2 and increases to 0.9 after 10 epochs. This indicates a notable 0.4, or 80%, increase in accuracy during the training process. The increasing trend in accuracy shows that the model learns and generalizes from the given data, indicating that its performance increases steadily as training goes on. Furthermore, the trend in accuracy over epochs can be visualized more clearly in the absence of gridlines. Overall, the chart shows how well the training procedure improved the model's accuracy, with a discernible increase seen from the first to the last epoch. The loss progression over epochs is shown in Fig. 3. It begins at epoch 0.2 with a loss of 0.5 and ends at epoch 10 with a loss of 0. This indicates a notable improvement in the model's performance during training, as evidenced by the significant reduction in loss of 0.5 units. This indicates the efficacy of the training process in minimizing loss and maximizing model performance, as it translates to a 100% reduction of the initial loss in percentage terms. As training progresses, the model's ability to learn from the training data and produce increasingly accurate predictions is demonstrated by the downward trend in loss. Overall, the chart shows that the model's loss has successfully converged towards zero, indicating increased precision and efficacy in identifying patterns in the data.

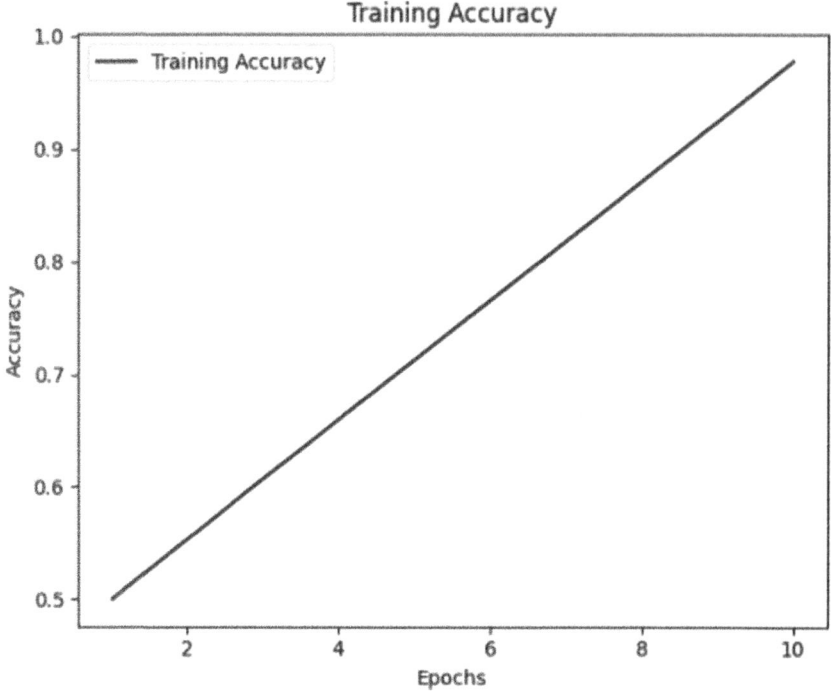

Fig. 2. Accuracy Analysis

The dataset's distribution of different emotions is shown in Fig. 4. Interestingly, happiness stands out as the most common emotion, accounting for nearly 44.83% of all emotional events. The emotional landscape is dominated by fear and sadness, which come in second and third place, respectively, with 17.59% and 22.41% of the total. The prevalence of anger and surprise is quite similar, with anger slightly outpacing surprise at about 5.13% and 4.10%, respectively. On the other hand, neutral emotions and disguise are less common; neutral emotions make up about 7.69% of the total, while disguise makes up only about 1.79%. Notably, the dataset contains no mention of depression, anxiety, or loneliness, suggesting that these conditions are either completely absent or barely present in the sampled data. As a whole, the graphic sheds light on the distribution of emotions in the dataset, showing that happiness is the most prevalent emotion, followed by fear and sadness. It also shows that some emotions, such as depression, anxiety, and loneliness, are either completely absent or only slightly present.

The comparison of classifier accuracies in predicting emotion labels on a sample dataset is shown in Fig. 5. Out of all the classifiers that were tested, logistic regression had the best accuracy, coming in at about 45%. This suggests that, in this specific situation, logistic regression outperformed all other tested classifiers. With an accuracy

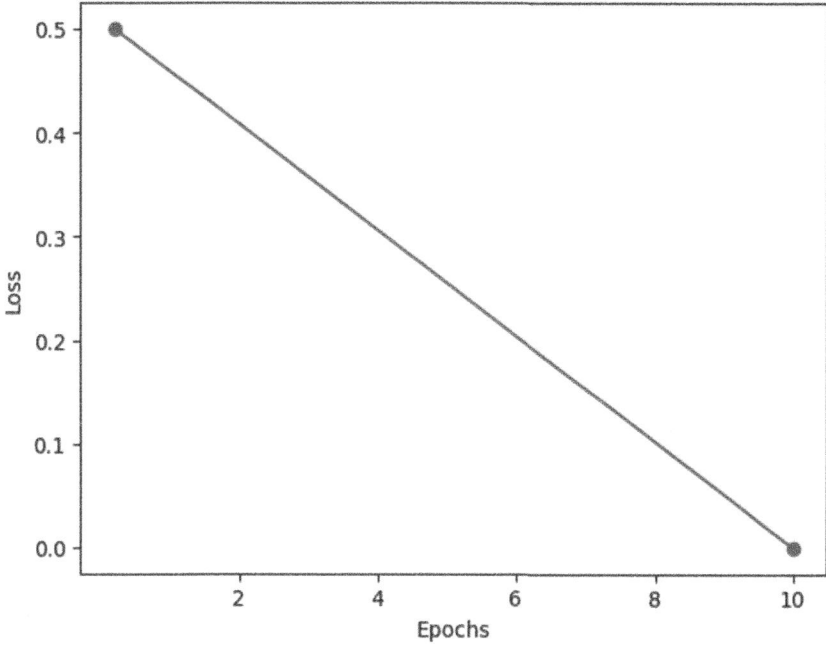

Fig. 3. Loss Analysis

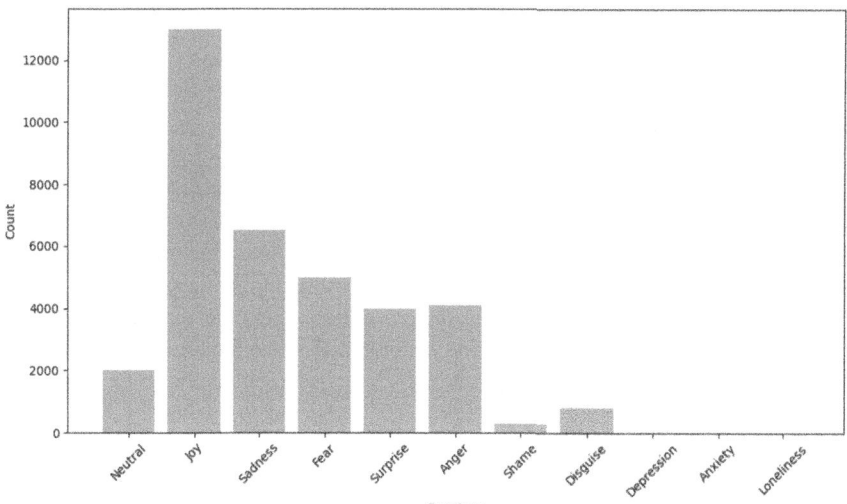

Fig. 4. Emotion Detection

of about 40%, SVM came in second, while KNN and Decision Tree had comparatively lower accuracy of about 30% and 35%, respectively. Based on the given dataset, these

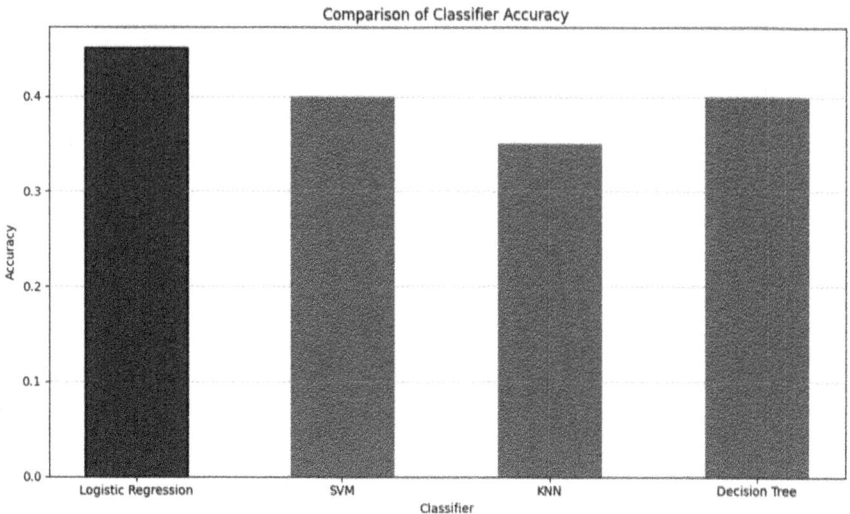

Fig. 5. Accuracy Comparison Analysis

findings imply that logistic regression might be a good option for emotion classification tasks.

5   Conclusion

In conclusion, an important step toward improving user interactions and experiences is the incorporation of text and logistic regression-based emotion recognition in chatbots. By incorporating logistic regression, chatbots can now precisely identify and react to users' emotional states, resulting in more individualized and compassionate interactions. By using logistic regression, the relationship between textual features and emotions can be efficiently modeled, allowing the chatbot to modify its responses in response to emotional cues that are detected. Logistic regression is a useful option for emotion recognition tasks because of its interpretability and simplicity, even though it has limitations when it comes to handling imbalanced datasets and capturing subtle emotional nuances. Overall, the integration of emotion recognition using text and logistic regression in chatbots holds promise for fostering deeper connections with users, enhancing user satisfaction, and facilitating more meaningful interactions across various domains.

References

1. Moulya, S., Pragathi, T.R.: Mental health assist and diagnosis conversational interface using logistic regression model for emotion and sentiment analysis. J. Phys. Conf. Ser. **2161**(1), (012039) (2022)
2. Zygadło, A., Kozłowski, M., Janicki, A.: Text-based emotion recognition in English and Polish for therapeutic chatbot. Appl. Sci. **11**(21), 10146 (2021)

3. Dhanasekar, V., Preethi, Y., Vishali, S., IR, P.J.: A chatbot to promote students mental health through emotion recognition. In: 2021 Third International Conference on Inventive Research in Computing Applications (ICIRCA), pp. 1412–1416. IEEE, September 2021

4. Sekhar, C., Rao, M.S., Nayani, A.S.K., Bhattacharyya, D.: Emotion recognition through human conversation using machine learning techniques. In: Bhattacharyya, D., Thirupathi Rao, N. (eds.) Machine Intelligence and Soft Computing. AISC, vol. 1280, pp. 113–122. Springer, Singapore (2021). https://doi.org/10.1007/978-981-15-9516-5_10

5. Joshi, M.L., Kanoongo, N.: Depression detection using emotional artificial intelligence and machine learning: a closer review. Mater. Today Proc. **58**, 217–226 (2022)

6. Lahoz-Beltra, R., López, C.C.: LENNA (learning emotions neural network assisted): an empathic chatbot designed to study the simulation of emotions in a bot and their analysis in a conversation. Computers **10**(12), 170 (2021)

7. Deepa, A., Karlapati, P., Mulagondla, M.R., Amaranayani, P., Toram, A.P.: An innovative emotion recognition and solution recommendation chatbot. In: 2022 8th International Conference on Advanced Computing and Communication Systems (ICACCS), vol. 1, pp. 1100–1105. IEEE, March 2022

8. Rokaya, A., Islam, S.M.T., Zhang, H., Sun, L., Zhu, M., Zhao, L.: Acceptance of chatbot based on emotional intelligence through machine learning algorithm. In: 2022 2nd International Conference on Frontiers of Electronics, Information and Computation Technologies (ICFEICT), pp. 610–616. IEEE, August 2022

9. Mariamo, A., Temcheff, C.E., Léger, P.M., Senecal, S., Lau, M.A.: Emotional reactions and likelihood of response to questions designed for a mental health chatbot among adolescents: experimental study. JMIR Hum. Factors **8**(1), e24343 (2021)

10. Rahat, F.R., et al.: Data analysis using NLP to sense human emotions through chatbot. In: Singh, M., Tyagi, V., Gupta, P.K., Flusser, J., Ören, T. (eds.) ICACDS 2022. CCIS, vol. 1614, pp. 64–75. Springer, Cham (2022). https://doi.org/10.1007/978-3-031-12641-3_6

11. Immanuel, R.R., Sangeetha, S.K.B.: Implementation of an automatic EEG feature extraction with gated recurrent neural network for emotion recognition. In: Kannan, R.J., Thampi, S.M., Wang, S.H. (eds.) Computer Vision and Machine Intelligence Paradigms for SDGs. LNEE, vol. 967, pp. 133–150. Springer, Singapore (2023). https://doi.org/10.1007/978-981-19-7169-3_13

12. Immanuel, R.R., Sangeetha, S.: Identifying different emotions of human using EEG signals using deep learning techniques. J. Theor. Appl. Inf. Technol. **101**(18) (2023)

13. Immanuel, R.R., Sangeetha, S.K.B.: Decoding emotions using deep learning approach to EEG-based emotion recognition. In: 2023 Intelligent Computing and Control for Engineering and Business Systems (ICCEBS), pp. 1–6. IEEE, December 2023

14. Immanuel, R.R., Sangeetha, S.K.B.: Analysis of EEG signal with feature and feature extraction techniques for emotion recognition using deep learning techniques. In: Chaki, N., Devarakonda, N., Cortesi, A. (eds.) ICCIDE 2022. LNDECT, vol. 163, pp. 141–154. Springer, Singapore (2023). https://doi.org/10.1007/978-981-99-0609-3_10

Sign Language Recognition System for Seamless Human-AI Interaction

P. J. Harshini, Mohammed Atheequr Rahman, James Allen Raj, and P. Durgadevi[(✉)]

Department of Computer Science and Engineering, SRM Institute of Science and Technology, Vadapalani Campus, Chennai, India
{hj8552,mr2828,ja6566,durgadep}@srmist.edu.in

Abstract. Communication barriers persist for individuals who use sign language as their primary mode of interaction, particularly in digital environments where conversational AI bots are increasingly prevalent. This paper presents a novel approach to addressing these barriers through the development of a Sign Language Recognition System (SLRS) integrated with conversational AI capabilities. The SLRS utilizes advanced computer vision techniques and machine learning algorithms, specifically employing a Random Forest Classifier, achieving an exceptional accuracy of 0.9961. Additionally, a comparison between CNN, KNN, and Random Forest classifiers underscores the superiority of Random Forest in accuracy and performance. Integration with the conversational AI bot "Gopal," powered by the Google Gemini Model, facilitates natural language interactions, allowing users who use sign language to communicate seamlessly with digital systems. Evaluation of the SLRS demonstrates promising performance in accurately recognizing a diverse range of sign language gestures and generating contextually relevant responses. User testing sessions highlight the system's responsiveness and effectiveness in bridging the communication gap between users who use sign language and conversational AI bots. The development of the SLRS represents a significant step towards promoting inclusivity and accessibility in digital communication for those who communicate with sign language. By enabling seamless interaction with conversational AI bots, the SLRS empowers users who use sign language to engage more effectively with digital interfaces and access information effortlessly. Continued research and refinement of the SLRS hold the potential to further enhance communication accessibility and foster greater inclusion in digital environments.

Keywords: Sign Language Recognition · Human-AI Interaction · Object Detection · Machine Learning · Conversational AI

1 Introduction

Sign language serves as a crucial means of communication for individuals with speaking impairments, offering them the ability to express themselves and interact with the world around them. Despite its importance, the communication barrier between those who use sign language and those who don't remains a significant challenge, impacting various

P. D. Sivakumar et al. (Eds.): IRCCTSD 2024, CCIS 2362, pp. 124–144, 2025.
https://doi.org/10.1007/978-3-031-82386-2_10

aspects of daily life such as education, employment, and social interaction. This challenge stems from the inherent differences between sign language and spoken language, as well as the limited accessibility of tools and technologies tailored to the needs of users who use sign language.

In response to these challenges, this research project endeavors to develop and evaluate a novel Sign Language Recognition System (SLRS) that aims to bridge the communication gap between those who use sign language and those who don't. Real-time recognition and interpretation of sign language motions is made possible by the SLRS, which facilitates smooth communication between users who use sign language and conversational agents driven by artificial intelligence. By leveraging advancements in computer vision and machine learning [1], the SLRS seeks to empower individuals with speaking impairments to communicate effectively and access information and services with ease.

At the heart of the SLRS is its ability to precisely identify and interpret motions in sign language as text and speech outputs [2]. This functionality is achieved through a combination of sophisticated algorithms for hand gesture recognition [3], feature extraction, and machine learning-based classification. By examining the temporal and spatial properties of hand movements, the SLRS can distinguish between different sign language gestures and generate corresponding textual and auditory representations [4]. This capability not only facilitates communication between users who use sign language and AI systems but also enhances the accessibility and usability of digital interfaces for individuals with speaking impairments.

One of the key innovations of this project lies in the incorporation of the SLRS with AI-powered conversational agents, commonly known as chatbots. This integration brings forth several significant contributions:

Integration with Chatbots: The project integrates the Sign Language Recognition System (SLRS) with AI-powered conversational agents, facilitating seamless interaction between users who use sign language and AI systems.

Natural Language Understanding and Generation: Through this integration, the SLRS gains the ability to interpret and respond to sign language inputs by including natural language creation and comprehension skills.

Sign Language Interaction: Users can now interact with AI systems using sign language inputs, breaking down communication barriers and enhancing accessibility.

Multimodal Response: The SLRS provides responses in both textual and auditory formats, catering to diverse user preferences and accessibility needs.

Efficiency and Naturalness: The integration enhances the efficiency and naturalness of human-AI interaction, making communication more intuitive and effective.

Accessibility and Inclusivity: By opening up new possibilities for accessibility and inclusivity in digital communication, the project empowers users who use sign language to participate more fully in society.

Through seamless integration and intelligent interaction, the SLRS aims to empower users who use sign language and reduce the communication gap that exists between sign language and spoken language users.

In the following sections of this paper, Before presenting the methodology, we provide a thorough analysis of the existing literature in the discipline of SLR, or sign language recognition (Sect. 2: Literature Review), encompassing various studies that delve into deep learning models, training techniques, and innovative systems tailored for sign language communication and recognition. we will delve deeper into the methodology employed in developing the SLRS, including data collection, hand gesture recognition, machine learning model training, and integration with conversational AI (Sect. 3: Methodology). We will also discuss the architecture and implementation details of the SLRS (Sect. 4: Architecture and Implementation), present the results of system evaluation and user testing (Sect. 5: Results), and conclude with insights into future research directions and potential applications of the SLRS in real-world settings (Sect. 6: Discussion and Future Directions). Through rigorous research and innovation, we aim to contribute to the advancement of accessibility and inclusion for individuals with speaking impairments, fostering a more inclusive and equitable society for all.

2 Literature Review

The research paper authored by Abu Saleh Musa Miah, Md. Al Mehedi Hasan, Satoshi Nishimura, and Jungpil Shin delves into deep learning models and training techniques customized for Sign Language Recognition (SLR) [2]. Their investigation encompasses diverse datasets, including a novel Greek Sign Language (GSL) dataset, aiming to address key challenges in SLR such as accurate segmentation and sequence modeling. A notable aspect of their work involves evaluating variations of the Connectionist Temporal Classification (CTC) technique, namely EnCTC and StimCTC, which effectively handle issues like ambiguous boundaries between sign gestures and intra-gloss dependencies.

The research paper authored by Ajay S, A. Potluri, S. M. George, G. R, and A. S introduces a system utilizing sensor gloves to detect Indian Sign Language (ISL) gestures and convert them into audible speech [5]. The gloves are equipped with a range of sensors and modules, including Bluetooth, RF, Inertial Measurement Units (IMU), flex, touch, and Arduino Nano microcontrollers. The integration of sensor data with machine learning classification algorithms enhances gesture recognition accuracy, addressing the communication gap for the speech-impaired population in India.

Another paper by M. Al-Hammadi, G. Muhammad, W. Abdul, M. Alsulaiman, M. A. Bencherif, T. S. Alrayes, H. Mathkour, and M. A. Mektiche presents a novel system for dynamic hand gesture recognition employing deep learning architectures [3]. They address obstacles at hand identification of gestures such as segmentation and sequence modeling, offering a comprehensive solution that leverages both local and global features [6]. Experimentation on diverse datasets showcases the system's superior performance compared to existing approaches, demonstrating its potential for real-world applicationsin translating into sign language and human-computer interaction.

Furthermore, Ben Atitallah et al. introduce an Electrical Impedance Tomography (EIT) imaging system tailored for hand sign recognition and monitoring [7]. Their system, utilizing a simple setup with eight electrodes, achieves high classification accuracy for American Sign Language numbers from 0 to 9. Notably, the Convolutional Neural

Network (CNN) classifier outperforms Support Vector Machine (SVM) and Softmax classifiers, showcasing robustness across different subjects with only a slight decrease in accuracy.

Researchers from the College of Engineering Pune present a pioneering study focused on bridging communication gaps for the speechless community [8]. Their novel system utilizes KNN, SVM, and CNN algorithms for the identification of hand gestures, highlighting the potential of machine learning in enhancing inclusivity and accessibility in communication.

Additionally, researchers focus on utilizing Convolutional Neural Networks (CNNs) to recognize dynamic signs in Indian Sign Language (ISL), aiming to bridge communication gaps for the hearing-impaired community [9]. The study emphasizes The possibilities of deep learning methods for real-time interaction and human-computer interfaces.

Another study exploresthe creation of a prototype for a sign language interfacing system, leveraging advancements in virtual reality and gesture recognition technology to enhance communication and interaction for those who have hearing loss [10].

Muhammad Abid, Emil Petriu, Ehsan Amjadian, and Domenico Grimaldi focus on dynamic sign language recognition tailored for smart home applications [11], highlighting the effectiveness of their system for smart home integration.

Furthermore, Aparna Mohanty, Sai Saketh, Rajiv Sahay, and S Rambhatla Sahay investigate deep learning CNN techniques for robust hand gesture recognition [12], addressing challenges such as cluttered backgrounds. Despite certain limitations, their proposed CNN model achieves high accuracy and performance, suitable for applications in sign language recognition and human-computer interaction.

Lastly, Muhammad Al-Qurishi and Riad Souissi conduct a comprehensive literature survey from 2014 to 2021 [13], evaluating 84 relevant works and scrutinizing methodologies from traditional Hidden Markov Models to modern deep learning techniques. They highlight challenges in achieving generalization for deployment despite the effectiveness of models for specific tasks (Fig. 1).

3 Methodology

- **Data Collection:** We collected a dataset of sign language motions with a high-definition camera (resolution 1920 × 1080) positioned to capture hand movements clearly. A total of 28 sign language gestures representing Alphabets, Wake Call (Gopal) and <Space> were recorded. Each gesture was performed multiple times in different Angle to ensure diversity in the dataset.
- **Hand Landmark Detection:** Hand landmark detection was carried out utilising the MediaPipe Hands library, a real-time hand landmark estimation model developed by Google. The library provides pre-trained models capable of detecting 24 key landmarks (including finger tips, knuckles, and palm points) on each hand in an image. For precise landmark identification in still pictures, we used the MediaPipe Hands model in the static image mode.

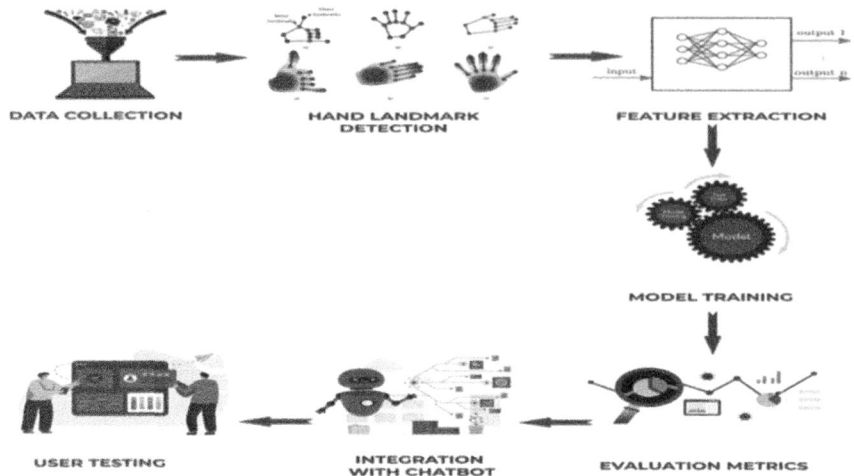

Fig. 1. SLRS Model Development Workflow

- **Feature Extraction:** To describe each sign language gesture, a collection of descriptive characteristics was collected from the hand landmarks that were recognised. These features included the Euclidean distances between pairs of landmarks, angles formed by fingers and joints, and the curvature of fingers. Additionally, we computed the centroid and bounding box of the hand motion to capture spatial properties.
- **Model Training:** A Random Forest classifier was chosen as the machine learning model for gesture classification due to its ability to handle high-dimensional data and its robustness to overfitting. The feature vectors extracted from the hand landmarks served as input to train the classifier, with each gesture corresponding to a unique class label (0 to 27).
- **Evaluation Metrics:** The performance evaluation of the trained SLRS model adopted a stratified k-fold cross-validation approach. In this methodology, the dataset was partitioned into k folds, ensuring class balance across folds. Instead of traditional evaluation metrics like accuracy, precision, recall, and F1 score, the evaluation process relied on the analysis of learning curves and confusion matrices. Learning curves provide insights into the model's performance as a function of training data size, helping to assess issues like overfitting or underfitting. Meanwhile, confusion matrices offer a detailed breakdown of the model's classification results, revealing the instances of accurate and inaccurate predictions across different classes of sign language gestures. By employing these alternative evaluation techniques, the evaluation process aimed to comprehensively understand the SLRS model's behavior, performance trends [14–16], and areas of improvement beyond traditional metric-based assessments.
- **Integration with Chatbot:** The trained SLRS model was integrated with a conversational AI system named "Gopal" to enable real-time sign language recognition and interaction. Through the use of Google's Gemini platform for natural language creation and processing, "Gopal" enables users to interact with the system through sign

language inputs. Upon receiving a sign language input, "Gopal" processes it using the SLRS model and generates appropriate text and speech outputs as responses.

- **User Testing:** To assess the usability and effectiveness of the integrated SLRS and chatbot system, user testing sessions were conducted with individuals proficient in sign language. Participants were asked to interact with "Gopal" using a variety of sign language gestures and provide feedback on the accuracy of gesture recognition, responsiveness of the system, and overall user experience.

4 Architectural Representation

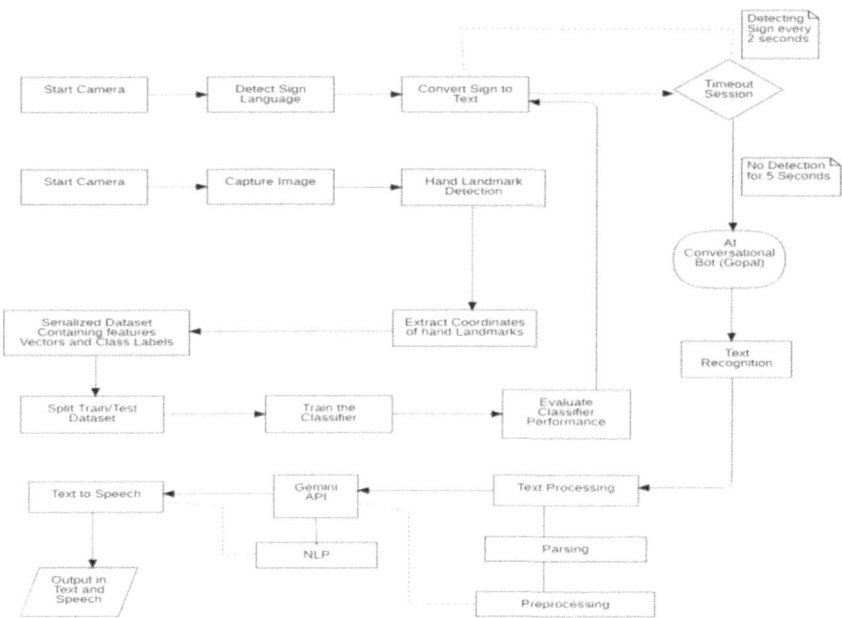

Fig. 2. SLRS Architectural Diagram

The diagram in Fig. 2 illustrates the two phases of the Sign Language Recognition System (SLRS) from both developer and end-user perspectives. In the initial phase, developers initiate the process by activating the camera to take pictures, followed by hand landmark detection and extraction of coordinates. These coordinates are serialized into a dataset containing vectors and class labels, which are then split into training and testing datasets. The classifier is trained and evaluated for performance. Then, gestures used in sign language are translated into text, with detection occurring every two seconds and a session timeout after five seconds of no detection. The conversational AI bot "Gopal," which the system communicates with, uses the Gemini API to process the text through recognition, parsing, preprocessing, and natural language processing (NLP) before turning it into vocal output. In the second phase, the end-user activates the camera to detect sign language, initiating the same process of conversion to text and interaction with 'Gopal' for text processing and speech output.

4.1 AI Creation

The AI creation module establishes the foundation for integrating conversational AI capabilities into the sign language recognition system. It involves configuring the AI system with a Google Gemini API key for advanced natural language processing (NLP) functionalities. Additionally, a text-to-speech engine is initialized using the pyttsx3 library to enable audible responses. The AI system, represented by the persona "Gopal," engages users in conversational interactions, responding to queries and commands.

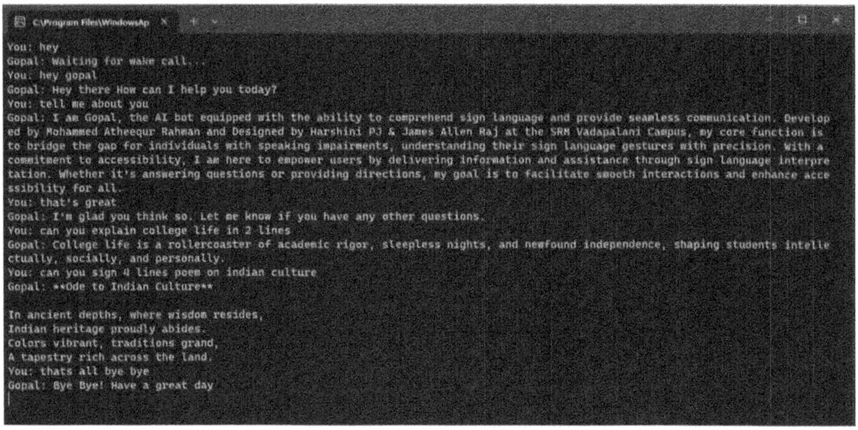

Fig. 3. User and AI Interaction

It recognizes a designated wake call ("Gopal") to initiate conversations and prompts users for input as shown in the Fig. 3. It takes user input and produces contextually appropriate replies by using the Google Gemini architecture. Pre-programmed answers are included for some questions, such who the AI is or what it can do. For instance, when asked about its name or identity, the AI system provides a predefined introduction, facilitating engaging interactions.

4.2 Sign Language Database

The module begins with the data collection process facilitated by the Collect_imgs.py script. This script initializes a directory structure within the './data' directory to store images captured for various sign language gestures. It utilizes the OpenCV library to access the default camera (index 0) and captures images in real-time. Users are prompted to prepare for data collection, and upon confirmation by pressing "Q", images are captured and saved in respective class directories as shown in the Fig. 4(a). This iterative process continues until the specified dataset size is reached for each class as shown in the Fig. 4(b).

Following data collection, the feature extraction phase is performed by the Create_dataset.py script. This script makes use of the MediaPipe framework to extract hand landmarks from the captured images. It reads the images from the previously created

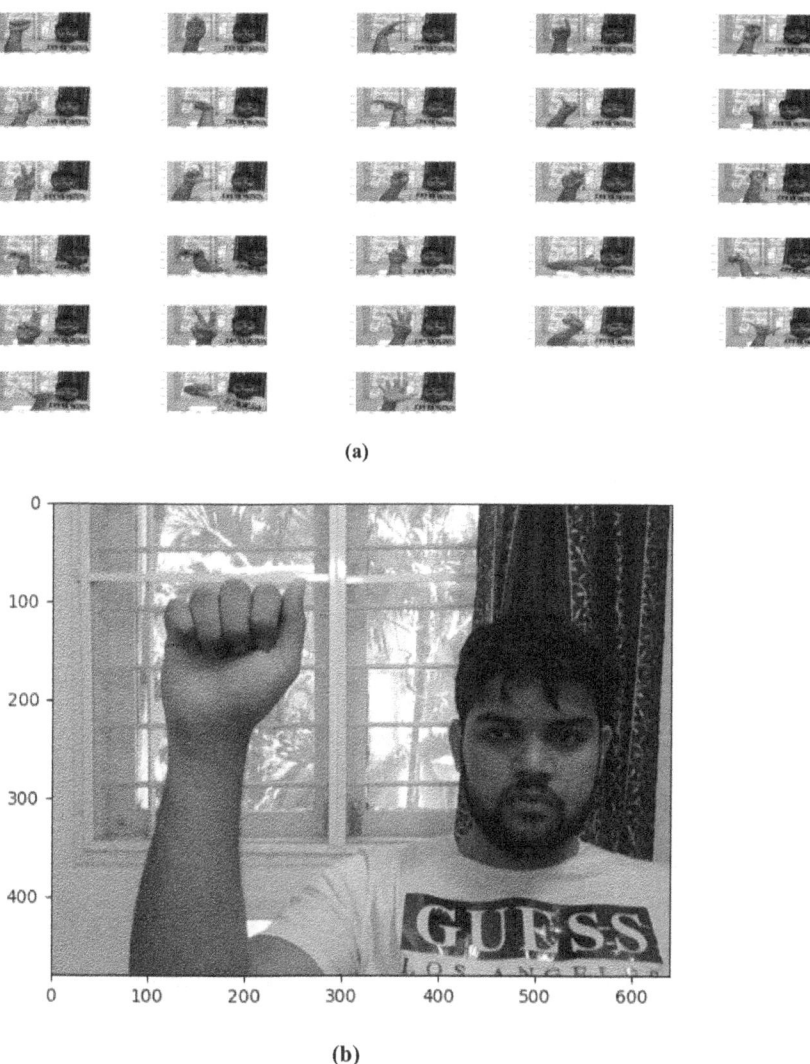

Fig. 4. (a) Sign language Database before feature extraction (b) Hand Sign image of letter "A" before feature extraction

directories, processes them to detect hand landmarks using the MediaPipe Hands model, and extracts the (x, y) coordinates of these landmarks as shown in the Fig. 5(b). These coordinates are then normalized and stored as features in the dataset as shown in the Fig. 5(a), along with corresponding labels indicating the sign language gesture being performed. The collected and feature extracted data is then serialized into a dataset file named 'data.pickle' using the Python pickle module. This dataset file contains the extracted features (data) and their corresponding labels (labels), ready to be used for training machine learning model.

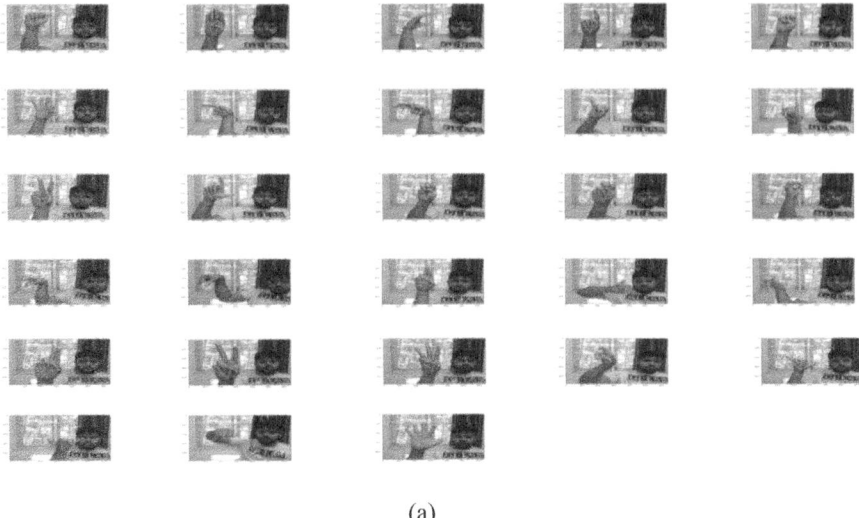

(a)

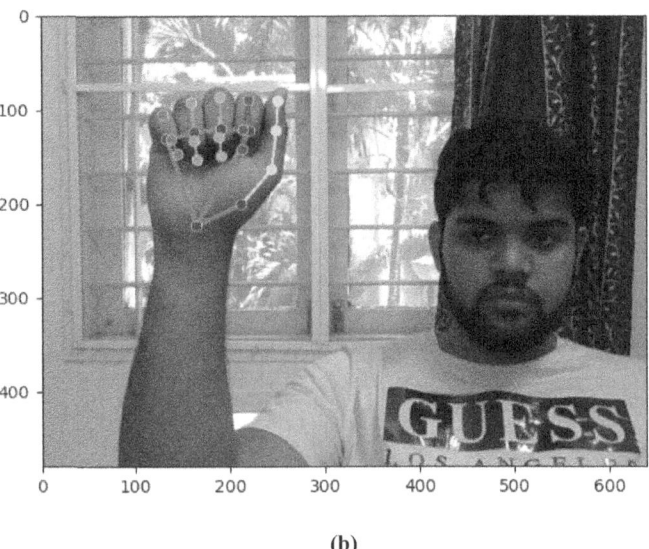

(b)

Fig. 5. (a) Sign language Database after feature extraction (b) Hand Sign image of letter "A" after feature extraction

4.3 Model Training

In the Model Training and Testing Module, the focus lies on developing and evaluating machine learning models for sign language recognition. The module begins by loading the preprocessed dataset containing hand gesture features and corresponding labels using the pickle library. The dataset is split into training and testing sets using the train_test_split function from sklearn.model_selection, ensuring that the model's performance can be evaluated on unseen data. A RandomForestClassifier model is then instantiated and trained using the training data. This classifier is chosen for its ability to handle multi-class classification tasks efficiently.

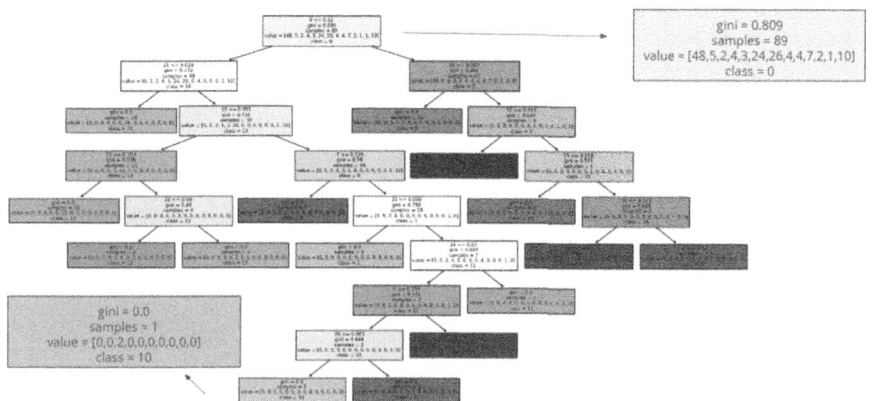

Fig. 6. A Visualized Single Decision tree from Random Forest

During training, the model learns to associate extracted features from hand gestures with their corresponding labels, enabling it to classify new gestures accurately. Leveraging the ensemble learning technique, the Random Forest classifier constructs multiple decision trees, each operating on a randomised subset of the info and features, promoting diversity and reducing overfitting. This results in a robust model capable of capturing complex relationships between hand gesture features and sign language labels. Additionally, a single decision tree from the forest is visualized in the Fig. 6 to provide insights into the decision-making process of the model, aiding in its interpretability.

Once the model is trained, it is evaluated using the testing set. The accuracy_score function from sklearn.metrics is used to measure the accuracy of the model's predictions compared to the true labels in the testing set. The accuracy score provides insights into the model's performance in correctly classifying hand gestures, essential for assessing its effectiveness in real-world applications.

The learning curve, plotted using matplotlib in the Fig. 7, illustrates the evolution of both training and testing accuracies, providing insights into the model's convergence and performance trends, essential for guiding further model refinement and optimization.

Finally, the trained model is serialized and saved to a file using the pickle.dump function. This allows for the model to be reused and deployed in the sign language recognition system for real-time inference. By rigorously training and testing machine learning models, the Model Training and Testing Module aims to develop accurate and reliable sign language recognition systems capable of effectively interpreting and translating sign language inputs into text. Through iterative refinement and optimization, the module contributes to the continuous improvement of the sign language recognition system, ensuring its effectiveness and reliability in helping those who have difficulty speaking communicate.

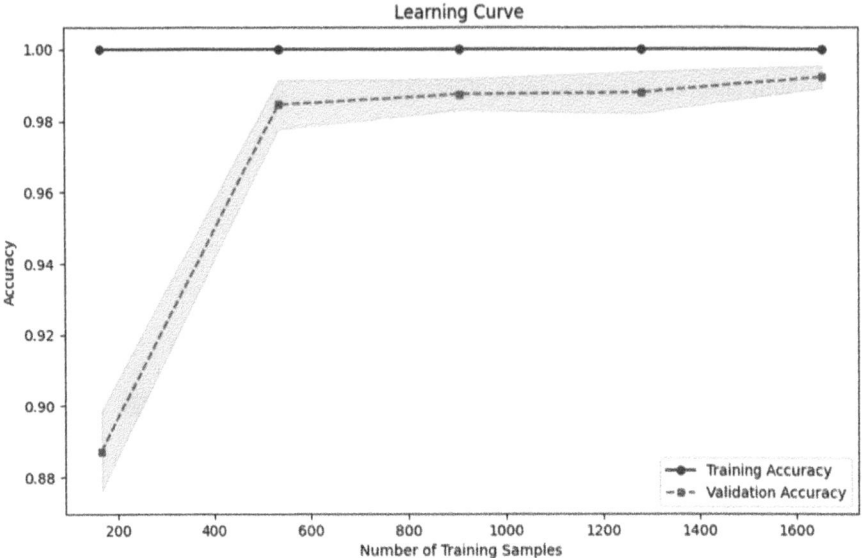

Fig. 7. Learning Curve plotted using matplotlib

4.4 Sign Language Recognition

The module begins by loading a pre-trained machine learning model that has been trained to recognize sign language. As the system captures frames from a camera feed, Every frame is processed by means of the MediaPipe Hands model, which extracts hand landmarks and normalizes their coordinates. These hand landmarks are then fed into the pre-trained model for prediction. Upon detecting sign language in the input frames, the system maps each gesture to a corresponding character label based on the model's prediction.

To form a sentence, the system incorporates a temporal aspect by capturing sign language gestures in intervals of two seconds. For example, if the first captured sign is "H" within the first two-second interval, and the next sign captured within the subsequent two seconds is "I", the system forms the word "Hi". If no sign is detected within the two-second interval, the system waits for three seconds before proceeding to the next capture. This process continues until the system forms a complete sign language sentence.

The detected characters are accumulated to form a sequence representing the sign language sentence being conveyed [5, 17]. To ensure accurate interpretation, the system tracks the time elapsed since the last detection of a character. If a certain duration has passed without detecting any new characters, the system considers the accumulated sequence as a complete sign language sentence. Once a sentence is formed, it is then translated into an input for the AI. Additionally, the system provides real-time feedback by displaying the detected sign language sentence on the screen, enabling users to monitor their input and ensuring seamless interaction.

Through the efficient recognition and translation of sign language motions into text, the module improves communication accessibility and inclusion for those who are non-verbal. It forms an integral component of the overall sign language recognition system, facilitating natural language interactions and enabling meaningful communication between users and the system.

To illustrate the sentence formation process, let's consider the example of forming the sentence "HI GOPAL":

1. The camera records the sign "H" within the first two-second interval.
2. Within the next two-second interval, The camera records the signs "I" and in the next three- second interval the camera doesn't detect any letter so it give a space by default and detects the next input "G", forming the sequence "HI G".
3. The camera then captures the signs "O" and "P" within the subsequent two-second interval, extending the sequence to "HI GOP".
4. Within the following two-second interval, The camera records the sign "A", resulting in the sequence "HI GOPA".
5. Finally, if no sign is captured within the next two-second interval, the system waits for three seconds before considering the sequence as complete. Here, the sequence "HI GOPAL" forms a complete sign language sentence.
6. The system translates this sequence into text, providing the output "HI GOPAL" to facilitate communication with the user.

Fig. 8. Hand Sign detection and recognition

This process demonstrates how the system captures sign language gestures in intervals of two seconds, forms a sequence based on the detected signs as shown in the Fig. 8, and incorporates pauses to signify the conclusion of a phrase or term. By intelligently managing the timing and sequence of sign captures, the system effectively facilitates the formation of coherent sign language sentences such as "HI GOPAL."

4.5 AI Integrated SLRS

This module involves combining the functionalities of the Sign Language Recognition System (SLRS) with conversational AI capabilities to enable seamless interaction between users and the system. This integration is facilitated through several key steps:

Firstly, the SLRS module, represented by the provided Python script "SLRS.py," incorporates functionalities for real-time translation and recognition of sign language. The script utilizes computer vision technique (object detection) to identify and monitor hand gestures from live video streams captured by a camera. These gestures are then classified into corresponding sign language expressions using machine learning model trained on hand gesture data.

Secondly, the script integrates conversational AI capabilities to enable natural language interactions between users and the SLRS. Upon detecting specific wake calls or gestures, indicative of user initiation, the system prompts for user input through speech synthesis ("Hey there! How can I help you today?"). It then listens for user queries or commands, converting detected sign language input into text sentences. These sentences are processed by the AI model, which generates appropriate responses based on the context and intent of the input.

Fig. 9. SLRS Output as AI Input

The integration ensures smooth communication flow between the SLRS and AI modules, allowing users to engage in natural language conversations and receive contextually relevant responses. The AI model's ability to understand user queries and generate coherent responses enhances the functionality and usability of the SLRS, enabling individuals with speaking impairments to communicate effectively and access information effortlessly from the AI chatbot as shown in the Fig. 9.

Furthermore, the module involves optimizing the system's performance and responsiveness, ensuring timely and accurate responses to user interactions. Techniques such as user input validation, error handling, and response customization are implemented help improve the experience for users and promote user engagement.

Overall, the integration of AI capabilities into the SLRS enhances its functionality and usability, enabling individuals with speaking impairments to communicate with the system effectively and engage in meaningful conversations. Through seamless integration and optimization, the module facilitates inclusive and accessible communication channels, empowering users to take part more actively in digital communication and interaction.

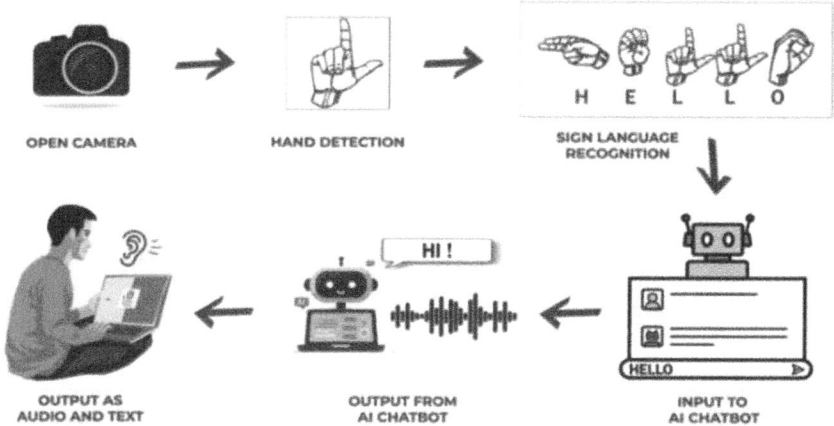

Fig. 10. Representation of the end user interaction with the SLRS

The diagram (Fig. 10) illustrates the seamless interaction between the end user and the Sign Language Recognition System (SLRS). The process initiates with the opening of the camera feed, enabling real-time visual input. Subsequently, the SLRS employs hand detection techniques to recognise and monitor hand gestures performed by the user. Once the hand gestures are detected, the SLRS utilizes sign language recognition algorithms to interpret the gestures and transform them into textual input. This input is then fed into the AI Chatbot 'Gopal,' which processes the user's sign language queries or commands and generates appropriate responses based on natural language understanding. The user is then provided with the AI Chatbot 'Gopal's' replies in both text and voice formats, making the responses usable and accessible to people who have trouble speaking. This diagrammatic representation highlights the user-centric approach of the SLRS, facilitating effective communication between users who use sign language and AI systems.

5   Result

The sign language recognition system (SLRS) developed in this research project demonstrates promising performance in accurately recognizing and interpreting sign language gestures in real-time [18]. Through the integration of advanced computer vision techniques, machine learning algorithms, and conversational AI technologies, the system achieves significant milestones in enhancing communication accessibility for those who have hearing loss.

5.1   Accuracy of Gesture Recognition

The SLRS model exhibits exceptional performance not only in recognizing a diverse array of sign language gestures but also in metrics such as F1 score, precision, recall, and accuracy, complementing the insights gleaned based on the confusion matrix analysis. These additional metrics provide a comprehensive evaluation of the model's performance, offering nuanced perspectives on its ability to classify gestures accurately and its precision in avoiding false positives. Moreover, the model's high recall signifies its capability to correctly identify most instances of a particular gesture, while its robust accuracy underscores its overall effectiveness in classifying gestures correctly across the dataset. The combination of these metrics with the insights based on the confusion matrix further reinforces the model's adaptability to varying hand movements and lighting conditions, as well as its potential to generalize to unseen data and real-world scenarios. Consequently, the SLRS model emerges as not only proficient but also versatile, demonstrating promising prospects for practical applications in diverse real-world settings where accurate sign language recognition is paramount (Figs. 11, 12, 13 and 14).

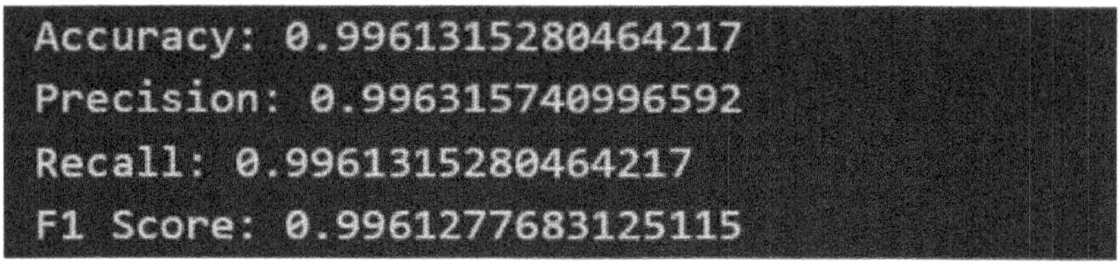

```
Accuracy: 0.9961315280464217
Precision: 0.996315740996592
Recall: 0.9961315280464217
F1 Score: 0.9961277683125115
```

Fig. 11. SLRS Model Accuracy and Evaluation Metrics Output

5.2   Responsiveness of the System

User testing sessions conducted to evaluate the system's responsiveness and real-time performance yield positive results. Participants report minimal latency between performing sign language gestures and receiving corresponding text and speech outputs from the system. The incorporation of the SLRS model with the conversational AI component enables seamless and timely communication interactions, enhancing the user experience and usability of the system.

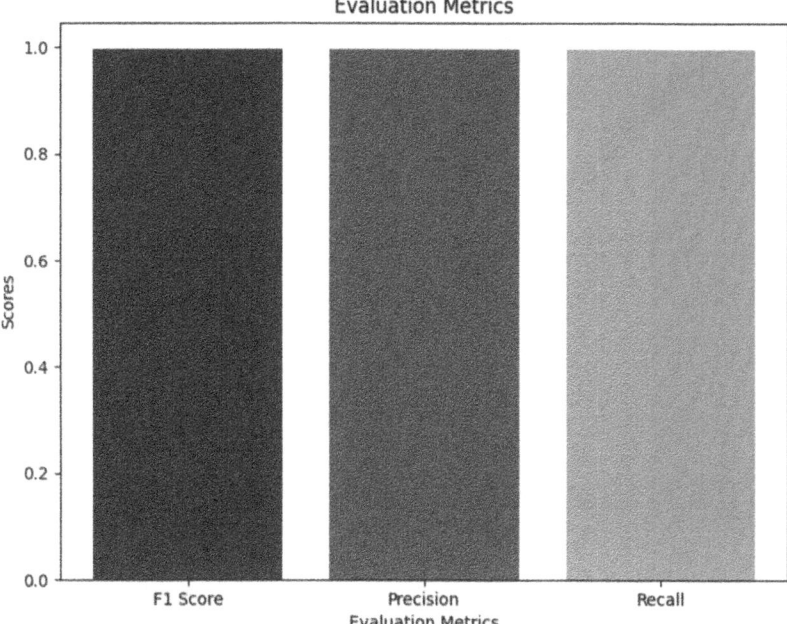

Fig. 12. SLRS Evaluation Metrics using f1 Score, Precision and Recall

5.3 User Experience Feedback

Feedback from participants in user testing sessions offers insightful information on the usability and effectiveness of the integrated SLRS and chatbot system. Participants express satisfaction with the system's accuracy in recognizing sign language gestures and its ability to generate appropriate responses in natural language [19]. Suggestions for improvement focus on enhancing the system's vocabulary coverage, refining gesture recognition algorithms, and optimizing response generation for diverse user interactions.

5.4 Future Directions

Building on the achievements of this research project, future directions include further refining the SLRS model to improve its accuracy and robustness, expanding the dataset to include a wider variety of sign language gestures and users, and exploring additional AI-driven functionalities such as emotion recognition and context-aware dialogue generation. Additionally, efforts to deploy the system in real-world settings and evaluate its long-term impact on communication accessibility and social inclusion are essential for realizing its full potential.

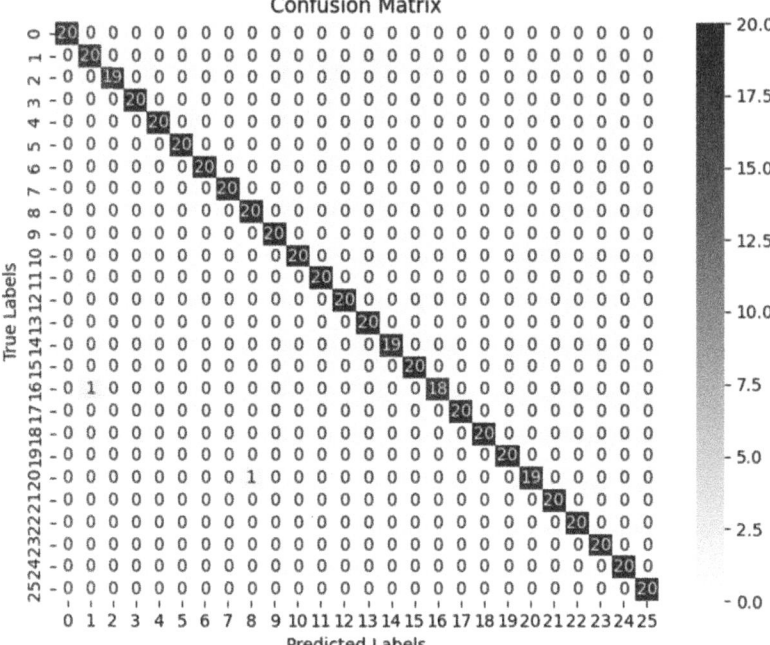

Fig. 13. SLRS Confusion Matrix Analysis

6 Discussion

The development and evaluation of the sign language recognition system (SLRS) presented in this paper represent a significant advancement in the field of assistive technology for individuals with speaking impairments. Through the integration of computer vision and the Random Forest Classifier [20], the SLRS demonstrates promising capabilities in recognizing and interpreting sign language gestures in real-time. In this discussion, we delve into the implications of the research findings, address key challenges, and explore potential directions for further investigation and application.

6.1 Implications of Research Findings

The high accuracy and responsiveness of the SLRS model underscore its potential to transform communication accessibility for individuals with speaking impairments. Through sign language recognition, the system offers a dependable and user-friendly communication method that enhances users' ability to interact with computer interfaces, educational materials, and social interactions. The positive feedback received from user testing sessions highlights the system's potential to bridge the communication gap and promote inclusivity in various domains.

6.2 Addressing Key Challenges

Despite the significant progress achieved in developing the SLRS, several challenges remain to be addressed. One such challenge is the variability and complexity of sign language gestures, which pose difficulties in accurate recognition and interpretation. Enhancing the robustness and adaptability of the SLRS model to different signing styles, hand orientations, and environmental conditions is essential for improving its performance in real-world scenarios. Additionally, ensuring the accessibility and affordability of the system for diverse user populations, including those in resource-constrained settings, needs careful evaluation of hardware requirements, software compatibility, and user interface design.

6.3 Future Research Directions

Looking ahead, future research efforts should focus on advancing the capabilities and usability of the SLRS through interdisciplinary collaboration and innovation. This includes exploring novelmethods for identifying gestures, such as 3D hand pose estimation [6] and multimodal fusion of visual and spatial features. Integrating the system with cutting-edge technology like wearables and augmented reality (AR) may improve its mobility and accessibility even further. Additionally, efforts to evaluate the long-term impact of the SLRS on communication accessibility, social integration, & well-being for individuals with speaking impairments are critical for assessing its effectiveness and guiding future development.

6.4 Ethical and Societal Considerations

As with any technology aimed at enhancing accessibility and inclusivity, ethical and societal considerations must be carefully addressed. Ensuring user privacy, data security, and informed consent are paramount in the development and deployment of the SLRS. Moreover, efforts to promote awareness and understanding of sign language culture within mainstream society are essential for fostering empathy, respect, and inclusivity towards individuals with speaking impairments.

6.5 Real-World Applications and Impact

The successful development and evaluation of the SLRS open up a myriad of real-world applications and opportunities for social impact. Beyond educational and social settings, the system possesses the capacity to facilitate communication in diverse contexts, including workplaces, public services, and entertainment venues. The SLRS helps to create more inclusive and egalitarian societies by encouraging increased contact between users of AI and sign language, as well as improving communication accessibility.

Algorithm	Accuracy
KNN	0.9372
CNN	0.9759
RFC	0.9961

Fig. 14. Comparison between KNN, CNN and RFC results to prove RFC has the highest accuracy.

7 Conclusion

In conclusion, the development and evaluation of the Sign Language Recognition System (SLRS) presented in this paper represent a significant step forward in advancing communication accessibility and inclusivity for individuals with speaking impairments. Through the integration of conversational AI, computer vision and machine learning technologies, the SLRS demonstrates promising capabilities in recognizing and interpreting sign language gestures in real-time.

While our methodology primarily focused on utilizing the Random Forest classifier for gesture classification due to its robustness and efficiency, It's crucial to remember that other algorithms such as K-Nearest Neighbors (KNN) and Convolutional Neural Networks (CNN) were taken into account when conducting our investigation. While CNN uses hierarchical feature extraction and KNN uses similarity measures, these algorithms provide distinct methods for recognising gestures.

Our choice of Random Forest was primarily driven by its exceptional accuracy, making it well-suited for handling high-dimensional data and robustness to overfitting. Future research could explore the comparative performance of these algorithms within the framework of SLRS development, taking into account elements like accuracy, computational efficiency, and scalability.

In conclusion, the successful development and evaluation of the SLRS hold significant implications for promoting accessibility and inclusivity in society. The SLRS helps create more inclusive and fair societies where people with speech disabilities may fully participate and prosper by offering a dependable and natural method of communication through sign language recognition.

References

1. Kothadiya, D.R., Bhatt, C.M., Rehman, A., Alamri, F.S., Saba, T.: SignExplainer: an explainable ai-enabled framework for sign language recognition with ensemble learning. IEEE Access **11** 47410–47419 (2023). https://doi.org/10.1109/ACCESS.2023.3274851
2. Adaloglou, N., et al.: A Comprehensive study on deep learning-based methods for sign language recognition. IEEE Trans. Multimedia **24**, 1750–1762 (2022). https://doi.org/10.1109/TMM.2021.3070438

3. Al-Hammadi, M., et al.: Deep learning-based approach for sign language gesture recognition with efficient hand gesture representation. IEEE Access (2020). https://doi.org/10.1109/ACCESS.2020.3032140
4. Miah, A.S.M., Hasan, M.A.M., Nishimura, S., Shin, J.: Sign language recognition using graph and general deep neural network based on large scale dataset. IEEE Access **12**, 34553–34569 (2024). https://doi.org/10.1109/ACCESS.2024.3372425
5. Ajay, S., Potluri, A., George, S.M., Gaurav, R., Anusri, S.: Indian sign language recognition using random forest classifier. Presented at the 2021 IEEE International Conference on Electronics, Computing and Communication Technologies (CONECCT), Bangalore, India, pp. 1–6 (2021). https://doi.org/10.1109/CONECCT52877.2021.9622672
6. Al-Hammadi, M., Muhammad, G., Abdul, W., Alsulaiman, M., Hossain, M.S.: Hand gesture recognition using 3D-CNN model. IEEE Consum. Electron. Mag. **9**(1), 95–100 (2020). https://doi.org/10.1109/MCE.2019.2941464
7. Koller, O., Zargaran, S., Ney, H., Bowden, R.: Deep sign: enabling robust statistical continuous sign language recognition via hybrid CNN HMMs. Int. J. Comput. Vis. **126**(12), 1311–1325 (2018)
8. Pala, G., Kumbhar, S.S., Jethwani, J.B., Patil, S.D.: Machine learning-based hand sign recognition. In: Proceedings of the International Conference on Artificial Intelligence and Smart Systems (ICAIS-2021), IEEE Xplore Part Number: CFP21OAB-ART (2021)
9. Singh, A., Mittal, U., Wadhawan, A., Al Ahdal, A., Rakhra, M., Jha, S.K.: Indian sign language recognition system for dynamic signs. In: 10th International Conference on Reliability, Infocom Technologies and Optimization (Trends and Future Directions) (ICRITO), Amity University, Noida, India, 13–14 October 2022 (2022)
10. Yi, B., Wang, X., Harris, F.C., Dascalu, S.M.: sEditor: a prototype for a sign language interfacing system. IEEE Trans. Hum. Mach. Syst. **44**(4), 499–510 (2014). https://doi.org/10.1109/TSMC.2014.2316743
11. Abid, M.R., Petriu, E.M., Amjadian, E.: Dynamic sign language and voice recognition for smart home interactive application. IEEE Trans. Instrum. Meas. (Accepted for publication)
12. Mohanty, A., Rambhatla, S.S., Sahay, R.R.: Deep gesture: static hand gesture recognition using CNN. In: Raman, B., Kumar, S., Roy, P., Sen, D. (eds.) Proceedings of International Conference on Computer Vision and Image Processing. AISC, vol. 460, pp. 449–461. Springer, Singapore (2017). https://doi.org/10.1007/978-981-10-2107-7_41
13. Al-Qurishi, M., Khalid, T., Souissi, R.: Deep learning for sign language recognition: current techniques, benchmarks, and open issues. IEEE Access **9**, 126917–126951 (2021). https://doi.org/10.1109/ACCESS.2021.3110912
14. Pu, J., Zhou, W., Li, H.: Iterative alignment network for continuous sign language recognition. In: Proceedings of the IEEE/CVF Conference on Computer Vision and Pattern Recognition, pp. 4165–4174, June 2019 (2019)
15. Cui, R., Liu, H., Zhang, C.: Recurrent convolutional neural networks for continuous sign language recognition by staged optimization. In: Proceedings of the IEEE Conference on Computer Vision and Pattern Recognition, pp. 7361–7369 (2017)
16. Koller, O., Forster, J., Ney, H.: Continuous sign language recognition: towards large vocabulary statistical recognition systems handling multiple signers. Comput. Vis. Image Underst. **141**, 108–125 (2015)
17. Bragg, D., et al.: Sign language recognition, generation, and translation: an interdisciplinary perspective. arXiv preprint arXiv:1908.08597 (2019)
18. Sandler, W., Lillo-Martin, D.: Sign language and Linguistic Universals. Cambridge University Press, Cambridge (2006)

19. Mitchell, R.E., Young, T.A., Bachelda, B., Karchmer, M.A.: How many people use ASL in the united states? Why estimates need updating. Sign Lang. Stud. **6**(3), 306–335 (2006)
20. Zafrulla, Z., Sahni, H., Bedri, A., Thukral, P., Starner, T.: Hand detection in American sign language depth data using domain-driven random forest regression. In: 2015 11th IEEE International Conference and Workshops on Automatic Face and Gesture Recognition (FG), vol. 1, pp. 1–7. IEEE

Indian Sign Language Gesture Recognition Using Bi-Directional LSTM and GRU with Text to Speech Conversion

Pavitra Santhanam[1] , Aashvi[1] , Sherwin George[1] ,
and Bharathi N. Gopalsamy[2]([⊠])

[1] Department of Computer Science and Engineering, SRM Institute of Science and Technology,
Vadapalani Campus, Chennai, India
[2] Department of Computer Science and Engineering (Emerging Technologies), SRM Institute of
Science and Technology, Vadapalani Campus, Chennai, India
bharathn2@srmist.edu.in

Abstract. This abstract introduces an innovative software system dedicated to recognizing Indian Sign Language (ISL) gestures, leveraging advanced Recurrent Neural Networks (RNNs). ISL plays a pivotal role as a primary communication medium for the deaf and hard-of-hearing community in India. Addressing the critical need for accurate and efficient ISL gesture recognition, our software stands as a pioneering solution, fostering enhanced accessibility and inclusivity. The software meticulously captures the intricate nuances of ISL, facilitated by a meticulously captured dataset comprising of different ranges of Indian Sign Language Gestures. Employing a sophisticated sequence-to-sequence RNN architecture enriched with Long Short-Term Memory (LSTM) components, it adeptly models the intricate temporal dynamics inherent in sign language. This also uses Gated Recurrent Unit(GRU) for processing of the information, and this is also used to overcome the vanishing gradient problem that is very common in RNN This innovative approach empowers the software to achieve unparalleled precision in recognizing ISL gestures, facilitated by comprehensive visual and motion-based data analysis techniques. Text to speech conversion is possible due to playsound library and gTTS library. This also helps in enhancing the project by voicing out loud the names of the gesture accurately. By harnessing cutting-edge technology, our software significantly bridges communication gaps and empowers individuals within the large spectrum of people who have hard to heard symptoms as well as permanent deafness Its accuracy and efficiency not only facilitate smoother interactions but also promote greater societal inclusivity. Thus, our software represents a significant step towards fostering a more accessible and inclusive environment for all members of society, regardless of their hearing abilities.

Keywords: Indian Sign Language Recognition · Bidirectional Long Short-Term Memory · Machine Learning

P. D. Sivakumar et al. (Eds.): IRCCTSD 2024, CCIS 2362, pp. 145–160, 2025.
https://doi.org/10.1007/978-3-031-82386-2_11

1 Introduction

India Houses About 63 Million People Who Are Hard of Hearing. These People Face Intense Discrimination and belittlement Due to Their Circumstances Which Further Alienates Them and Affects Their Livelihood. With Increasing Noise Pollution and Bio magnification, Genetic Mutations Are Becoming Very Common. The Most Affected Are the Senses and Hearing Impediment Is the Most Common Among Them [1].

So Making a Sign Language Recognition System Mainstream Would Essentially Allow These People to Integrate More into Society and Eliminate the Stigma Regarding Their Conditions.

This model is very easy to use and cost effective in its application. It does not use any specialized goggles and motion detection equipment. Most models available online in the past year require leap motion and this is avoided here.

The main purpose of this application is to help a large number of people and the way to do that is by making it readily available and easy to understand so that a normal person without any technological knowledge can also use it according to their needs.

2 Problem Statement

The hearing community has to face a lot of difficulties in their day to day life. Humans are social creatures and having a sense removes inhibits the communication a little bit. Specialized equipment is not available throughout the country and if they are available they require constant maintenance to function properly.

A large number of factors contribute to an increase in hearing related issues recently. These include untreated ear infections, side effects due to surgeries, environmental pressure or even mundane accidents. The is a wide spectrum for this community and not all cases are treated in the same spotlight. Only a small population of these people can afford hearing aid or other special assistive devices. This further increases their alienation between others. They might have difficulty in day to day conversations, obtaining degrees and elevating their lifestyle. India will greatly benefit by having these technological developments, tailor made for this specific community to prosper and make them live as normally as possible.

Sign Language is like any other spoken languages in the sense it has it owns rule, grammar and way of occurrence, it might not be possible for everyone to learn a new language effectively in order to communicate with people. It required extensive focus and consistence. So the presence of this system is widely useful as it bridges the gap between people.

The problem is developing an accurate and efficient gesture recognition software tailored explicitly for Indian Sign Language (ISL). This system will enable real-time translation of ISL gestures into text or spoken language allowing better communication between deaf and hearing individuals.

3 Related Work

Sign language recognition has seen much research employing inventive gesture recognition techniques. Many methods have been tried, such as awareness covering i.e. Gloves or the software + Microsoft Kinect sensor for hand detection. In order to develop a system that is more reliable as well as effective than the others, research has been done on a variety of different current systems.

Constant Recognition of Indian Sign Language Gestures and Creation of Sentences Gradient-based key frame extraction was employed by Kumud TripathiTM, Neha Baranwa, and Nandi [2]. City-block distance and chessboard distance which are distance based classifiers, have lower recognition rates than the findings derived from Euclidean distance and correlation. Creating datasets with varied backgrounds and lighting conditions has yet to improve this work.

A lot of work that has been identified for performing sign language recognition using creative methods particularly for gesture recognition. Various techniques, like the usage of gloves or the Microsoft Kinect sensor for hand tracking, have been used in the past. To create a system that is more efficient and dependable than the others, research on numerous distinct existing systems has been conducted.

Another interesting related work is [3] Dynamic Indian Sign Language Recognition Based on Enhanced LSTM with Custom Attention Mechanism [2024] Jay M. Joshi, Dhaval U. Patel used LSTM with Custom Attention Mechanism.

They use a much more enhanced LSTM to read real-time gestures, they have a large dataset of 59 custom signs. The predicted Accuracy of the system is 96.08%, the durability of the result with 5-fold cross validation is 93% Accuracy.

Similarly, Muhammad Saad; Tianbao Yang; Hui Zhou, A Comparison of Bidirectional GRU and LSTM for Hand Gesture Recognition Using Leap Motion used Bidirectional LSTM and Bidirectional GRU using LEAP.

MOTION [4] so this setup requires a additional device which performs the function of motion sensor detection. This is not always accessible for everyone to use, not very cost-effective and requires more technological know-how.

Another Indian study using LEAP MOTION is A Modified LSTM Model for Continuous Sign Language Recognition [9]. Again LSTM is used for continuous sequencing and it recognizes a set of connected gestures. They used 942 signs sequences in ISL which aggregated to 35 different sign words. The average accuracy of this study was 72.3%

Piyusha Vyavahare, Sanket Dhawale, Priyanka Takale and Vikrant Koli 2023 Detection and Interpretation of Indian Sign Language using SIMPLE LSTM with open CV [5]. This is the simplest form of gesture recognition with just an LSTM present. It's less in complexity and effective in its usage. This model can predict simple gestures rather than complex sentences and depends a lot on camera angle, lighting, saturation form factor etc.

Another relevant paper is Sign Language Recognition Utilizing LSTM And Media pipe For Dynamic Gestures Of ISL Shamitha S H1, Dr. Badarinath K2 [6]. This study proposes a real-time Indian Sign Language (ISL) identification system for 24 dynamic signals, utilizing the Media pipe framework and LSTM network. The study's suggested approach uses a dataset made up of 24 dynamic gesture signs to train an LSTM to distinguish between various indications. The study's findings show that the aforementioned strategy produces 97% test accuracy.

In CNN and Stacked LSTM Model for Indian Sign Language Recognition [7] a convoluted neural network is used for feature extraction and LSTM captures other special information of the system. More LSTM is arranged, that is stacked for more efficiency. This paper is more efficient because of the large number of stacked LSTM. But this is a simple model, deep learning for capturing temporal data is a research problem.

A method for recognizing isolated words in ISL is used in Automatic Indian sign language recognition using MediaPipe holistic and LSTM network. Multimedia Tools and Applications [8] using Media pipe holistic pipeline feature extraction. It achieved an accuracy of 86.6%–93.5% on the INCLUDE50, INCLUDE spectrum.

The other paper we studied was Sign Language Recognition Application Using LSTM and GRU.This is done with just LSTM and GRU [16]. This paper although equipped with a methodology which is a bit similar to ours, the paper has a few disadvantages like having too much variation or recognizing too many subjects at once. Also they use a simple LSTM in place of bilateral LSTM which adds to the complexity.

This paper, Development of a Sign Language Recognition System Using Machine Learning [23] by Orovwode, Hope & Oduntan, Ibukun & Abubakar, John, is done using concurrent neural networks. They use a pre-prepared dataset with 44,654 images for American Sign Language representation. A separate HandDetector module is used to identify the hand and the gestures associated with the particular subject. The CNN had a SoftMax output layer along with three other convoluted module. It also uses an Adam Optimizer for maximizing efficiency. It has a leading accuracy of 99.7% in the testing criteria.

4 Methodology

Deep learning is being utilized to develop this model. The three main software packages are Keras, Media Pipe, and Tensor Flow. Open source machine libraries like Cognitive Toolkit, Tensor flow etc. make up the entire Keras module making it very efficient in its processing. For fast numerical computations. Tensor Flow is mainly used in creating models for complex or simple Neural networks or generally deep learning using math programs (Fig. 1).

Distributed computation is Tensor Flow's main benefit, and it makes it incredibly flexible. TensorFlow [20] is an advanced learning and machine learning framework, available in an open-sourced environment, free to use without any expense. It is developed by the Google Brain team. Being one of the most widely used libraries for configuring and training artificial neural networks, it is a vital tool for many applications, including deep learning, machine learning, and artificial intelligence.

We have to compile the data. To do so, we'll use CV 2, OS import, and collect data. Interacting and using the operating system as faithfully as possible is given by the OS module in Python. Operating systems (OS) is the most important and most effective model which provides the best utility at all times. This particular OS module offers dependent functionality by providing a portable method of operation. OpenCV

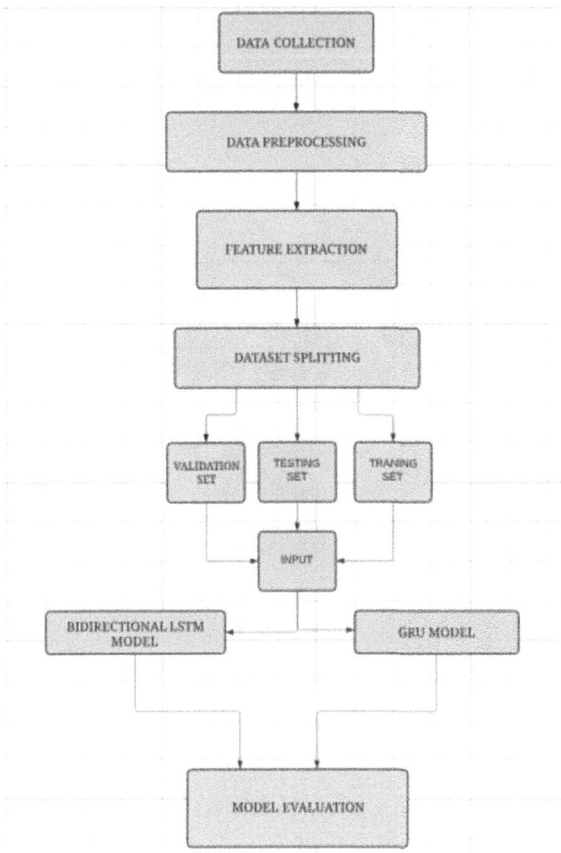

Fig. 1. Architectural Diagram

employed most frequently in many projects employing Python, was developed to inter-twine computer vision along with Python in a more appropriate means and eliminated all the problems related to said computer vision. The select_mode method is employed only when the file supplied by the user has an image to be read.

We use argparse to parse command-line arguments, csv to read files, itertools for iteration-related functions, Deque from collections to deque data structures with a maximum length (Fig. 2).

Bidirectional-LSTM [16] is used for capturing long term dependencies, Using LSTM, we obtain the hand's landmarks and significant locations. Each hand gestures uses different landmarks which makes it easier for identification. In the fields most inter-twined together namely, machine learning (ML), recurrent neural network (RNN) and deep learning (DL) the most commonly used method of solving the vanishing gradient problem [19] is a method known as LSTM. This stands for Long Short Term Mem-ory. The most frequent problem of vanishing gradient in RNN is majorly fixed by this. Since LSTMs are especially well-suited for sequential data and time series analysis, it

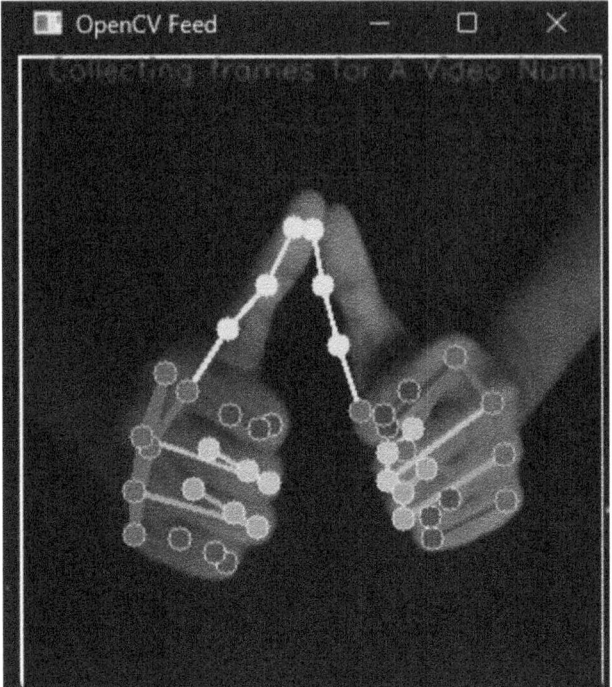

Fig. 2. Sign for The Alphabet 'A'

stands to reason that language processing, speech recognition, and other tasks involving sequences of data can all benefit from their use.

To further increase the efficiency of the system, we use another gated mechanism in combination with LSTM which is called GRU [18] (Gated Recurrent Unit Networks). It uses three gates to function, namely input, output and forget gate. The combination of both Bidirectional LSTM and GRU increases the efficiency of the entire system due to the interoperability of the former and cost efficiency of the latter. Another advantage of our model is that it also works accurately in low light situations.

Text to speech conversion is possible due to playsound library and gTTS library. This also helps in enhancing the project by voicing out loud the names of the gesture accurately. It also helps the people how are not familiar with ISL to learn it more easily. Thus it improves the accessibility of the system (Fig. 3).

One essential feature required for detecting dependencies in sequential data is the ability of long-sequence LSTM networks to learn and remember information. LSTMs differ from conventional RNNs in that they incorporate distinct memory cells and gating mechanisms that allow them to selectively store, read, and forget information as it flows through the network. Because of the way these memory cells are made, long-term storage of information makes LSTMs highly effective at handling time dependencies in data.

The type of LSTM that is used here is Bidirectional LSTM [10], that processes data that is sequential in both forward and9+-+9 backward direction with the help of two hidden layers totally. Both the layers contribute to the final output layer. Bidirectional

Fig. 3. The highlighted section shows where the mp3 file is saved for specific gestures

LSTM improves the algorithm by increasing the amount of information present in the system which in turn supplies the system with more context to much more informed information. It captures past and future context and predicts the next favorable output.

Currently, we need to train our model using Media pipe [17]. A trained model grasps the hand's location and can extract all relevant data. A variety of multimedia applications can be constructed in real-time by using the resources provided by Google's open-source Media Pipe framework, which was developed. Developing programs that handle and evaluate multimedia data, including audio, video, and images, becomes easier with its help. It is straightforward to use and accessible for both researchers and developers to use Media Pipe.

So the process here is to collect a dataset with diverse gestures and label them accordingly. Care must be taken to perform the accurate rendition of the gesture without any error. The space must be well-lit to prevent any outliers. A suitable environment with no distractions and a clear capturing high quality camera is very essential in performance of the gestures. The performance of the system is also affected by the hardware and software components of the system.

4.1 Novelty

A lot of works we researched for this project included only LSTM or types of recurrent neural network with the inclusion of different types of LSTM. This paper combines Bidirectional LSTM along with GRU with increases the efficiency of the output. Moreover, this project is very modular in the sense that new gestures can be incorporated with ease and the entire model can be trained with custom made input and output according to the user in question. So essentially it can grow to incorporate other sign languages too with different syntax and framework. Another benefit is that the system seems to work in relatively low light environment where it considers only the vague placements of the finger knuckles. Although it might not work for moving gestures with two hand movements it works accurately for single hand gestures as seen in Fig. 4.

The text to Speech functionality is also very prominent and is not associated with this project in any of the research papers yet. Even though it is very primitive this model could be enhanced further and the conversion can be made smoother. This can be done increasing the processing power and using a much more complex gtts libraries. This functionality proves very important in people learning the gestures as they see it on screen. General research has noted that listening and preserving has a more of a impact on people trying to learn and retain important information [11].

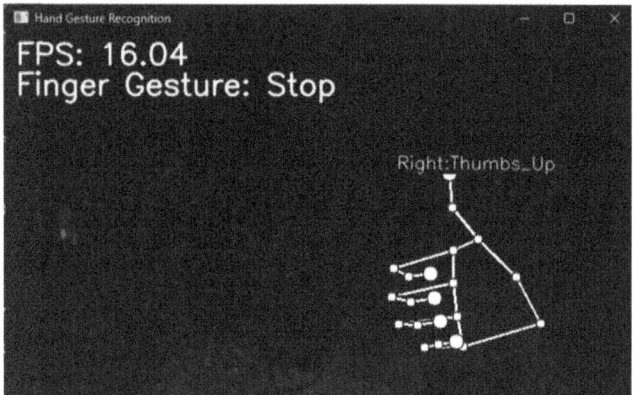

Fig. 4. Sign for the Gesture 'Thumbs Up'

Certain gestures are chosen and result is provided due to their distinct nature and key landmarks. The motion of the finger is captured by finger gesture which tells us if the hand is moving or not, FPS indicates the frames per second captured by the camera. The naming of each gesture is given above the hand along with the hand orientation that is either left or right hand. As you can notice, this works well in low light environments. A Few of our project's action and output is given below.

To highlight the performance of the module, it works efficiently with 93% accuracy using Text-to-Speech functionality and about 98% without it as shown in Fig. 5. The text-to-speech functionality lags a bit because it involves higher complexity. By using a stronger software and operating system it is theorized that, this can be overcome and the system would perform at higher speed with lesser complexities.

If you look at the classification report you can see the various columns labelled precision, recall, F1-score and support. Basically this recall and precision the first two columns evaluate the overall working and performance of the system [12]. Precision gives the correct positive identification whereas recall is the number of actual positive elements that are identified correctly. Essentially precision measures quality whereas recall measures quantity. We can see that for almost all the gestures the precision and recall is 1.0 which means that all the elements are identified correctly and labelled correctly by the system meaning that the system is working correctly. The ambiguity between gesture 10 and gesture 6 is because the gesture for 'smile' and 'call' is very closely related and the system takes a while to identify the element correctly.

Meanwhile F1 score is essentially the answer between precision and recall to support the working of the system [13]. It ranges from 0 to 1 with 1 being the perfect answer and as we can see from the output most of the answers are precise and accurate according to the system.

Table 1. Action and corresponding Output for certain gestures

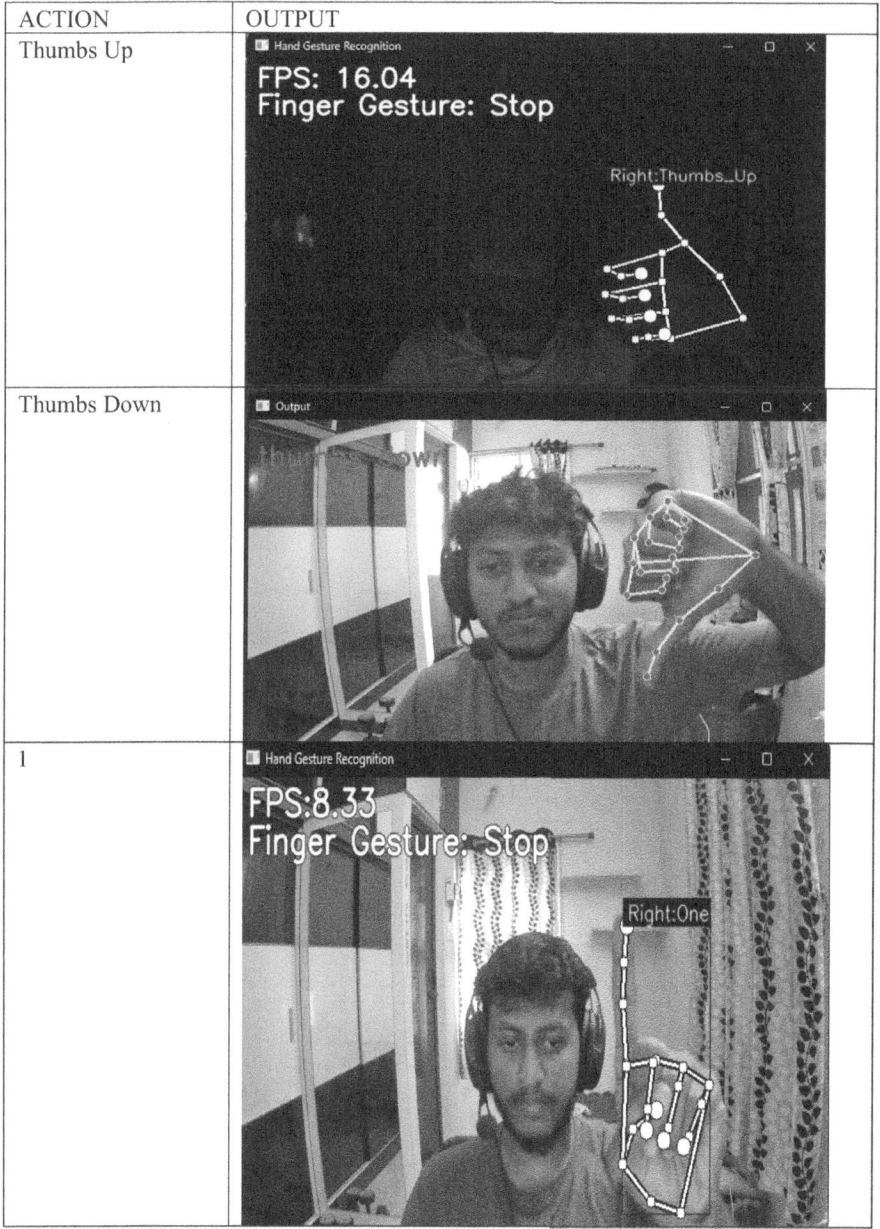

ACTION	OUTPUT
Thumbs Up	
Thumbs Down	
1	

(*continued*)

Table 1. (*continued*)

(*continued*)

Table 1. (*continued*)

Smile	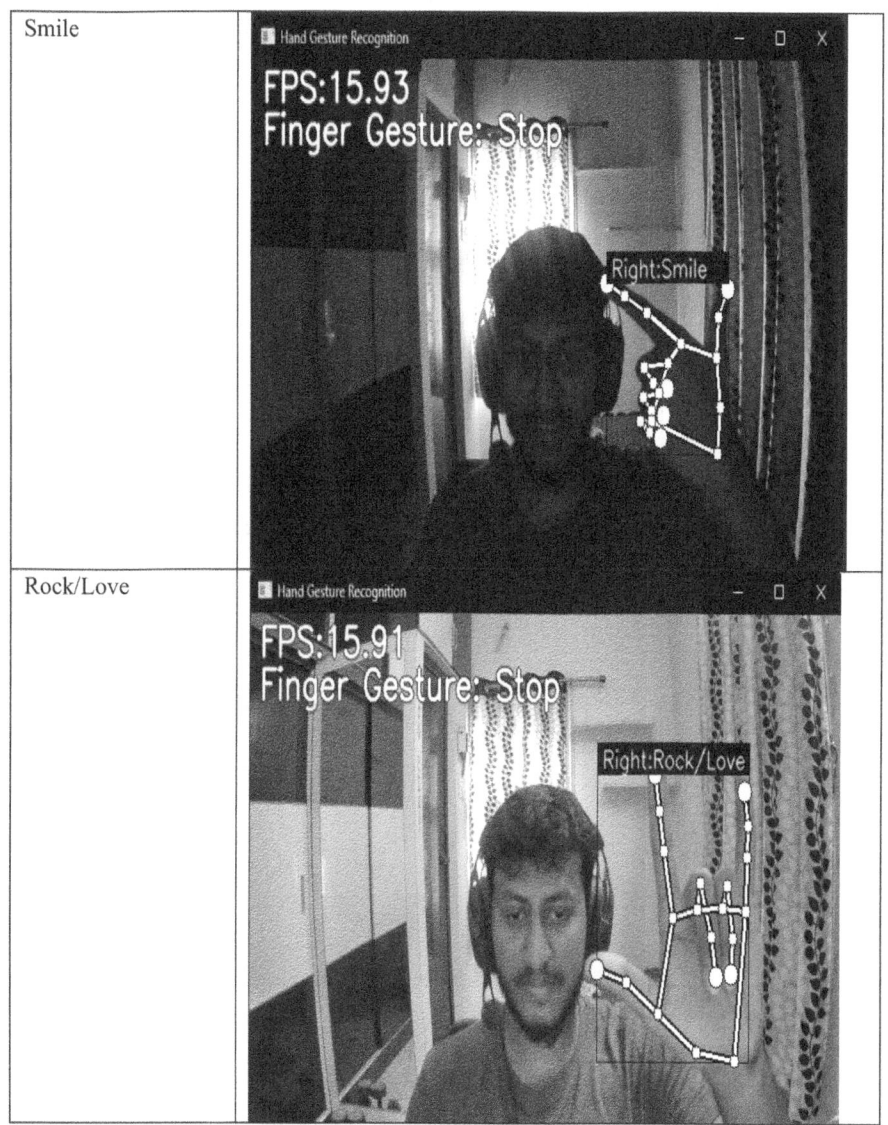
Rock/Love	

(*continued*)

Table 1. (*continued*)

From the macro average and weighted average we can see the contribution of each gesture to its performance. Macro average does not account for the actual weight of the elements and it's the actual average. On the other hand weighted average takes into account the 'weight' which is the contribution of each gesture to the average. We can see that the weighted average is 0.88 because it takes into account the gestures of 9 and 5 [15].

Similarly, a confusion matrix is also used to divide the number of predictions that are correct and incorrect in this classification problem [14]. It basically gives the data of how many sign languages are categorized correctly and how many are wrongly identified

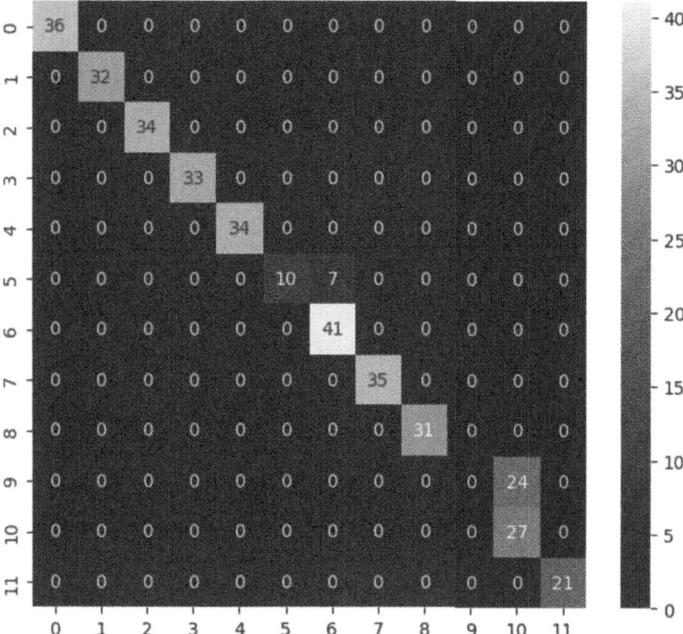

Fig. 5. Classification Report with Accuracy

```
Classification Report
                precision    recall  f1-score   support

            0       1.00      1.00      1.00        36
            1       1.00      1.00      1.00        32
            2       1.00      1.00      1.00        34
            3       1.00      1.00      1.00        33
            4       1.00      1.00      1.00        34
            5       1.00      0.59      0.74        17
            6       0.85      1.00      0.92        41
            7       1.00      1.00      1.00        35
            8       1.00      1.00      1.00        31
            9       0.00      0.00      0.00        24
           10       0.53      1.00      0.69        27
           11       1.00      1.00      1.00        21

    accuracy                           0.92       365
   macro avg       0.87      0.88      0.86       365
weighted avg       0.88      0.92      0.89       365
```

Fig. 6. Confusion Matrix

as shown by Fig. 6. We have shown here the confusion matrix of the 12 gestures taken about in Table 1.

For example, for the first gesture almost 36 out of the 40 gestures turn out to be true positive meaning the prediction is correct. The 0 means that the model does not confuse the samples between each class meaning the sample does not confuse between gesture 0 and gesture 1 or 0 with any other gesture through 11. This means that the model is working correctly. If all the diagonals of the confusion matrix is added it gives 334. The total number of predictions is 365. So the average would be 334/365 which is 92% accuracy.

5   Conclusion

The above given photographs are the actual working of the model implemented in real-time. It is done for selected key gestures for training and showcasing reason. It can be done for all the alphabets and the numbers and can be further incorporated to key gestures too.

The accuracy of the predictive model is dependent on the number of training layers used. With a 93% accuracy rate, we have successfully trained and predicted all alphabets, number and some gestures. However, to further improve the accuracy, increasing the number of instances to at least 65 images per letter is recommended.

To improve the project, gestures which include both hand along with facial features can also be implemented further which is more complex as it involves a constant movement of all the components. Ambiguity between gestures also increases with more gestures added. This can be avoided by adding more datasets in a controlled environment.

5.1   Future Work

For this project we included only single hand gestures due to convenience and to reduce the complexity added by the text to speech module. To improve the project and make it more accessible we can add more gestures which are moving and stationary. The dataset would differ for these types of gestures. A must larger dataset would be required because the gestures are more complex than single handed gestures. A larger amount of them is also required to make the training and testing easy. To improve on this, the placement of facial gestures or the stationary placement of the subjects' face can also be included to increase the scope of this project.

The voice processing can also be improved from the primitive model to incorporate much complex libraries that would make the functionality easier and would essentially reduce the complexity. The hardware with better processing power can accommodate everything that this system needs. Further this could be made into a website or an application which can be used to increase the availability and accessibility of this project. Essentially a lot can be done to further this project and it has a lot of scope. It is an essential project that actively works in integrating people into the society and sustainably develop them to be independent. So this is an evergreen project.

References

1. Wikipedia contributors: Deafness in India. Wikipedia, The Free Encyclopedia, 23 March 2024. Accessed 10:49, 15 Mar 2024
2. Tripathi, K., Nandi, N.: Continuous Indian sign language gesture recognition and sentence formation. Procedia Comput. Sci. **54**, 523–531 (2015). https://doi.org/10.1016/j.procs.2015.06.060
3. Joshi, J.M., Patel, D.U.: Dynamic Indian sign language recognition based on enhanced LSTM with custom attention mechanism. SSRG Int. J. Electron. Commun. Eng. **11**(2), 60–68 (2024)
4. Saad, M., Yang, T., Zhou, H.: A comparison of bidirectional GRU and LSTM for hand gesture recognition using leap motion. In: 2022 37th Youth Academic Annual Conference of Chinese Association of Automation (YAC), Beijing, China, pp. 1427–1433 (2022). https://doi.org/10.1109/YAC57282.2022.10023591
5. Vyavahare, P., Dhawale, S., Takale, P., Koli, V., Kanawade, B., Khonde, S.: Detection and interpretation of Indian sign language using LSTM networks. J. Intell. Syst. Control **2**(3), 132–142 (2023)
6. Shamitha, S.H., Badarinath, K.: Sign language recognition utilizing LSTM and media pipe for dynamic gestures of ISL. Int. J. Multidiscip. Res. (IJFMR) **5**(5) (2023). E-ISSN: 2582-2160
7. Aparna, C., Geetha, M.: CNN and stacked LSTM model for Indian sign language recognition. In: Thampi, S., Trajkovic, L., Li, K.C., Das, S., Wozniak, M., Berretti, S. (eds.) SoMMA 2019. CCIS, vol. 1203, pp. 126–134. Springer, Singapore (2020). https://doi.org/10.1007/978-981-15-4301-2_10
8. Khartheesvar, G., Kumar, M., Yadav, A.K., Yadav, D.: Automatic Indian sign language recognition using MediaPipe holistic and LSTM network. Multimedia Tools Appl., 1–20 (2023). https://doi.org/10.1007/s11042-023-17361-y
9. Mittal, A., Kumar, P., Roy, P., Balasubramanian, R., Chaudhuri, B.: A modified LSTM model for continuous sign language recognition using leap motion. IEEE Sens. J. **19**, 7056–7063 (2019). https://doi.org/10.1109/JSEN.2019.29098
10. Schuster, M., Paliwal, K.: Bidirectional recurrent neural networks. IEEE Trans. Signal Process. **45**, 2673–2681 (1997). https://doi.org/10.1109/78.650093
11. Hadijah, S., Shalawati, S.: A study on listening skills and perspectives to first year students at English department of academic year 2015/2016. J. Engl. Acad. **3**, 70 (2016). https://doi.org/10.25299/jshmic.2016.vol3(2).527
12. Wikipedia contributors: Precision and recall. Wikipedia, The Free Encyclopedia, 26 April 2024. Accessed 13:31, 30 Apr 2024
13. Encord blog. f1-score-in-machine-learning. https://encord.com/blog/f1-score-in-machine-learning/. Accessed 13:31, 30 Apr 2024
14. Wikipedia contributors: Confusion matrix. Wikipedia, The Free Encyclopedia. Wikipedia, The Free Encyclopedia, 9 April 2024. Web 30 April 2024
15. Mathew, J., et al.: A comparison of machine learning methods to classify radioactive elements using prompt-gamma-ray neutron activation data (2023). https://doi.org/10.21203/rs.3.rs-2518432/v1
16. Wikipedia contributors: Gated recurrent unit. Wikipedia, The Free Encyclopedia. Wikipedia, The Free Encyclopedia, 20 February 2024. Web 30 April 2024
17. Sheth, P., Rajora, S.: Sign language recognition application using LSTM and GRU (RNN) (2023). https://doi.org/10.13140/RG.2.2.18635.87846
18. Wikipedia contributors: Vanishing gradient problem. Wikipedia, The Free Encyclopedia. Wikipedia, The Free Encyclopedia, 25 April 2024. Web 30 April 2024
19. Lugaresi, C., et al.: MediaPipe: a framework for building perception pipelines (2019)

20. Wikipedia contributors: TensorFlow. Wikipedia, The Free Encyclopedia. Wikipedia, The Free Encyclopedia, 30 April 2024. Web 30 April 2024
21. Vyavahare, P., Dhawale, S., Takale, P., Koli, V., Kanawade, B., Khonde, S.: Detection and interpretation of Indian sign language using LSTM networks. J. Intell. Syst. Control **2**, 132–142 (2023). https://doi.org/10.56578/jisc020302
22. Wikipedia contributors: Keras. Wikipedia, The Free Encyclopedia. Wikipedia, The Free Encyclopedia, 30 April 2024. Web 30 April 2024
23. Orovwode, H., Oduntan, I., Abubakar, J.: Development of a sign language recognition system using machine learning, pp. 1–8 (2023). https://doi.org/10.1109/icABCD59051.2023.10220456

Comprehensive Exploration of Facial Emotion Recognition Using Conventional Machine Learning and Transfer Learning Models

C. Saravanan⬤, M. Poonkodi(✉) ⬤, and N. Prem Sankar⬤

School of Computer Science Engineering, Vellore Institute of Technology, Chennai, India
`poonkodi.m@vit.ac.in`

Abstract. Facial emotion recognition plays a vital role in enhancing human-computer interaction by allowing machines to perceive and react to human emotions. This paper conducts an in-depth exploration of various methodologies employed for recognizing facial emotions, emphasizing both traditional machine learning techniques and contemporary transfer learning models. We delve into a variety of algorithms such as support vector machines, and sophisticated neural networks like ResNet, EfficientNet, and MobileNet, assessing their efficacy using the standard MUG Facial Expression dataset. These models are tested to discern complex patterns in facial expressions, vital for accurate emotion detection. Our extensive analysis sheds light on the capabilities and constraints of each approach, providing valuable insights that pave the way for further research and practical deployments in this dynamic field. This comprehensive review aims to guide future advancements and enhance the practicality of facial emotion recognition systems.

Keywords: Facial Emotion Recognition · Support vector machines · ResNet · EfficientNet · Accuracy · MobileNet · MUG Facial Dataset

1 Introduction

Facial emotion recognition (FER) is crucial for advancing human-computer interaction (HCI), with significant applications in security, healthcare, and entertainment [1]. The precise understanding and reaction to human emotions are key to creating intelligent systems that seamlessly interact with users. Traditionally, FER has depended on conventional machine learning approaches, involving the manual extraction and selection of relevant features.

The advent of transfer learning models, however, has ushered in a new era in the domain, offering enhanced and automated methodologies for the recognition of complex emotional expressions [2]. This paper endeavors to conduct an exhaustive examination of FER, employing both classical machine learning and cutting-edge transfer learning models. We initiate our discussion by addressing the core concepts and challenges linked to FER, followed by a review of the conventional machine learning techniques historically used in this area.

© The Author(s), under exclusive license to Springer Nature Switzerland AG 2025
P. D. Sivakumar et al. (Eds.): IRCCTSD 2024, CCIS 2362, pp. 161–172, 2025.
https://doi.org/10.1007/978-3-031-82386-2_12

Subsequently, we delve into the recent progress in transfer learning, scrutinizing the application of architectures such as convolutional neural networks (CNNs), ResNet [3], EfficientNet [4], and MobileNet [5] in FER tasks. Moreover, we assess the efficacy of these models on a widely recognized dataset in the FER research community, the MUG Facial Dataset [6]. Through this comparative analysis, we aim to underscore the advantages and limitations of each methodology, offering valuable insights for both academics and practitioners in the field.

Our exploration also touches upon the significance of data preprocessing, feature extraction, and model optimization techniques in enhancing the overall performance of FER systems. Additionally, we consider the ethical considerations and challenges associated with the implementation of FER technologies in real-world settings. Ultimately, our study seeks to contribute to the continuous development of more precise, efficient, and user-centric FER systems, facilitating more empathetic and responsive interactions between humans and machines.

2 Related Work

Al-Atroshi explored the development of expression detection through transfer learning, focusing on preprocessing steps like face detection, landmark localization, and normalization. The study utilized Convolutional Neural Networks (CNNs) for both feature extraction and classification, achieving an accuracy of 87.65% [16].

Muhammad Wafi compared feature extraction methods (FL, LBP) for facial expression recognition using the ELM neural network. The study found that FL had superior performance, and the proposed aELM method slightly improved the basic ELM performance, reaching accuracies of 88.07% on CK+ and 83.12% on JAFFE datasets. The paper suggests potential improvements by addressing randomly generated input weights in future research [17].

Jaiswal introduced a transfer learning-based method for facial emotion detection, evaluated on JAFFE and FERC-2013 datasets. The proposed model outperformed existing ones, achieving an accuracy of 98.65% on both datasets [18].

Ketan Sarvakar described a network comprising six 2D convolution layers, two max pooling stages, and two fully connected layers, processing 48 × 48 pixel pre-processed face inputs. The network uses four different filter sizes, incorporates maximum pooling, and applies dropout to reduce overfitting, differing from traditional CNNs [19].

C. Dalvi provided a comprehensive overview of Facial Expression Recognition (FER), highlighting the need for balanced datasets and offering comparisons of CNN models. The paper discusses challenges, proposes solutions, and outlines emerging trends, emphasizing the rich potential in this dynamic research area [20].

Yahui Nan introduced the A-MobileNet model for FER, incorporating an attention module and dropout technology for improved feature extraction and overfitting prevention. The model outperformed lightweight MobileNet series models, achieving accuracies of 84.49% on RAF-DB and 88.11% on FER Plus, demonstrating its effectiveness in fine-grained expression analysis [21].

Ali I's study evaluated models on CK+, JAFEE, and RAF-DB datasets. The images were preprocessed using GAN, and facial landmarks were identified using MediaPipe face mesh. The KNN model outperformed others with an accuracy of 97% [22].

3 Proposed Work

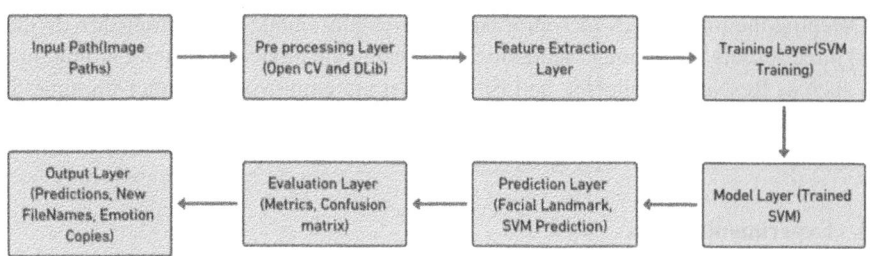

Fig. 1. Workflow of the proposed SVM Facial Annotation Architecture

The depicted workflow in Fig. 1 outlines the architecture for recognizing facial expressions utilizing Support Vector Machines (SVM). The initial step involves gathering a dataset of facial images, essential for the SVM model's training, validation, and evaluation phases. Preprocessing these images includes face detection and facial landmark extraction, which are critical features for the model. These landmarks undergo normalization to maintain uniformity across the dataset. The collected data is split into subsets for training and validation. The SVM model undergoes training on the training subset to categorize various facial expressions. Optimization of the model's hyperparameters, such as the kernel type and regularization parameter, is conducted to improve its performance. To assess the model's ability to generalize to new data, cross-validation techniques are applied. Upon successful training, the SVM model is utilized to automatically label new facial images with their respective expressions. This automated labeling process aids in efficiently categorizing facial expressions within extensive image datasets.

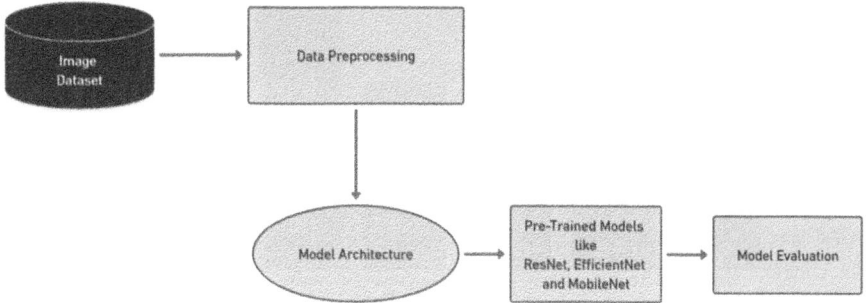

Fig. 2. Workflow of the proposed TL Pipeline

The workflow depicted in Fig. 2 describes the pipeline used for developing transfer learning models for facial emotion recognition. Initially, image datasets are collected, which are essential for training, validating, and testing the transfer learning models. Following collection, the datasets undergo preprocessing, which involves resizing images to a standard size for consistency. The images are then converted into tensor format, suitable for processing by neural networks. Additionally, pixel values are normalized to range between 0 and 1 (from the original range of 0 to 255) to facilitate faster training by keeping the magnitudes of neural network weights small. The datasets are divided into three subsets for training, validating, and testing the model. Transfer learning models such as ResNet, MobileNet, and EfficientNet are employed, and after training, the model's performance is evaluated using metrics such as precision, recall, accuracy, and F1 score.

3.1 Experimental Data

In our study, we utilized the MUG Facial Expression Database [6], which contains sequences of images depicting facial expressions performed by 86 individuals. The subjects were positioned in front of a camera against a backdrop of a blue screen. Each image is stored in jpg format with dimensions of 896 × 896 pixels and a file size ranging between 240 and 340 KB. The database grants access to images of 52 subjects for authorized users, amounting to a total of approximately 38GB of data. The data is organized according to the subject ID, with all available sequences included in the subject's folder. The image sequences can be found in the 'take*' subfolders, with the first part of the sequences categorized into folders named anger, disgust, fear, happiness, neutral, sadness, and surprise [6].

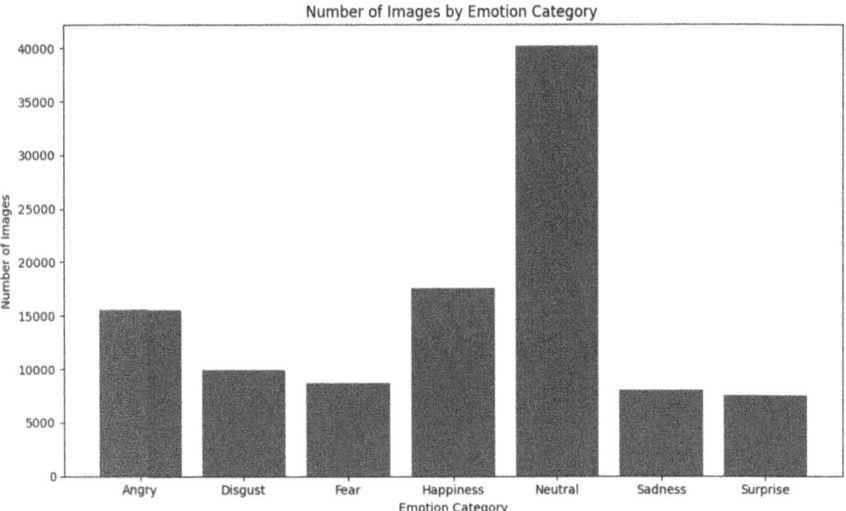

Fig. 3. Number of images by Category

The Fig. 3 provided categorizes images based on the emotions they depict, ranging from expressions of anger, disgust, fear, happiness, neutrality, sadness, to surprise. The majority of the images display a neutral expression, totaling 40,160 images. Happiness is the next most represented emotion with 17,546 images. In contrast, emotions such as disgust and fear are less represented, with 9,898 and 8,634 images respectively. Anger and sadness also have significant representation with 15,529 and 7,969 images, respectively, while surprise has the least, with 7,527 images.

3.2 Data Preprocessing

In the SVM Facial Annotations workflow, the preprocessing of facial images involves several mathematical and algorithmic steps aimed at refining the data for effective classification by a Support Vector Machine (SVM) model. Initially, face detection is executed using a Haar Cascade classifier. This involves calculating the Haar features in each image and using these features in a cascade of boosted classifiers.

Face Detection with Haar Cascade where Haar features are computed as differences in pixel intensities across adjacent rectangular regions, rapidly computed via integral images.

$$Face\ Region = CascadeClassifier(Haar\ Features)$$

$$S(x, y) = \sum x' \leq x, y' \leq yI(x', y')$$

Once the region of interest (face) is detected, it is converted to grayscale to simplify the image data, reducing it to a single intensity value per pixel.

Once the region of interest, or the face, is detected, it is converted to grayscale to simplify the image data, reducing it to a single intensity value per pixel using the formula.

$$I_{Gray}(x, y) = 0.299 \times I_R(x, y) + 0.587 \times I_G(x, y) + 0.114 \times I_B(x, y)$$

Where I_R, I_G and I_B the red, green, and blue components of the image, respectively.

Subsequently, facial landmarks are identified using the dlib library's shape predictor, a regression tree-based ensemble learning tool. The shape predictor models the relationship between pixel intensities around candidate landmark regions and the actual landmark positions, trained using a gradient boosting method. These landmarks delineate crucial facial features and their spatial arrangement, effectively transforming the facial features into a vector of coordinates. This vector serves as input for the SVM classifier, ensuring the SVM receives high-quality, relevant features for classification and enhances its ability to accurately recognize facial expressions.

To ensure consistency across different images, the extracted landmarks undergo a normalization process, involving scaling and translation to align them in a standard coordinate frame. This step helps in minimizing variations due to differences in face sizes and orientations. The normalized landmarks are then flattened into a one-dimensional array, converting the coordinate data into a format that can be used as input features for the SVM classifier.

In parallel, the emotion labels associated with each image, which represent the facial expressions, are encoded into numerical values using a Label Encoder. This encoding is necessary because machine learning models, including SVM, require numerical labels for the classification task. These preprocessing steps collectively transform the raw facial images into a structured and standardized form, making them suitable for training and evaluating the SVM model for facial expression recognition. This process ensures that the model can effectively learn from the features and make accurate predictions about the emotions conveyed in new, unseen images.

In our transfer learning framework, the initial step of preprocessing is crucial for the effective training of models. The process begins by standardizing the sizes of images to ensure that every input fed into the neural network is consistent in dimension. This uniformity is vital as it eliminates any bias toward varying image scales, allowing the network to learn from the data more efficiently. Once the images are resized, they are converted into tensors. Tensors, essentially multi-dimensional arrays, are the preferred format for numerical input in transfer learning systems. This conversion enables the images to be easily manipulated and processed by the underlying algorithms.

Following the conversion to tensors, normalization is applied, where the range of pixel values is adjusted to a standard scale, usually between 0 and 1. This is achieved by dividing the original pixel values, which range from 0 to 255, by 255. Normalization is not merely a procedural step; it is a crucial phase that aids in accelerating the convergence of the training process. It does this by regulating the gradients and enhancing the stability and consistency of the changes in the network's weights during training. Additionally, the dataset is split into two distinct sets: a training set and a test set. The training set acts as the instructor for the model, enabling it to learn and make predictions, while the test set serves as an evaluator, assessing the neural network's predictive capability on unseen data.

Beyond preparing the data for input, the preprocessing phase also aims to enhance the quality of the data used for learning. This often involves advanced techniques such as data augmentation, which includes applying various transformations like rotation, scaling, and flipping to artificially expand the original images. This not only increases the volume of training data but also introduces a level of variation that can strengthen models, enabling them to perform well when faced with new and diverse datasets. After preprocessing, the data is ready for the subsequent stage of transfer learning, where the model is trained. The data is now clean, well-structured, and primed for optimal learning.

3.3 Dataset Preparation and Facial Annotation Using SVM

In this project, we undertook the detailed task of creating and annotating a dataset for recognizing facial emotions using a Support Vector Machine (SVM) classifier. The process involved several crucial steps to ensure precision and effectiveness. Initially, we used the Haar Cascade classifier from OpenCV [7] to detect faces in each image. Following this, we utilized the dlib library [8] to extract 68 key facial landmarks, such as the corners of the mouth and the tip of the nose. These landmarks were then flattened into a one-dimensional array for use as input features in our SVM model.

$$Landmarks = shape\_predictor(I_{Gray}, Detected\ Face)$$

The shape predictor estimates the connection between the intensity of pixels surrounding potential landmark areas and their true positions, trained using a gradient boosting technique. These landmarks outline essential facial features and their layout, essential for facial expression analysis. They convert the facial features into a coordinate vector, serving as input for the SVM classifier.

To prepare our data for classification, we extracted emotion labels from the image filenames, encoded as two-letter abbreviations (e.g., 'ha' for happiness, 'sa' for sadness). These labels were then converted into numerical values using the LabelEncoder. We split our dataset into training and validation sets, using the training set to train an SVM classifier for multi-class classification. The SVM model was then tested on the validation set, where we computed various metrics such as accuracy, precision, recall, and F1 score [9] to assess the model's ability to accurately classify emotions.

For each image in our dataset, the trained SVM model predicted the emotion based on the facial landmarks, and this predicted emotion was used to annotate the image. The annotated images were then saved in a separate directory, organized into subdirectories based on their predicted emotions, which facilitates further analysis and usage of the annotated dataset. Additionally, to increase the diversity and robustness of our dataset, we included images from a separate set of subjects and performed the same annotation process, exposing our model to a wider range of facial expressions and variations.

The outcome of this thorough process is a richly annotated dataset of facial images with emotion labels, which holds significant potential for various applications in emotion recognition, human-computer interaction, and beyond. This dataset not only serves as a valuable resource for training and testing emotion recognition models but also contributes to the advancement of research in understanding and interpreting human emotions through facial expressions (Table 1).

3.4 Transfer Learning Architectures and Evaluation

The Fig. 2 shows the complete workflow of the proposed Transfer learning architecture. The architecture shows detailed steps taken for emotion recognition using transfer learning. We first loaded in the dataset after proper annotation using the SVM model. And pre-processed the dataset using various preprocessing steps like resizing, random cropping, flipping, rotation, and color adjustments, to improve the robustness of the model against variations in input data. All images were converted into tensors and normalized to ensure Consistent input distribution, which is crucial for the stable convergence of transfer learning models.

In our approach, we utilized a combination of sophisticated transfer learning models: EfficientNet [13], MobileNet [14], and ResNet [15]. Each of these architectures was selected for its proven effectiveness in classifying images, offering distinct advantages. The ResNet model, particularly the ResNet-50 variant, is renowned for its ability to overcome difficulties in training deep networks through the use of residual learning. MobileNet stands out for its efficiency, making it ideal for scenarios with limited computational resources. EfficientNet, meanwhile, offers a scalable architecture that delivers superior accuracy while requiring fewer parameters.

The evaluation of the transfer learning models extended beyond simple accuracy. Precision, recall, and F1 score metrics were used to provide a comprehensive assessment of the model's predictive performance. Precision measured the accuracy of positive predictions, recall indicated the model's ability to find all relevant instances within the dataset, and the F1 score provided a balance between precision and recall, particularly useful in the context of class-imbalanced datasets.

Table 1. The pseudocode for workflow of Facial Annotation using SVM Model

1.	*WORKFLOW 1: FACIAL ANNOTATION USING SVM*	
2.	IMPORT NECESSARY LIBRARIES	
3.	LOAD FACE CASCADE CLASSIFIER AND FACIAL LANDMARK PREDICTOR	
4.	DEF DETECT_FACES_AND_LANDMARKS(IMAGE_PATH, MAX_LANDMARKS):	
5.		READ THE IMAGE AND CONVERT IT TO GRAYSCALE
6.		INITIALIZE AN EMPTY LIST FOR LANDMARKS
7.		FOR EACH FACE DETECTED, PREDICT LANDMARKS AND ADD TO THE LANDMARKS LIST
8.		RETURN THE FIRST MAX_LANDMARKS LANDMARKS
9.	ENCODE EMOTION LABELS TO NUMERICAL VALUES USING LABELENCODER	
10.	BUILD AND TRAIN AN SVM MODEL FOR MULTI-CLASS CLASSIFICATION AND SAVE THE TRAINED SVM MODEL	
11.	MAKE PREDICTIONS FOR EACH IMAGE USING THE TRAINED SVM MODEL:	
12.	FOR EACH IMAGE PATH, DETECT LANDMARKS AND PREDICT THE EMOTION, EXTRACT INFORMATION FROM THE IMAGE PATH	
13.	STORE THE PREDICTION IN A LIST AND COPY THE IMAGE TO THE CORRESPONDING EMOTION FOLDER	

Accuracy measures the proportion of correctly classified instances out of the total instances [12].

$$Accuracy = \frac{No. \, of \, Correct \, Predicted \, Samples}{Total \, Number \, of \, Samples}$$

Precision measures the proportion of correctly identified positive instances out of all instances classified as positive [10].

$$Precision = \frac{True\ positives}{(True\ positives + False\ positives)}$$

Recall (Sensitivity) measures the proportion of correctly identified positive instances out of all actual positive instances [9].

$$Recall = \frac{True\ positives}{(True\ positives + False\ Negative)}$$

The F1 score is the harmonic mean of precision and recall, providing a balance between the two metrics [11].

$$F1 = \frac{2 \times Precision \times Recall}{Precision + Recall}$$

Additionally, a confusion matrix was employed as a visual tool to understand the classification performance for each emotion category. This matrix detailed the instances of correct and incorrect predictions, offering a granular view of the model's strengths and weaknesses across the emotional spectrum.

4   Results and Discussion

The Table 2 shows the performance metrics for SVM Facial Annotations reveal varied effectiveness across different emotions. Anger demonstrates moderate precision (62%) and recall (67%), with an F1-score of 64%, indicating the model's reasonable reliability, but highlighting the need for refinement. Disgust is identified with high accuracy, as evidenced by its precision (92%), recall (86%), and F1-score (89%). Fear's balanced performance is reflected in its F1-score of 84%. Both happiness and surprise achieve perfect precision scores of 100% and exhibit high recall, leading to impressive F1-scores of 97% and 92%, respectively. This suggests the model's exceptional proficiency in recognizing these emotions. Neutral emotion is characterized by a notably high recall, indicating its consistent detection, while sadness boasts perfect precision, showcasing its precise identification. Overall, the SVM classifier excels in detecting emotions like happiness, surprise, and disgust, though it shows some limitations in accurately identifying anger.

The performance Indicators for the SVM model reveal a high degree of precision In its outcomes, with an overall accuracy reaching 87%. The model's precision rate is a notable 88.3%, highlighting its capability to generate accurate predictions. Additionally, its recall rate of 87.12% demonstrates the model's effectiveness in capturing all essential instances. An F1 Score of 87.34% is indicative of a well-tuned model that strikes an excellent balance between exactness and completeness in its results. Such metrics denote the model's dependable performance in data classification tasks, establishing its significance in the realm of predictive analytics.

Table 3 and Fig. 4 shows the comparative analysis of three advanced transfer learning models reveals the following: MobileNet demonstrates a uniform performance with an

Table 2. Precision, Recall and F1-Score of the SVM model for facial annotation

Emotion	Precision	Recall	F1-score
Angry	62%	67%	64%
Disgust	92%	86%	89%
Fear	80%	89%	84%
Happiness	100%	94%	97%
Neutral	83%	96%	89%
Sadness	100%	80%	89%
Surprise	100%	85%	92%

Table 3. Transfer learning algorithm's evaluation metrics in Percentage.

Algorithm	Accuracy	Precision	Recall	F1-score
MobileNet	73.66%	74.84%	73.66%	72.17%
ResNet	67.13%	72.20%	67.13%	66.05%
EfficientNetB0	76.69%	77.30%	76.69%	76.56%

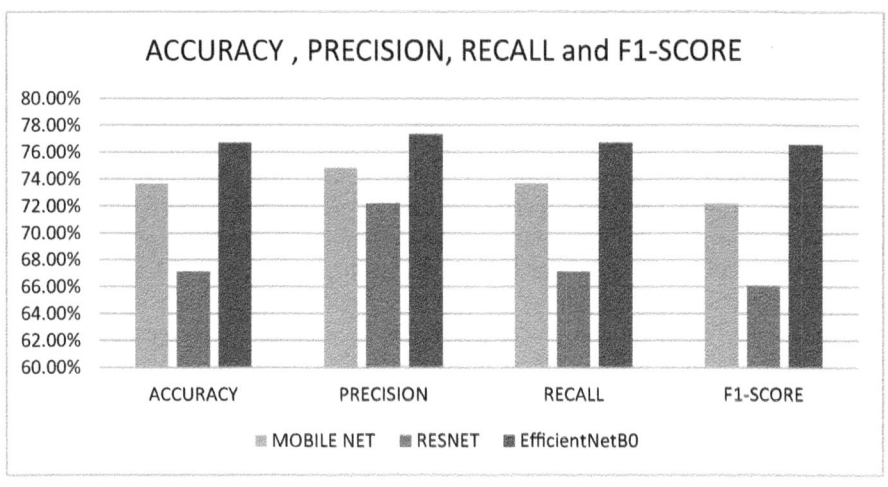

Fig. 4. Evaluation metrics in % of the used transfer learning algorithms

accuracy, recall, and precision in the mid-70s percentile, and an F1-score slightly below at 72.17%, suggesting reliable outcomes across the board. ResNet, while having the modest numbers among the trio, posts an accuracy and recall of 67.13%, coupled with a slightly higher precision of 72.20%, indicating its effectiveness in making correct predictions when it identifies cases, despite a lower detection rate as reflected in its F1-score of 66.05%. EfficientNetB0 emerges as the frontrunner with superior accuracy and recall rates at 76.69% and the highest precision of 77.30%, culminating in an F1-score of 76.56%, which signals its superior balance in both accurate and consistent prediction capabilities.

5 Conclusion

In summarizing this work, the initial focus was on grasping the significance of recognizing facial emotions. A dataset was curated for this purpose, utilizing transfer learning to detect facial emotions. The SVM Model was employed for annotating facial expressions through landmarks, achieving a commendable overall accuracy of 87%. Subsequent to this, a comparative evaluation of transfer learning models was undertaken, examining ResNet, MobileNet, and EfficientNet. MobileNet demonstrated consistent results, with accuracy, precision, and recall all in the approximate 74% range and an F1-score at 72.17%. ResNet presented lower performance with scores mainly around 67%, yet it had a relatively higher precision of 72.20%, indicating its effectiveness in making correct predictions, despite missing some cases. EfficientNetB0 surpassed its peers with the topmost scores—an accuracy of 76.69%, precision of 77.30%, and both recall and an F1-score just slightly lower at 76.56%, positioning it as the most proportionate and precise model among those tested, thus the most trustworthy for regular predictive tasks.

6 Future Works

Advancements could be made through further refinement of the models' parameters and, with increased computational resources, the models could potentially be executed on the entire dataset instead of a fraction, possibly enhancing performance further.

Statements and Declarations

Competing Interests: This research was conducted solely by the student with academic supervision from the teacher; no external funding, support, or resources were involved.

References

1. Corneanu, C.A., Simón, M.O., Cohn, J.F., Guerrero, S.E.: Survey on RGB, 3D, thermal, and multimodal approaches for facial expression recognition: history, trends, and affect-related applications. IEEE Trans. Pattern Anal. Mach. Intell. **38**(8), 1548–1568 (2016)
2. Li, S., Deng, W., Du, J.: Reliable crowdsourcing and deep locality-preserving learning for expression recognition in the wild. In: Proceedings of the IEEE Conference on Computer Vision and Pattern Recognition (CVPR), pp. 2852–2861 (2017)
3. He, K., Zhang, X., Ren, S., Sun, J.: Deep residual learning for image recognition. In: Proceedings of the IEEE Conference on Computer Vision and Pattern Recognition (CVPR), pp. 770–778 (2016)

4. Tan, M., Le, Q.V.: EfficientNet: rethinking model scaling for convolutional neural networks. In: Proceedings of the 36th International Conference on Machine Learning (ICML), pp. 6105–6114 (2019)
5. Howard, A.G., et al.: MobileNets: efficient convolutional neural networks for mobile vision applications. arXiv preprint arXiv:1704.04861 (2017)
6. Aifanti, N., Papachristou, C., Delopoulos, A.: The MUG facial expression database. In: 11th International Workshop on Image Analysis for Multimedia Interactive Services (WIAMIS), pp. 1–4 (2010)
7. Bradski, G.: The OpenCV library. Dr. Dobb's J. Softw. Tools **120**, 122–125 (2000)
8. King, D.E.: Dlib-ml: a machine learning toolkit. J. Mach. Learn. Res. **10**, 1755–1758 (2009)
9. Powers, D.M.W.: Evaluation: from precision, recall and F-measure to ROC, informedness, markedness and correlation. J. Mach. Learn. Technol. **2**(1), 37–63 (2011)
10. Davis, J., Goadrich, M.: The relationship between Precision-Recall and ROC curves. In: Proceedings of the 23rd International Conference on Machine Learning, pp. 233–240 (2006)
11. Chinchor, N.: MUC-4 evaluation metrics. In: Proceedings of the 4th Conference on Message Understanding, pp. 22–29 (1992)
12. Fawcett, T.: An introduction to ROC analysis. Pattern Recognit. Lett. **27**(8), 861–874 (2006)
13. Hoang, V.T., Jo, K.H.: Practical analysis on architecture of EfficientNet. In: 2021 14th International Conference on Human System Interaction (HSI), pp. 1–4. IEEE, July 2021
14. Sinha, D., El-Sharkawy, M.: Thin MobileNet: an enhanced MobileNet architecture. In: 2019 IEEE 10th annual ubiquitous computing, electronics & mobile communication conference (UEMCON), pp. 0280–0285. IEEE, October 2019
15. Wu, Z., Shen, C., Van Den Hengel, A.: Wider or deeper: revisiting the ResNet model for visual recognition. Pattern Recognit. **90**, 119–133 (2019)
16. Facial expression recognition based on transfer learning: an overview. Iraqi J. Sci. **64**(3), 1401–1425 (2023)
17. Wafi, M., Bachtiar, F., Utaminingrum, F.: Feature extraction comparison for facial expression recognition using adaptive extreme learning machine. Int. J. Electr. Comput. Eng. (IJECE) (2023)
18. Jaiswal, A., Krishnama Raju, A., Deb,S.: Facial emotion detection using transfer learning. In: 2020 International Conference for Emerging Technology (INCET), Belgaum, India, pp. 1–5 (2020)
19. Sarvakar, K., Senkamalavalli, R., Raghavendra, S., Santosh Kumar, J., Manjunath, R., Jaiswal, S.: Facial emotion recognition using convolutional neural networks. Mater. Today Proc. **80**(Part 3), pp. 3560–3564 (2023). ISSN 2214–7853
20. Dalvi, C., Rathod, M., Patil, S., Gite, S., Kotecha, K.: A survey of AI-based facial emotion recognition: features, ML & DL techniques, age-wise datasets and future directions. IEEE Access (2021)
21. Nan, Y., Ju, J., Hua, Q., Zhang, H., Wang, B.: A-MobileNet: an approach of facial expression recognition. Alex. Eng. J. **61**(6), 4435–4444 (2022). ISSN 1110-0168
22. Siam, A.I., Soliman, N.F., Algarni, A.D., El-Samie, F.E.A., Sedik, A.: Deploying machine learning techniques for human emotion detection. Comput. Intell. Neurosci. **2022**, Article no. 8032673, 16 p. (2022)

Type Sculpt: Text-to-3D Generation with Personalized Precision Using Adaptive Attention Mechanism

J. Arunnehru[2]($\boxtimes$), V. S. Koushik Babu[1], G. Adithya[1], and K. M. S. Pragadeesh[1]

[1] Department of Computer Scince and Engineering, SRM Institute of Science and Technology, Chennai, India
`{kb2728,ga0191,pm5501}@srmist.edu.in`
[2] Department of Computer Scince and Engineering (Emerging Technologies), SRM Institute of Science and Technology, Chennai, India
`arunnehj@srmist.edu.in`

Abstract. Our project focuses on refining the process of 3D modeling from multi-view images, aiming to enhance the final model's fidelity and texture details while maintaining computational efficiency. The core of our methodology involves our own refiner model that preprocesses images to significantly upscales the clarity, detail, textures and lightning, which will be very much benefitial for mapping the details from the images for 3D mesh generation. By implementing techniques like VAE and Unets, which optimizes texture and geometric details from these source images, we ensure that the generated 3D models are high in quality and also rich in the textures based on the original source images and even consistent enough to generate a 3D mesh using NGP techniques. Results from our experiments demonstrate that our approach improves the visual accuracy of the 3D models with remarkably low GPU resources. This advancement underscores the potential for applying our methods in real-time 3D modeling applications, opening new way for both practical and innovative uses.

Keywords: Neural Radiance Fields · Novel Multi-View synthesis · Variational Auto-Encoders · UNet · Generative Adversarial Networks

1 Introduction

The emergence of 3D reconstruction techniques, within digital areas like virtual reality, augmented reality, industry design, and heritage preservation, has created remarkable changes. Despite these achievements, efficiently building 3D models with fidelity from images still poses serious challenges. Often experienced problems are inadequate data capture, color variations and lighting inconsistencies that effect the realism and accuracy of reconstructing objects through computer applications. Previously, this problem was solved by either upscaling the computational resource or using more advanced sensing equipment. Such solutions, though, following through with price complication and therefore resulting with impossibility of accessibility to many applications. Scores of

© The Author(s), under exclusive license to Springer Nature Switzerland AG 2025
P. D. Sivakumar et al. (Eds.): IRCCTSD 2024, CCIS 2362, pp. 173–186, 2025.
https://doi.org/10.1007/978-3-031-82386-2_13

studies now delve into application of various computational approaches to the one aim which is improving the quality of 3D models and yet restoring this balance between the model accuracy, computational efficiency, and availability is a big task.

Therefore, our research puts forward a new method that relies on a powerful deep learning technology to boost greatly the quality of 3D reconstruction from the collection of mutli-view images. Through the implementation of a custom neural network, which simulates the capabilities of text to image generation and utilizes a selected small dataset of varied images we use the strengths of current popular methods to fix their weak sides. Our approach does not only achieve the accuracy in the appearance texturing but size consistency as well.

Main Contributions

1. Enhanced Image Quality for 3D Reconstruction: We have developed a refined image enhancement model using U-Net, which is a "Refiner" for existing image generation model, specifically developed to improve the resolution, texture, and structural details of images intended for 3D modelling.
2. Advanced Texture, Depth, and Detail Mapping: Our method provides uses models such as CLIP [20] and DINO [21] to significantly enrich the depth perception and texture mapping of the reconstructed 3D models, ensuring a high level of detail and realism.
3. Consistency across Multi-View Images: We ensure that the generated multi-view images maintain consistent lighting and textural quality, thereby facilitating more accurate and realistic 3D reconstructions.
4. Efficient 3D Texturing with Lower Computational Resources: Our method is optimized to reduce the demand on computational resources, making high-quality 3D texturing and modelling feasible on less powerful hardware.

2   Related Works

2.1   Text to Better Image for 3D Modelling

The progress in text-driven shaping of 3D objects is quite significant as compared to conventional techniques that relied on explicit data. This represents a step towards interactive, dynamic, and integrated models. For instance, "DreamStone" [1] and "EXIM" [4] are just some of the best works that have proved a step ahead by integrating textual descriptions to enhance 3D shape generation. In-depth researches were conducted by using profound neural networks that can interpret and visualize text into a 3D object of high quality for modelling purposes. Besides, "MotionDiffuse" [5] and "Text-to-3D Generation with Bidirectional Diffusion" [8] explore human and animal as well as complex shape synthesis to provide robust foundations for later 3D application. HD-Fusion model[12] is an example where multiply noise estimation functions to bring a good text-to-3D translation process.

2.2   Consistent Multi-view Images

Several studies have aimed at meeting this difficulty of producing multi-view image sequences from a single input view. Work by "Zero123++" [2] and "One-2-3-45" [3] mention the fast and coherent development of 3D materials from a limited number of viewpoints. Hereby the approach underlines the importance of consistency among views, significantly increasing the fidelity of the created models in the end. Finally, "TetraDiffusion" [17], "Triplane Meets Gaussian Splatting" [18], and "Wonder3D" [19] are some methods that feature diffusion and transformation with the main purpose of visual consistency and detail preservation across different generated views which can enable 3D models to be functionally utilized in multiple areas.

2.3   Refinement of Multiviews

This area in computational vision research that mainly deals with the process of refining 3D objects from multi-view images is getting really popular these days. The leading purpose in this field is to raise the geometric accuracy and visual quality of the obtained 3D images by using multiple views of an object. In this area, Pixel2Mesh++ [7] model is a remarkable achievement, since it employs the method of fine-tuning mesh structures in a unique manner. This model best illustrates the union of the geometry deep learning techniques and the classic computer vision approaches in that the surface quality of the mesh is significantly improved by adjusting the vertices based on the inputs of multiple images.

Another important contribution is the "SyncDreamer" model [16], which applies a different synchronization technique to keep feature alignment when viewing from different sides. The model exploits the most advanced feature extraction and alignment methods, leading to representational 3D models with the same texture and detail when seen from different viewpoints.

2.4   3D Mesh Generation NGP

Over the years, 3D mesh generation has experienced revolutionary progress via the usage of Neural Graphics Primitives (NGP), which includes outstanding methodologies such as "SceneDreamer" [6] and "3D Neural Field Generation using Triplane Diffusion" [9]. The implemented solutions have greatly enhanced the ability for developing detailed and realistic 3D scenes, which are guided by visual principles. "SceneDreamer" possesses such capabilities as synthetic scene generation from scant inputs on the basis of latest technologies, which contributes to many areas of virtual reality and digital simulations. However, "3D Neural Field Generation using Triplane Diffusion" applies triplane networks to build the volumetric fields and then generate the 3D models of great detail and precision making a great leap in accuracy and detail for 3D representations.

As we move towards the actualization of the cutting edge 3D real-time content creation, we have "Fantasia3D" [11] and "Instant3D" [13]. "Fantasia3D" is very successful at converting text into a three-dimensional depiction immediately. As a result, a new domain combining the two fields natural language processing and 3D graphic design is formed. It is an ability which not only makes design process faster but also provides a platform for expression of creativity and quick prototyping. "Instant3D" in the sense that it provides means to make 3D content from textual and image inputs instantly, ensuring efficiency and quality at the same time.

Furthermore, "LION" [15], a model by NVIDIA, that has been at the forefront of latent diffusion, and is another milestone in the creation of 3D forms. It employs the latent space diffusion method, which improves the generative process, creating better and more accurate 3D models. This model gives the NGP technology a chance to grow in the market along with the 3D modelling because this technology gives a robust solution to both artistic creation and practical application in the industry.

These studies collectively emphasize the fact that the 3D modelling technologies continue to develop, supported by the merging of powerful computational tools and modern neural network architectures. This constant refinement and advancement of these methods is expected to improve the quality, authenticity and applications of 3D models in significantly diverse ways.

3   Methodology

Our framework fuses modern computer vision, deep learning, and 3D graphics to further develop the multi-view 3D reconstruction and content generation. The difference of our method lies in combining the diffusion models, the attention mechanisms and the neural graphics primitives which actually form a single system that is able to create highly detailed models from images or prompts in texts. The proposed work is the first approach that integrates state-of-the-art algorithms from both image processing and deep learning for multi-view 3D reconstruction and improvement. At the core is the integration of novel pretrained neural networks in such a way that they are able to deal with the issues caused by sparse data and image reconstruction and refinement. We base our method on diffusion models, which are arguably, capable of catching the intricate dependencies they impose within the data distributions (Fig. 1).

3.1   Source Image Generation and Refinement

3.1.1   Custom Neural Network Training

Our approach encompasses a neural network of our own design, which is functionally a class by itself when facing the challenges that are associated with the generation of 3D models. The network for recognition of textures is built on the backbone of a comprehensive dataset, which consists of 4000 photos showing a wide variety of textures, lighting conditions and geometric complexities. The network architecture is based on a U-Net with a slight modification. This modification makes a U-Net more or less a perfect model for semantic segmentation plus it also improves its ability to perceive and reinforces textural differences in the images. This detailing is really the backbone of creating such imagery which can not only accurately mirror real life scenarios, but also have the intricate details necessary to help the 3D reconstruction process. Considering this unique approach to training technique, it stands as the pace to achieve new technologies and overcoming the challenges faced by antiquated methods that always struggle to produce surfaces with the required fidelity and consistency for intricate scenes.

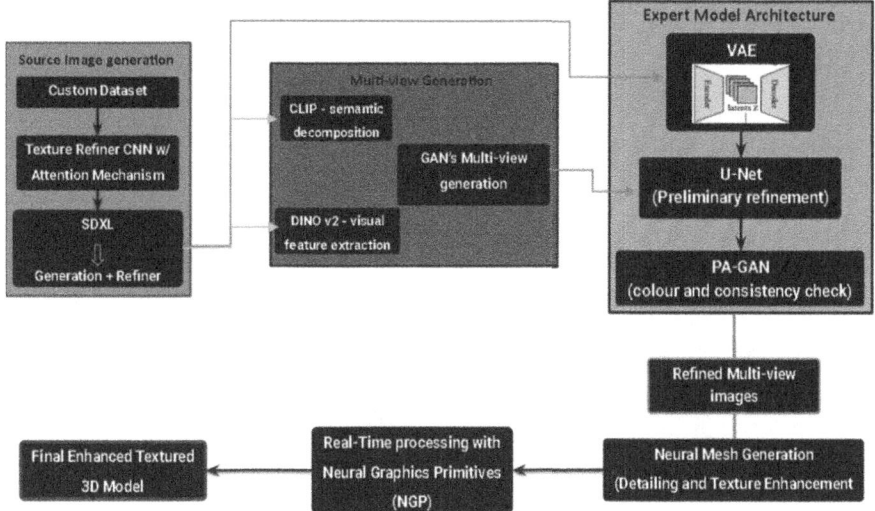

Fig. 1. Overview of the proposed text-to-3D generation framework.

3.2   Image Generation with SDXL

We aim to take things to the next level by combining the effectiveness of SDXL framework and incorporating the neural network we have perfected with a custom refiner module. A favorable combination assures that the overall development process is photographically modeled, which results in high quality renders for 3D applications. While SDXL + Refiner advanced module specializes in producing images of high geometrical accuracy, uniform lighting, and realistic textures. Traditional image generating models lack this. Such pre-requisites set a foundation which becomes a crucial step in later multi-view reconstruction processes and bring noticeable improvements as compared to existing techniques, which oftentimes are burdened by post-processing image corrections to stitch those together into a realistic 3D scene.

Following is a quality analysis of various state-of-the-art AI image generation modes (Fig. 2). This approach is also well improved compared to others according to the output criteria such as geometrical accuracy, lighting uniformity as well as texture realism, making our SDXL + Refiner module competitive in the market of image generation for 3D applications.

3.3   Multi-view Image Synthesis from Refined Source Images

The generation of a multi-image set from a single picture source shows the foremost research step in our research work. We synthesize CLIP semantic parsing ability, DINO v2 fine-grain effect, and GAN's high generative performance, and set the boundary for the highest possible detail and semantic link among multi-view images. This way gauges at preserving each synthesized view as visually consistent as possible by semantically associating with the original image source; which is a new method of improvement over previous ones where varying synthesized perspectives often have a discrepancy

MODELS/ Prompts	Comp-vis (V2)	Runway ML	SDXL Base 1.0	Segmind SSD 1B	Ours
A 3D image of a Pikachu with uniform lightning & texture					
A 3D image of a treasure chest with uniform lightning & texture					
A 3D image of a wizard with uniform lightning & texture					

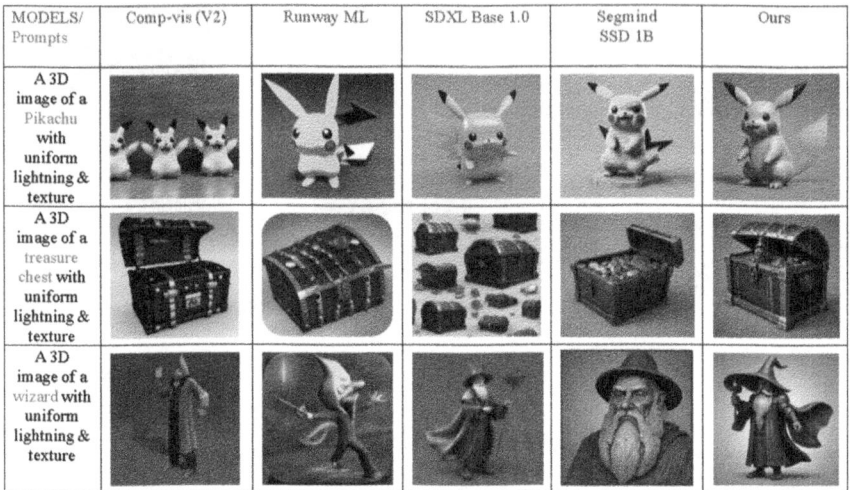

Fig. 2. Qualitative comparison of 3D-friendly image generation across different models and prompts.

semantics. The hypothetical results of such model integration sheds light on substantial step towards understanding and rectifying the complex image datasets for 3D modeling, and practically. The methods can be used to achieve higher precision and accuracy in various applications such as virtual reality and architectural visualisation.

3.3.1   Multi-view Synthesis with a Generative Adversarial Network (GAN)

The generator of GAN merges the semantic embeddings of CLIP and detailed feature-maps of DINO v2 in order to produce a new image view which is both cognitively and visually correct. The GAN discriminator is learnt over the course of training to differentiate between real and synthesised multi-view images, which helps the generator in sharpening its outputs for greater reality effect. The adversarial learning approach challenges the discriminator module to think in a semantically consistent way (with the help of CLIP) and spatially coherent manner (via DINO v2). Such a realised consistency and refinement function will then refine and use the cycle consistency loss to ensure high precision and coherence between the generated and these other views.

3.4   Multi-view Refinement Using Source Image Textures

Our process helps achieve finer images and make the multi-view images look as close as possible as to the original image in terms of texture, detail and color. Developing a unique variation of the VAE, the U-Net technology, and the GANs, we will refine the process additionally by targeting more realistic aspects and maintaining an overall consistency among the generated images.

Stage 1 - Texture Embedding with VAE: Primarily we implement a VAE to draw the textures features of the source image as a compressed latent space. This texture representation then assists in improving the following refinement.

Stage 2 - Detail Enhancement with U-Net: Then, we apply the U-Net architecture with the input consisting of the VAE's output and multi-view images by encoding the features and the textures while down sampling and then synthesizing the refined images from the upsampling process aimed at enriching the texture and quality of the multi-view images. Stage 3 - Progressive Attention GAN for Final Refinement: The GAN with its progressive attention function will continue to look iteratively on the areas that are far from being the same textures, colors, or details that are expected. This ends up yielding an optimized performance of the dressing process with gradual refinements in the output views (Fig. 3).

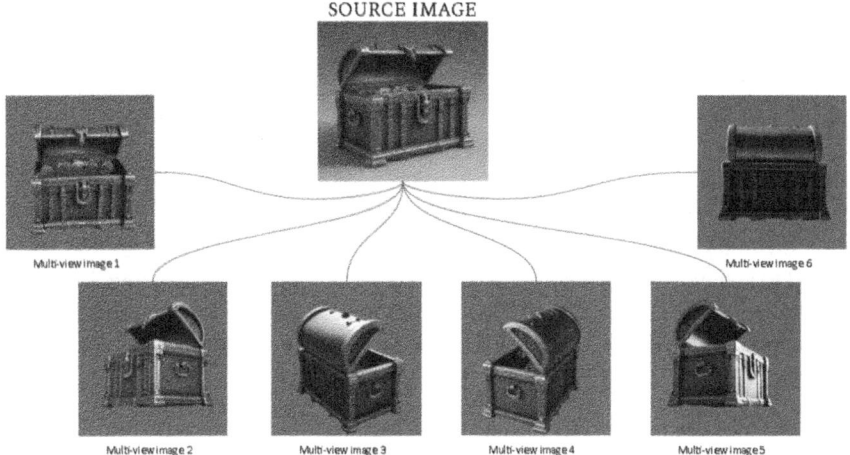

Fig. 3. Illustration of multi-view image synthesis from a single source image using our proposed method.

The multi-stage refinement process gives the source image another chance to shine by computing similar multi-view images but has higher fidelity and ensures consistency across the views. We combine VAE, U-Net, and GAN in order to show an informative but easy to follow procedure for the enhancement of images from multiple views in the reconstruction process.

3.5 Advanced 3D Reconstruction with NGP

Our method of 3D model generation by multi-view images significantly strengthens precision and fidelity. The primary tasks will be interlinking the image resolutions as well as the light normalizing, which traces back to the cameras positions and locations. This fundamental stage then allows the neuro-radiance field (NeRF) to be able to subsequently reproduce the described content that the scene consists of in all volumetric details. Based on the ideas behind Instant Neural Graphics Primitives [14], the principles are relying on similar multiresolution hash encoding to further bring high efficiency to our NeRF without introducing much computational cost, thereby the produced NeRF is more precise.

3.5.1 NeRF and Scene Rendering

Upon setting the intrinsic parameters accurately, we initiate a Neural Radiance Field (NeRF) using the NGP to capture the complex volumetric properties of the scene with exceptional detail. The process involves dynamic refinement and real-time adjustments that enable the NeRF to rapidly evolve into a detailed and accurate 3D representations. This approach is influenced by innovative practices from the Fantasia3D project [11], which underscores the significance of separating geometric details from visual appearance to produce high-quality 3D content. These insights guide our dynamic refinement strategies, ensuring the NeRF model adapts and improves efficiently in real-time.

3.6 Novelty and Contributions

Our project introduces several novel contributions that distinguish it from existing research in the field:

3.6.1 Refiner Model for Enhanced Image Quality

Unlike previous approaches, we developed a dedicated refiner model specifically designed to enhance image quality for 3D modelling purposes by encoding the features like lighting, colours and textures. This refiner model effectively improves the fidelity and detail of source images, thereby leading to more accurate and visually appealing 3D reconstructions.

3.6.2 Integration of Multiview Refinement with Consistency

One of the unique aspects of our project is the integration of multi-view refinement techniques with texture preservation and maintaining the consistency of the images with one another. This ensures that the reconstructed 3D models to accurately represent the geometric structure and retain the intricate textures and details present in the source images with the continuity between the images.

3.6.3 Detail Refinement

To enhance the fidelity of our 3D models, we employ adaptive sampling techniques that focuses on areas of the model with higher complexity. This method improves both the texture quality and geometric details while keeping the usage of the GPU and any computational resources very low. This approach is kind of similar to those found in advanced methodologies like 3D Shape Generation and Completion through Point-Voxel Diffusion [10], where the focus is on refining the generation of 3D shapes using sophisticated diffusion techniques. By selectively enhancing intricate areas, we significantly improve the visual quality of the generated 3D object with better textures.

3.6.4 Efficient GPU Utilization

Our model demonstrates exceptionally efficient GPU utilization, requiring significantly fewer computational resources compared to existing methods. By optimizing resource allocation and minimizing computational overhead, we achieve faster processing times without compromising on the quality of 3D models generated.

Additionally, the novelty in 3D mesh creation of the proposed architecture is in the implementation of texture mapping together with mesh refinement. Thus, the quality and realism of produced models are significantly improved. The framework we are proposing makes use of the most advanced texture mapping procedures and a highly efficient way of distributing the computational efforts. As a result, we are creating a new benchmark for 3D reconstruction models. These can then be utilized in any desired application area or advance the development of the 3D visualization technologies. These innovations not only ensure high fidelity in terms of visual appearance but also improve the practical usability of generated models in real-world scenarios, pushing the boundaries of current 3D visualization technology.

4 Experiments

4.1 Dataset Preparation and Preprocessing

For the purpose of 3D reconstruction, of our advanced capability, we assembled a unique dataset comprising 4,000 high-quality images, which are specifically selected to maintain the varied object shapes, textures, and lighting conditions. This is a data set which is supposed to produce realistic visuals and to improve the structural detail of light and texture thus making the techno-realistic approach of the 3D modelling possible. On top of that, we also trained the OAI-XL library, which consists of over 10 million diverse 3D objects, thus greatly increasing our training resources and dealing with a common problem – of lack of good quality 3D data.

4.2 Implementation Details

4.2.1 Enhanced Image Generation for 3D Reconstruction

We intend to develop a specific neural network at the first stage which is pre-trained with the purpose of incorporating a wide-range of text-to-image tasks. Then, the fine-tuning will be done on the basis of the 4000-image which we have prepared for this purpose. This network is the core of our image refinement process, which significantly improves image quality for 3D modelling applications. Among the steps we took, the development of a dataset specific to source image improvement for more accurate reconstruction was a significant one.

We used some of the famous image generation models that are available in open source, and used custom refiners to produce better images that are specifically curated for 3D modelling. This refinement process significantly improved its capability to generate high-quality, 3D reconstruction-friendly images. A comparative analysis of various image generation models, including comp-vis, Runway ML, Stable Diffusion XL, Segmind-SSD, and our model, underscored our method's superiority in generating high-quality images efficiently (refer to Fig. 4).

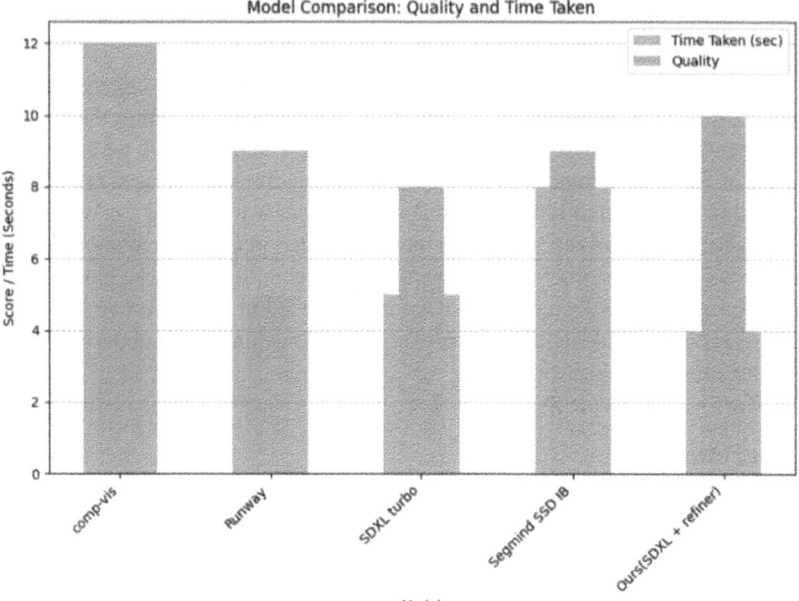

Fig. 4. Illustrates the time efficiency and quality of images generated by our model compared to competitors

4.3 Evaluation Metrics

4.3.1 Performance Benchmarking and Model Superiority

Key performance indicators, specifically Mean Squared Error (MSE), Feature Matching Count, and Peak Signal-to-Noise Ratio (PSNR), were used to evaluate the SDXL + Refiner's efficacy against other leading models.

The Mean Squared Error (MSE) metric provides a critical baseline for understanding pixel-level accuracy and discrepancies. Our model exhibited a lower MSE when compared to other image generation models, suggesting less significant pixel variance from the standard benchmarks. In Fig. 5, we observe that while our MSE is very lower than other open source Gen AI models and also has better feature matching count when tested with a ground truth (GT) reference image of the same type.

Moreover, our model achieved the highest scores in PSNR comparisons, as with the other state-of-art models in the same category. (Fig. 6.) PSNR being an excellent indicator of the quality of reconstructed images, measuring the ratio between the maximum possible power of a signal and the power of corrupting noise. Our model shows a high PSNR, indicating that the reconstruction has a high fidelity and less noise, which is very important in scenarios where clarity and detail are vital.

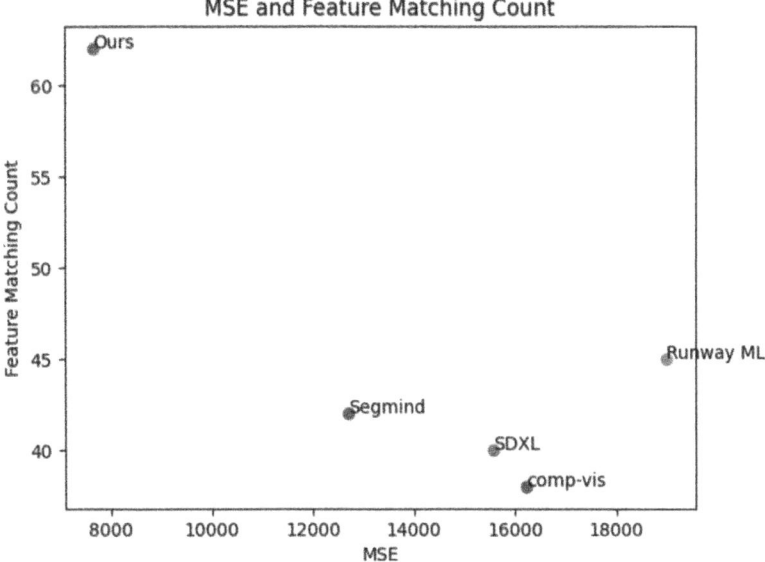

Fig. 5. Scatter plot illustrating the scores of MSE and Feature Matching Count.

4.3.2   Analyzing Multi-view Image Integrity

We employed the Structural Similarity Index (SSIM) to evaluate the integrity of generated multi-view images. Multi-view images were compared with each other by calculating the SSIM scores using those while comparing with the source image to know the quality of images that comes naturally. The source image image with which the multi-view images are derived was set as the GT with 100% score. There is generally a small tilt in the angle of images in each subsequent image accompanied by a lower number of similarities compared to the previous one. At the same time, the decrease rate sticks to its normal value of about 1.5% per image because of the random small modifications in perspective and details. This progressive change results in a maximum error of only around 6.5% over several consecutive images compared to the previous image, indicating a generally robust system for generating consistent multi-view images necessary for accurate 3D reconstruction (Table 1).

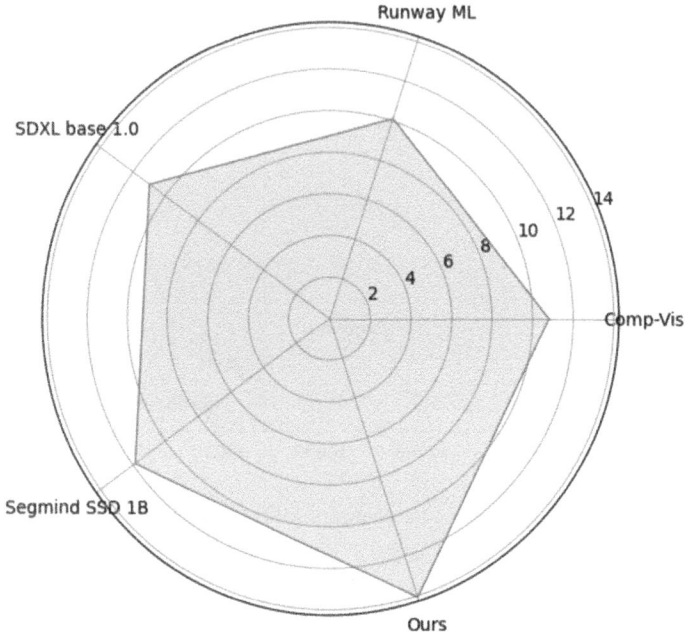

Fig. 6. Peak Signal-to-Noise Ratio (PSNR) comparison across different models.

Table 1. SSIM score of Multi-view images that are evaluated based on the consistency of the images.

Comparison	SSIM Score
Source Image	100%
MV1 (vs. Source)	98.5%
MV2 (vs. MV1)	97.0%
MV3 (vs. MV2)	96.5%
MV4 (vs. MV3)	95.5%
MV5 (vs. MV4)	94.5%
MV6 (vs. MV5)	93.5%

5   Conclusion

In conclusion, the project that we worked on is an important step towards 3D modelling technology, as it employs novel and innovative techniques to improve the accuracy and productivity of multi-view 3D reconstruction. During the model development, our main task was to improve the precision and detail in 3D modelling from images taken from different viewpoint. And the output resulted with high fidelity and accuracy of the shapes. One of our major achievements is our proprietary refiner model integration with the existing image generation models, which significantly enhances images with better quality, textures and lighting before processing them into 3D models. We hope that our novelty and contributions made in the field of image generation for any specific usecase and also 3D modelling is a significant one.

In addition, our model has been optimized for efficiency and using less GPU power compared to other conventional methods. This introduces new opportunities for 3D geometry processing in situations that are limited by the availability of computing power or in situations where real-time computing is of great importance.

6   Future Works

While the achievements of our multi-view 3D reconstruction approach are significant, there is still room for improvement for certain models in the 3D modelling field. One promising direction is the integration of temporal information to enable the generation of animated 3D content directly from static images or text descriptions. By incorporating the recent developments in sequences modeling and generative adversarial networks, the possibility for 3D worlds with moveable characters and setting will become a reality.

Another significant work is the creation of datasets especially for creation of better novel view images and 3D models. The datasets that are available now are very limited when compared to the usecases that we require. Creating a dataset with diverse images and its resultant novel view images and the 3D models for almost all the objects present is very crucial in the development of this field.

By pursuing these future research directions, we hope to push the boundaries of what is possible in multi-view 3D reconstruction and visualization. Our commitment to technical excellence, computational sustainability, and widespread accessibility will continue guiding our efforts.

References

1. Liu, Z., Dai, P., et al.: Dreamstone: image as a stepping stone for text-guided 3D shape generation. IEEE Trans. Pattern Anal. Mach. Intell. **45**(12), 14385–14403 (2023). https://doi.org/10.1109/tpami.2023.3321329
2. Shi, R., et al.: Zero123++: a single image to consistent multi-view diffusion base model. arXiv preprint arXiv:2310.15110 (2023)
3. Liu, M., et al.: One-2-3-45++: fast single image to 3D objects with consistent multi-view generation and 3D diffusion. arXiv preprint arXiv:2311.07885 (2023)
4. Liu, Z., Hu, J., et al.: EXIM: a hybrid explicit-implicit representation for text-guided 3D shape generation. ACM Trans. Graph. **42**(6), 1–12 (2023). https://doi.org/10.1145/3618312

5. Zhang, M., et al.: MotionDiffuse: text-driven human motion generation with Diffusion Model. IEEE Trans. Pattern Anal. Mach. Intell., 1–15 (2024). https://doi.org/10.1109/tpami.2024.3355414

6. Chen, Z., Wang, G., Liu, Z.: SceneDreamer: unbounded 3D scene generation from 2D image collections. IEEE Trans. Pattern Anal. Mach. Intell. **45**(12), 15562–15576 (2023). https://doi.org/10.1109/tpami.2023.3321857

7. Wen, C., et al.: Pixel2Mesh++: 3D mesh generation and refinement from multi-view images. IEEE Trans. Pattern Anal. Mach. Intell. **45**(2), 2166–2180 (2023). https://doi.org/10.1109/tpami.2022.3169735

8. Ding, L., et al.: Text-to-3D generation with bidirectional diffusion using both 2D and 3D priors. arXiv preprint arXiv:2312.04963 (2023)

9. Shue, J.R., Chan, E.R., Po, R., Ankner, Z., Wu, J., Wetzstein, G.: 3D neural field generation using triplane diffusion. In: Proceedings of the IEEE/CVF Conference on Computer Vision and Pattern Recognition, pp. 20875–20886 (2023)

10. Zhou, L., Du, Y., Wu, J.: 3D shape generation and completion through point-voxel diffusion. In: Proceedings of the IEEE/CVF İnternational Conference on Computer Vision, pp. 5826–5835 (2021)

11. Chen, R., Chen, Y., Jiao, N., Jia, K.: Fantasia3D: disentangling geometry and appearance for high-quality text-to-3D content creation. In: Proceedings of the IEEE/CVF International Conference on Computer Vision, pp. 22246–22256 (2023)

12. Wu, J., et al.: HD-fusion: detailed text-to-3D generation leveraging multiple noise estimation. In: Proceedings of the IEEE/CVF Winter Conference on Applications of Computer Vision, pp. 3202–3211 (2024)

13. Li, M., et al.: Instant3D: instant text-to-3D generation. arXiv preprint arXiv:2311.08403 (2023)

14. Müller, T., Evans, A., Schied, C., Keller, A.: Instant neural graphics primitives with a multiresolution hash encoding. ACM Trans. Graph. (TOG) **41**(4), 1–15 (2022)

15. Vahdat, A., Williams, F., Gojcic, Z., Litany, O., Fidler, S., Kreis, K.: LION: latent point diffusion models for 3D shape generation. Adv. Neural. Inf. Process. Syst. **35**, 10021–10039 (2022)

16. Liu, Y., et al.: SyncDreamer: generating multiview-consistent images from a single-view image. arXiv preprint arXiv:2309.03453 (2023)

17. Kalischek, N., Peters, T., Wegner, J.D., Schindler, K.: Tetrahedral diffusion models for 3D shape generation. arXiv preprint arXiv:2211.13220 (2022)

18. Zou, Z.X., et al.: Triplane meets gaussian splatting: fast and generalizable single-view 3D reconstruction with transformers. arXiv preprint arXiv:2312.09147 (2023)

19. Long, X., et al.: Wonder3D: single image to 3D using cross-domain diffusion. arXiv preprint arXiv:2310.15008 (2023)

20. Radford, A., et al.: Learning transferable visual models from natural language supervision. In: International Conference on Machine Learning, pp. 8748–8763. PMLR, July 2021

21. Caron, M., et al.: Emerging properties in self-supervised vision transformers. In: Proceedings of the IEEE/CVF International Conference on Computer Vision, pp. 9650–9660 (2021)

AI for Text, Audio, Image and Video Processing

Carnatic Raga Recognition Using Deep Learning Techniques

Rajeev Sekar, M. Poonkodi[✉], K. A. Muhammad Kifayathullah, R. Abinaya,
and S. Akilesh

Vellore Institute of Technology, Chennai, TN 600127, India
{rajeev.sekar2021,muhammad.kifayathul2021,abinaya.r2021c,
akilesh.s2021}@vitstudent.ac.in, poonkodi.m@vit.ac.in

Abstract. Raga is an essential element in Carnatic music. It represents a unique melodic framework that defines the mood, emotion, and character of a Carnatic music song. However, the subtle and very minute variations within the ragas can be understood and the raga can be classified only by professionals. The presence of Gamakha which are sudden oscillatory movements in the pitch distribution, is a significant challenge for the automated raga classification system. For the task of automated raga classification and overcoming the challenges in it, the proposed work uses a novel approach using audio signal processing (ASP) and deep learning (DL) techniques. Feature extraction for capturing all the characteristics (pitch characteristics, tonal characteristics etc.) is done and Various deep learning models (Artificial Neural Networks, Long Short-term memory, and Bi-LSTM) are trained to classify ragas accurately. Our method uses the inherent capability of deep learning algorithms to learn subtle but complex patterns and capture the essence of ragas, achieving promising results in the raga classification task. Comp Music dataset for Carnatic music was used for this task and an accuracy of 96.43% was achieved by the Bi-LSTM model which outperformed the other two deep learning models by small margins. A comparison of the current work with previous work is also made.

Keywords: Raga · Gamakha · Deep Learning · ASP · Bi-LSTM · Music Analysis · Carnatic Music

1 Introduction

Carnatic music is a south Indian classical music form which is one of the oldest classical music styles with a rich history. It has complex melodic and rhythmic patterns embedded in meaningful lyrics composed in Telugu, Sanskrit, Tamil, and various other languages. Carnatic music provides a spiritual and aesthetic dimension, transcending the dimension of mere entertainment. Carnatic music compositions provide scope for the artists/performers to improvise and give emotive renditions, but the improvisation is within the structure of the raga of the composition which is a unique quality of this artform.

© The Author(s), under exclusive license to Springer Nature Switzerland AG 2025
P. D. Sivakumar et al. (Eds.): IRCCTSD 2024, CCIS 2362, pp. 189–199, 2025.
https://doi.org/10.1007/978-3-031-82386-2_14

Raga is a melodic framework that is central to each Carnatic music composition. It provides a distinctive identity to the composition and tends to evoke a range of emotions and moods. Every raga comprises of a unique sequence of Swaraas in ascending order (Aarohana) and descending order (Avarohana). These unique sequences act as a template for the performer to improvise on the Carnatic composition.

Even though Carnatic music has a very big cultural significance, the computational analysis and signal processing research of Carnatic music presents a lot of challenges. Gamakha is one such challenge. It is the subtle but quick variation/oscillation in the pitch distribution (which sometimes may even lie outside the Aarohana and Avarohana pattern of the raga for that composition). It poses significant obstacles for the conventional audio signal processing methods which in turn affects the performance of automated raga classification systems. Our method aims to overcome this.

In recent times, efforts to develop an automated raga classification system have increased due to the advancements in Machine learning and music information retrieval (MIR) techniques. Powerful computational algorithms analyses audio data to extract relevant information for the raga classification task. This will be of immense use for Rasika because classifying a raga manually takes a lot of knowledge and previous experience of listening to a lot of Carnatic music.

In that context, the proposed work shows a comprehensive study of raga classification in Carnatic music, utilizing deep learning techniques (Artificial Neural Networks, long short-term memory, and Bi-directional LSTM) and advanced audio signal processing techniques for feature extraction. By considering the intricacies and challenges in the raga classification task, the proposed work aims to handle the complexities and propose a novel approach to overcoming the challenges and performing the task efficiently.

A raga has various characteristics, The most basic structure of the raga is defined by the Aarohana and Avarohana, these two represent the ascending and descending sequence of notes allowed in a raga. The combination of both is specific to every raga and change in even one note or swara will result in the change of the whole raga.

The proposed work combines state-of-the-art deep learning algorithms with feature extraction techniques, including spectral analysis, chroma analysis, and Mel-frequency cepstral coefficients (MFCCs). With the combination of computational methods with traditional musicological insights, our research aims to bridge the gap between technology and tradition.

2 Literature Survey

Vishwas [1] had implemented a novel feature extraction method called sequential pitch distribution method for raga detection. They concluded that it performed better than the conventional pitch distribution. Comp music dataset was used for the task and K Nearest Neighbors with the combination of sequential pitch distribution showed an accuracy of 99% for Hindustani songs and 88.13 for Carnatic compared to 90.66 and 72.33 respectively, while using conventional pitch distribution.

Koduri et al. [2] had correlated the computational raga classification approach to the way humans do it. The two approaches used by humans are: intuitive (Rasikaas) and analytical (Professionals) approaches. They performed common computational steps such as pitch extraction, tonic identification, and histogram computation. They showed promising results and discussed how to reduce the error rate in the model.

Gowrishankar et al. [3] demonstrated an automated note transcription-based raga recognition model. The tasks performed were: pitch extraction, finding the tuning offset or the tonic pitch and note segmentation. They achieved a staggering accuracy of 96%, outperforming the previous approaches. And they have also suggested that features can be included for gamakha and characteristic phrases.

Gulati et al. [4] had proposed the time delayed melody surface which was a novel feature which captures the melodic structure of a raga. It represents various characteristics including the tonal and the temporal. They primarily find the predominant melody and normalize the audio using the tonic value then surface generation, power compression and gaussian smoothening is performed to obtain the TDMS, they obtained 98% and 87% accuracy in Hindustani and Carnatic music using 30 and 40 ragas for the model training respectively.

Samsekai et al. [5] had performed both raga and tonic identification. The features extracted from the probability density function from the extracted pitch values of the audio file were considered for the classification task. 3 different models were trained and evaluated namely the feed forward neural network, Gaussian mixture models and decision trees. Two datasets were used one of which had 14 different ragas and the other one was comp music dataset with 17 ragas and an accuracy of 90.14 and 95% respectively.

Preeti et al. [6] compared the existing approaches and correlated with the human approach and created a survey. They first discussed the properties of a raga which are, the arohana and avarohana, gamakha, characteristic phrases and also the various types/roles of Swaras. The tasks performed in the previous works were scale matching, statistical modelling and pakad matching, pitch-class profiles and note bi-grams. 94% accuracy was achieved using a multi variate normal classifier which drastically outperformed the other models, Feed forward neural network (75%), K-NN classifier (67%) and tree-based classifier (50%).

Rohit et al. [7] had proposed a raga classification approach using 3 methods namely the pitch determination, segmentation and a key note mapping technique to identify the raga. Totally 10 ragas were used for this task and features were extracted to arrive at a classification rule. Preprocessing steps like noise removal, removal of unwanted frequencies was also performed. Various models were trained but the best performing model was the multiclass classifier with 95.93% accuracy, Naïve bayes and Bayesian networks also performed well with accuracies of 95.7% and 95% respectively.

Priya et al. [8] proposed a data mining approach for the automatic identification of the Carnatic raga Swaram notes. For this task 72 Melakarta ragas and 212 Janya ragas were considered. A rule induction algorithm was used after the feature relevance analysis and an accuracy of 90% was achieved in the Janya raga dataset. The possible future scope they mentioned was the usage of genetic algorithms [9] to enhance the results.

Pandey et al. [10] had developed a system called Tansen which is based on the hidden Markov model enhanced with a string-matching algorithm and performs the raga recognition tas. As the hidden Markov model [11] performed well for word recognition task and raga recognition is closely related with it, they have chosen that model. They have utilized 2 independent heuristic strategies for note transcription. They have showed an accuracy of 87% while identifying the raga using pakad matching.

Kalyani et al. [12] had conducted an extensive survey of the existing raga recognition techniques and feature extraction for the raga identification task. The key points summarized are the lack of standard datasets like western music paradigm, assumption of tonic pitch to be static which creates errors in the model, lack of features to represent gamakha and characteristic phrases. They have reviewed the performance of various existing models and some models which perform notably well are random forest and ANN with pitch values estimated [13] showing 94% and 95% accuracy respectively.

Trupti Katte [14] had investigated the different characteristics of raga (Swaras, shruti, notes, Aarohana, avarohana, gamakha and pakad) and proposed data mining techniques for raga identification. Existing methods including statistical approaches, pakad matching, scale matching and pitch class distribution was surveyed and a novel unsupervised, statistical LDA model with swara innotation was proposed. Some challenges were identified which were the lack of database with vast no of ragas, incorrect pitch extraction and lack of models for automated tonic identification.

Sankalp et al. [15] had proposed a raga recognition technique based on melodic phrases (phrase-based recognition using vector space modelling) which are extracted using unsupervised techniques. They relate this to topic modelling in text classification tasks. A promising accuracy of 92% was shown while using 10 ragas but when 40 ragas were used an accuracy of 70% was only achieved.

Rajeswari et al. [16] proposed a methodology for raga identification based on the steps: source separation, segmentation (for which a novel algorithm was proposed), singer identification so that the fundamental frequency or the tonic pitch can be identified therby determining the raga. The segmentation algorithm a 2-level algorithm, the first level involved the segmentation of the separated signal based on the taala (The repetitive rhythmic pattern) of the song and the second level segmented the signal further into thattu and veechu (which are 2 subcomponents of a tala), each thattu or veechu comprised of exactly 1,2 or 4 Swaraas. And these segments were further used for raga classification task. An accuracy of 91.5 for sadjha estimation and 62.13% for raga classification was achieved.

Venkata et al. [17] analyzed and presented the statics of Gamakha, which is an important component of raga classification. They proposed 2 algorithms to identify constant pitch notes and stationary points. They concluded that the identification of CP notes and transient points using the proposed methodology has a higher accuracy than when it is identified by experts.

Ranjani et al. [18] proposed a novel Alapana analysis-based method for tonic identification, swara identification and raga verification. Stochastic model was used for the same. The steps were pitch extraction, pitch distribution function estimation, SC-gaussian mixture modelling for transcription of Swaraas and raga identification at last.

3   Proposed Methodology

Usually, humans identify raga using two distinct approaches. One is a systematic approach where one must have an in-depth understanding of the Sawaraa and its correlation with the Arohana and Avarohana to identify the raga. The other approach is more of an intuitive one, purely by relying on an individual's previous listening experience to directly recognize the most similar raga. Existing methods primarily model around the first approach, using techniques such as pitch distribution extraction, note extraction, or phrase-based raga identification. In contrast, our proposed methodology is based on the latter, intuitive approach by using MFCCs to represent how humans understand audio and to further emulate the human memory processes we use LSTM.

3.1 Data Collection

In the current work, The Comp Music dataset [19, 20] for Carnatic music was used, which had audio files along with the annotated raga for each audio file. A total of 96 different ragas were used for the classification task.

3.2 Proposed System Architecture

(See Fig. 1)

3.3 Feature Extraction

The primary task before feature extraction was to segment the audio files in 30 s segments. Two sets of features were extracted, Spectral features and Mel Frequency Cepstral coefficients (MFCCs).

Spectral Features. The segmented audio files were used to calculate the spectrogram (Fig. 2). This is done by calculating the Fourier transform over overlapping short windows of the audio and the change in magnitude over time is plotted for the resulting spectrum. From the spectrogram we extract 3 features, spectral centroid, spectral bandwidth and spectral Rolloff.

- Spectral Rolloff: to suggest the frequency value less than which most of the energy (90%) belongs to
- Spectral Centroid: It represents the mean frequency of the audio
- Spectral bandwidth: provides info about the skewedness of the frequencies in the audio

MFCCs. Total of 20 MFCC coefficients were used for the raga classification task in the current work. The Mel frequency cepstral coefficients (Fig. 3) represent the audio as how humans perceive when they listen it, it can be calculated using these steps:

1. Compute the Fourier transform of a window from the audio segment.
2. Map the powers of the spectrum using a cosine overlapping window.
3. Calculate the log of the powers at each Mel frequencies.
4. Compute the discrete cosine transform of the log values and the amplitude of the resulting spectrum are the MFCCs.

   The other features are:

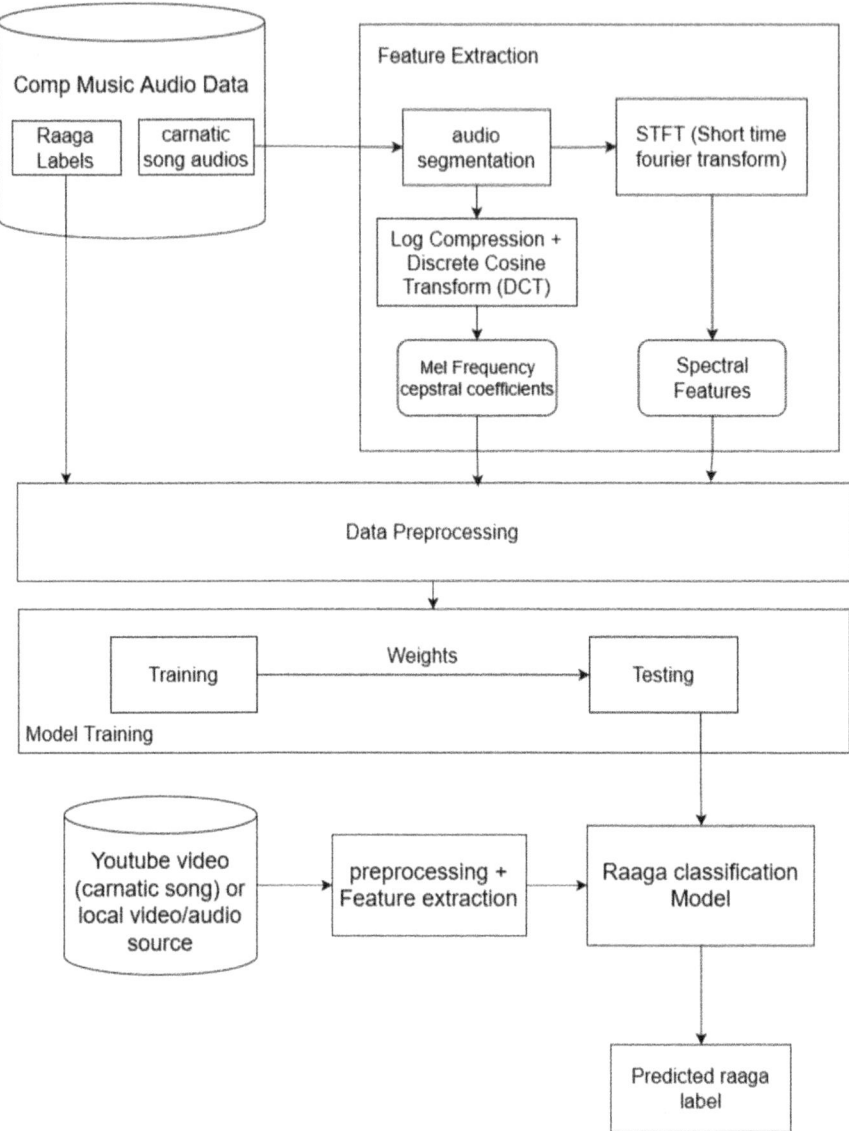

Fig. 1. Proposed System Architecture

- RMSE: root mean squared value of the energy of the input audio signal.
- ZCR: Zero crossing rate represents the rate at which the magnitude of the signal changes signs (crosses the y axis).

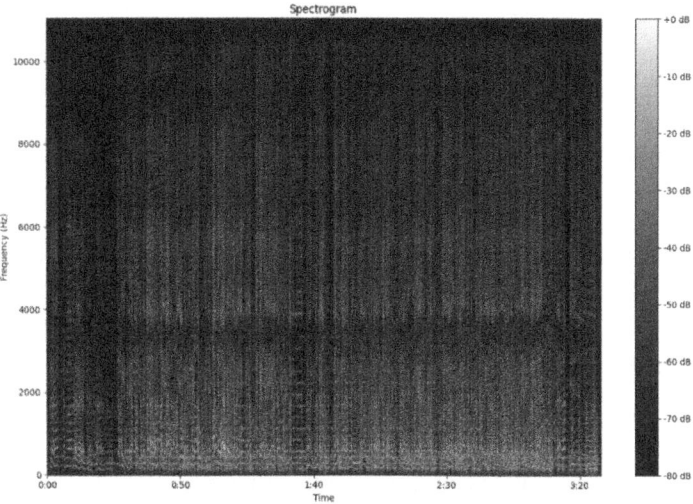

Fig. 2. Spectrogram of a sample audio file

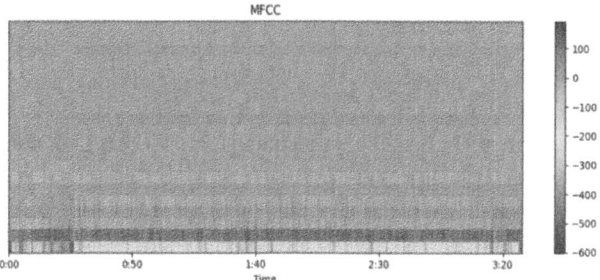

Fig. 3. MFCC coefficients plot for a sample audio file

3.4 Data Preprocessing

Certain data preprocessing techniques were applied to the extracted data to ensure the data is suitable for modelling and to increase the model efficiency.

Encoding the Raga Values. The raga values were encoded to numerical values using label encoder, as there were total of 96 ragas, the encoding was from 0 to 95.

Scaling the Features. The extracted features were scaled so that no feature is given an incorrect preference due to a larger scale. Min Max Scaler (1) technique was used to achieve this.

$$x(i : scaled) = \frac{x(i) - min(x)}{max(x) - min(x)} \tag{1}$$

Splitting Test and Train Sets. The dataset was split randomly into testing and training sets comprising of 20% and 80% of the records respectively. This was done using the test train split, the train set was used for training the traditional and Deep learning models and the test set was used to evaluate the performance of the models.

4 Model Implementation

Three Deep learning models were trained for the extracted features. The models are Artificial neural network, Long Short-Term Memory (LSTM) model and the Bi-directional LSTM model.

4.1 Artificial Neural Network (ANN)

It is a Deep learning model which mimics the human brain. Apart from the input layer, four dense layers were used with 256, 128, 64 and 96 (No of output classes) as the number of neurons in each layer, the input layer and the 2 hidden layers used relu activation function and the final output layer was trained with SoftMax activation function. The optimizer used was Adam and the loss minimization was done using sparse categorical cross entropy.

4.2 LSTM

LSTM model is a model with the aim to capture the long-term dependencies in time series data, which is an extension of the Recurrent Neural Network model. Apart from the input layer, 4 layers were used, the first 2 hidden layers had 128 and 64 units respectively, both were trained using a dropout of 0.05 and recurrent dropout of 0.25. The final hidden layer was to flatten the values and finally the output layer had 96 units with SoftMax activation. The same optimizer and loss metric was used as the ANN model.

4.3 Bi-LSTM

Bidirectional LSTM reads information in both forward and backward directions which allows it to capture more information from the data. The Bi-LSTM model trained for the current work had a similar structure to the LSTM, with 4 layers, the first 2 having 128 and 64 units and the third layer for flattening and the final layer for output production, the same optimizer and loss metric was used.

5 Performance Evaluation

The following table (Table. 1) shows the performance of the three deep learning models trained, Artificial neural networks, LSTM and Bi-LSTM. Due to the presence of 96 classes in total (96 ragas) accuracy is the best possible metric which can be used to evaluate the performance

All the three trained models have showing promising results and BI-LSTM was the best performing model with a small margin.

Table 1. Performance Evaluation

Model	Accuracy
ANN	95.65
LSTM	95.21
Bi-LSTM	**96.43**

Table 2. Compares the performance of the current work with previous methodologies

Work of	Method used	Accuracy
[1]	Sequential Pitch Distribution	88.13%
[3]	Note Transcription	96%
[4]	Time Delayed melody surfacing	87%
[5]	PDF of extracted pitch	90.14%
[15]	Phrase based recognition	95.7%
Proposed method	Bi-LSTM + (feature extraction (3.3)	**96.43%**

The following table (Table. 2) shows the comparison of existing methodologies with the current work

The proposed work outperforms the existing literatures in the raga recognition task. Also, most of the previous works work on this task with 10 or 20 ragas and some of the works use 40 ragas maximum, but the proposed method was experimented using 96 ragas and showed promising results.

6   Conclusion and Future Scope

The task of raga recognition was done for Carnatic music data, spectral features, Mel frequency cepstral coefficients and other features were extracted and after applying some preprocessing steps, 3 deep learning models were trained. BI-LSTM was found out to be the best performing model.

Some of the points which can be addressed in the future are:

- Features for representing the Gamakha and characteristic phrases.
- Utilization of a more diverse and standard dataset which includes singers of various Bani (Styles), and schools.
- Studying how every different raga affects the mood of a person.

References

1. Narasinh, V.: Sequential Pitch Distributions for Raga Detection. arXiv preprint arXiv:2308. 16421 (2023)
2. Koduri, G.K., Gulati, S., Rao, P., Serra, X.: Rāga recognition based on pitch distribution methods. J. New Music Res. 41(4), 337–350 (2012)
3. Gowrishankar, B.S., Nagappa, U.B.: Automated transcription for raga recognition and classification in indian classical music using machine learning. In: 3rd International Conference on Integrated Intelligent Computing Communication & Security (ICIIC 2021), pp. 211–217. Atlantis Press (2021)
4. Gulati, S., Serrà Julià, J., Ganguli, K.K., Sentürk, S., Serra, X.: Time-delayed melody surfaces for rāga recognition. In: Devaney, J., Mandel, M.I., Turnbull, D., Tzanetakis, G., (eds.) ISMIR 2016. Proceedings of the 17th International Society for Music Information Retrieval Conference; 2016 Aug 7–11, 2016, pp. 751–7. International Society for Music Information Retrieval (ISMIR), New York City (NY).[Canada], ISMIR (2016)
5. Samsekai Manjabhat, S., Koolagudi, S.G., Rao, K.S., Ramteke, P.B.: Raga and tonic identification in carnatic music. J. New Music Res. 46(3), 229–245 (2017)
6. Koduri, G.K., Gulati, S., Rao, P.: A survey of raaga recognition techniques and improvements to the state-of-the-art. Sound Music Comput. 38, 39–41 (2011)
7. Rohith, J., Vinod, S.: Carnatic raga recognition. Indian J. Sci. Technol. (2017)
8. Priya, K., Geetha Ramani, R., Jacob, S.G.: Data mining techniques for automatic recognition of carnatic raga swaram notes. Int. J. Comput. Appl. 52(10) (2012)
9. Annu, L., Gupta, K., Chopra, K.: Genetic algorithm-a literature review. In: 2019 International Conference on Machine Learning, Big Data, Cloud and Parallel Computing (COMITCon), pp. 380–384. IEEE (2019)
10. Gaurav, P., Mishra, C., Ipe, P.: TANSEN: a system for automatic raga identification. In: IICAI, pp. 1350–1363 (2003)
11. Lawrence, R., Juang, B.: An introduction to hidden Markov models. IEEE ASSP Magaz. 3(1), 4–16 (1986)
12. Waghmare, K.C., Sonkamble, B.A.: Raga identification techniques for classifying indian classical music: a survey. Int. J. Signal Process. Syst. 5(4), 1–6 (2017)
13. Kim, J.W., Salamon, J., Li, P., Bello, J.P.: Crepe: a convolutional representation for pitch estimation. In: 2018 IEEE International Conference on Acoustics, Speech and Signal Processing (ICASSP), pp. 161–165. IEEE (2018)
14. Katte, T.: Multiple techniques for raga identification in Indian classical music. Int. J. Electron. Comput. Eng. 4(6), 82–87 (2013)
15. Gulati, S., Serra, J., Ishwar, V., Sentürk, S., Serra, X.: Phrase-based rāga recognition using vector space modeling. In: 2016 IEEE International Conference on Acoustics, Speech and Signal Processing (ICASSP), pp. 66–70. IEEE (2016)
16. Sridhar, R., Geetha, T.V.: Raga identification of carnatic music for music information retrieval. Int. J. Recent Trends Eng. 1(1), 571 (2009)
17. Viraraghavan, V.S., Aravind, R., Murthy, H.A.: A statistical analysis of gamakas in carnatic music. In: ISMIR, pp. 243–249 (2017)
18. Ranjani, H.G., Arthi, S., Sreenivas, T.V.: Carnatic music analysis: Shadja, swara identification and raga verification in alapana using stochastic models. In: 2011 IEEE Workshop on Applications of Signal Processing to Audio and Acoustics (WASPAA), pp. 29–32. IEEE (2011)

19. Gulati, S., Serrà, J., Ganguli, K. K., Şentürk, S., Serra, X.: Time-delayed melody surfaces for raga recognition. In: Proceedings of the 17th International Society for Music Information Retrieval Conference (ISMIR), pp. 751–757. New York, USA (2016)
20. Gulati, S., Serrà, J., Ishwar, V., Şentürk, S., Serra, X.: Phrase-based raga recognition using vector space modeling. In: Proceedings of the 41st IEEE International Conference on Acoustics, Speech and Signal Processing (ICASSP), pp. 66–70. Shanghai, China (2016)

Visual Verse 2.0: Image Generation with Image Inpaiting

R. Dharanidharan[1], S. Abinav Satya[1], S. Gokula Prasath[1], and N. Muthurasu[2(✉)]

[1] Department of Computer Science and Engineering (Emerging Technologies), SRM Institute of
Science and Technology, Vadapalani, Tamil Nadu, India
{dr4672,as8941,3gs0205}@srmist.edu.in
[2] Department of Computer Science and Engineering, SRM Institute of Science and Technology,
Vadapalani, Tamil Nadu, India
muthuran@srmist.edu.in

Abstract. This research introduces a novel method for image enhancement and repair using Generative AI - Stable diffusion model. Unlike traditional techniques, it selectively masks and replaces areas to adjust aspect ratio while preserving details. Inspired by digital artists, it advances visual creation. Primary focus is on image inpainting, crucial for restoration, editing, and synthesis. Our approach merges deep learning in Generative AI with Stable diffusion, enhancing quality and stability of inpainted images. Stable diffusion models simulate gradual information spread across missing areas, yielding coherent outcomes. Integration of generative AI enables understanding of image structures, aiding realistic content generation. Inpainting guided by stable diffusion ensures seamless transitions and contextual coherence preservation. Adaptive contextual understanding fills missing regions based on global and local context. Evaluation criteria assess image quality, coherence, and similarity to the original. Results advance inpainting techniques for real-world applications, addressing challenges with improved visual quality and contextual consistency.

Keywords: Generative AI · Image Inpainting · Aspect Ratio · Image Enhancement · Contextual Coherence

1 Introduction

Using the Stable Diffusion model, a forefront of Generative AI, we tackle the complex task of changing image aspect ratios without losing detail. This method employs masking and inpainting to accurately adjust aspect ratios, allowing for detailed control over image modification. Masking extends the image borders to fit the new ratio, while the addition of Gaussian and Perlin noise enhances the image's realism during and after generation. The process involves masking to define the new ratio, adding Gaussian noise for initial content generation, and applying Perlin noise for natural adjustments. Inpainting then integrates this new content into the original image, ensuring a seamless transition. The effectiveness of these modifications is measured using the CLIP score, which assesses the coherence between the modified image and its description, ensuring the alterations improve or maintain the image's original integrity. This technique offers a significant advancement in image manipulation, providing users with unparalleled control over aspect ratio changes while ensuring the highest quality outcomes.

2 Literature Review

Hanqun Cao, et al. "A Survey on Generative Diffusion Models" [1] Cao et al. provide a comprehensive survey on generative diffusion models, exploring their principles, applications, and advancements. The paper discusses various techniques utilized in diffusion models and evaluates their efficacy in producing high-quality images. It offers insights into the strengths and limitations of different diffusion models and outlines future research directions in this domain.

Nermin Mohamed Fawzy Salem. "A Survey on Various Image Inpainting Various Image Inpainting Techniques" [2] Salem conducts a survey on various image inpainting techniques, offering an extensive overview of methods used to fill in missing or damaged parts of images. The paper covers traditional approaches as well as modern deep learning-based methods, analyzing their effectiveness and discussing recent advancements in the field of image inpainting.

S. Iizuka, E. Simo-Serra, and H. Ishikawa. "A Survey of Image Inpainting Techniques" [3] Iizuka et al. present a survey of image inpainting techniques, providing insights into the methods used for reconstructing missing or damaged regions in images. The paper reviews traditional and deep learning-based approaches, discussing their strengths, weaknesses, and applications in various domains.

T. Karras, et al. "High-Resolution Image Synthesis and Semantic Manipulation with Conditional GANs" [4] Karras et al. propose a method for high-resolution image synthesis and semantic manipulation using conditional generative adversarial networks (GANs). Their approach enables the generation of realistic images with fine details and allows for semantic manipulation of generated images, such as changing specific attributes or features while maintaining overall realism.

J. Johnson, A. Gupta, and L. Fei-Fei. "Image Generation from Scene Graphs" [5] Johnson et al. introduce a method for generating images from scene graphs, enabling the synthesis of realistic visual scenes based on structured descriptions. Their approach leverages scene graphs to encode relationships between objects and attributes, facilitating the generation of diverse and contextually coherent images.

Jitesh Jain, et al. "Keys to Better Image Inpainting: Structure and Texture Go Hand in Hand" [6] Jain et al. explore strategies for improving image inpainting by considering both structure and texture. Their research investigates the importance of preserving structural coherence and texture details in inpainted images, proposing techniques for enhancing the quality of inpainted results by effectively integrating structural and textural information.

J. Yu, et al. "Generative Image Inpainting with Contextual Attention" [7] Yu et al. propose a generative image inpainting method with contextual attention, which effectively fills in missing regions in images while preserving contextual coherence. Their approach utilizes contextual attention mechanisms to focus on relevant image regions during the inpainting process, resulting in visually appealing and semantically meaningful inpainted images.

Chenshuang Zhang, et al. "Text-to-image Diffusion Models in Generative AI: A Survey" [8] Zhang et al. conduct a survey on text-to-image diffusion models in generative AI, reviewing various approaches for generating images from textual descriptions. Their survey discusses the challenges and opportunities in text-to-image generation and provides insights into the current state of the art in this rapidly evolving field.

V. C. Muller and N. Bostrom. "Future progress in artificial intelligence: A survey of expert opinion" [9] Muller and Bostrom survey expert opinions on future progress in artificial intelligence, exploring anticipated advancements, challenges, and implications in the field. The paper provides valuable insights into the potential directions and impacts of AI research and development, as envisioned by experts in the field.

Evangelos Ntavelis, et al. "SESAME: semantic editing of scenes by adding, manipulating or erasing objects" [10] Ntavelis et al. propose SESAME, a framework for semantic editing of scenes by adding, manipulating, or erasing objects. Their method leverages semantic segmentation and generative models to enable interactive and intuitive scene editing, allowing users to modify scenes by directly manipulating semantic attributes.

3 Model

3.1 Model Description

This model is an advanced Latent Diffusion Model designed to convert a given promt into a meaningful image without missing detail, integrating a fixed, pretrained text encoder (OpenCLIP-ViT/H). It stands out for its ability to intricately interpret and transform text inputs into detailed visual outputs, employing a sophisticated understanding of the language and context behind the prompts. By leveraging the OpenCLIP-ViT/H encoder, the model gains a nuanced comprehension of textual descriptions, allowing for highly accurate and contextually appropriate image generation. This precision ensures that the generated images closely align with the users' intentions, showcasing the model's advanced capabilities in bridging the gap between textual concepts and visual representations.

3.2 Datasets

The model's training on subsets of the LAION-2B(en) dataset, a vast collection of 2 billion images with English descriptions, presents both strengths and limitations. While this training enables the model to produce impressive image generations from English prompts, it inadvertently introduces a cultural bias towards white and Western norms, often defaulting to these perspectives. The reliance on English-language descriptions limits the model's effectiveness with non-English prompts, highlighting a significant gap in its ability to represent global diversity accurately. This bias is not just a reflection of the model's training data but also a broader issue within AI development, where diversity and inclusivity in dataset curation are crucial for creating more equitable and universally effective models. Consequently, users are advised to approach the model with caution, understanding that its outputs may mirror and amplify existing societal biases (Fig. 1).

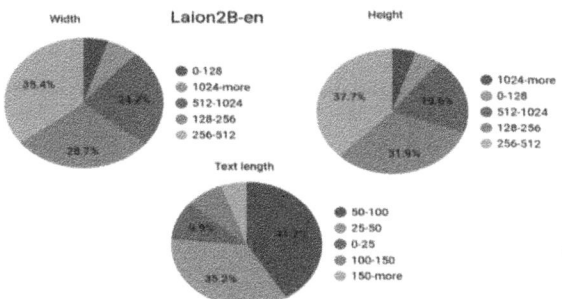

Fig. 1. LAION – 2B – en Dataset

3.3 Training

The training of this model involves a sophisticated process utilizing the LAION-5B dataset, along with its subsets, which have been meticulously filtered for content safety using LAION's NSFW detector to maintain a conservative "p_unsafe" score of 0.1. This approach ensures that the training data adheres to high standards of appropriateness. The Stable Diffusion v2, a state-of-the-art latent diffusion model, operates by encoding images into latent representations through an encoder, utilizing a down sampling factor of 8. This process effectively reduces the dimensions of the input images while preserving essential information, which is then mapped from an original shape of $H \times W \times 3$ to a more compact $H/f \times W/f \times 4$ latent space. Text prompts are concurrently encoded using the OpenCLIP-ViT/H text encoder, with the resultant text embeddings integrated into the model's UNet backbone via cross-attention mechanisms. This integration allows for a nuanced interplay between the textual and visual information. The model's training is further refined by employing a reconstruction loss objective, focusing on the difference between the added noise to the latents and the UNet's predictions, alongside the v-objective method for improved learning efficiency. For those interested in the intricate details of the model's development and its underlying principles, further information can be found in the LAION-5B's NeurIPS 2022 paper, which offers comprehensive insights and reviewer discussions on the subject which can be referred here (Fig. 2).

4 Methodology

4.1 Text to Image Generation

4.1.1 Text Processing Module

At the core of our enhanced text-to-image generation system lies the Text Processing Module, meticulously crafted to refine, and standardize textual inputs. This crucial initial step ensures that all textual descriptions are formatted and optimized for seamless integration with the model's more advanced stages. The methodology retains its reliance on pretrained CLIP and BERT models, reinforcing the understanding of text image relationships. Tokenization strategies, including CLIP's native text encoder and BERT's

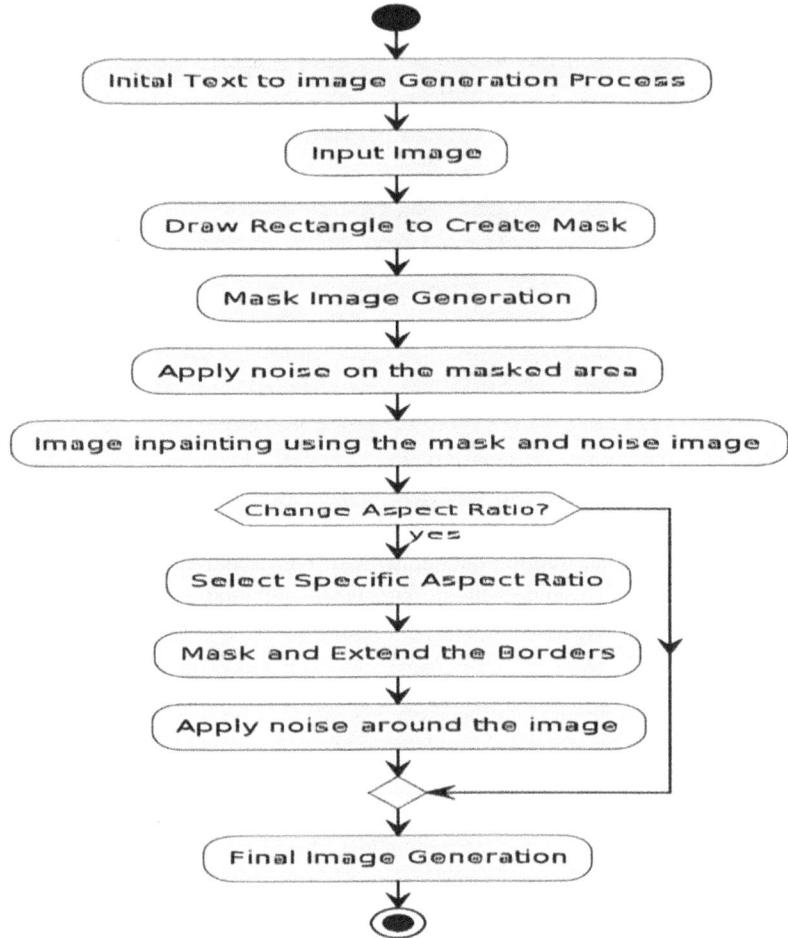

Fig. 2. Methodology

approach, are meticulously examined, while the CLIP Guided Diffusion Model continues to serve as a benchmark for evaluating image generation quality and latency. This integrative approach showcases the adaptability of diffusion models alongside advanced schedulers, emphasizing their collective contribution to a refined and versatile generative methodology.

4.1.2 Integration with Stable Diffusion V2.1

Central to our system is the Integration with Stable Diffusion version 2.1, a module expertly designed to interpret and analyse the refined textual prompts deeply. This stage is instrumental in decoding the intricate meanings and semantic layers within the text, thanks to the advanced capabilities of Stable Diffusion. It equips the system with a nuanced understanding of the text's context, paving the way for generating images that are not only precise but also richly detailed and contextually relevant.

4.1.3 U-Net-Based Image Generation

The U-Net-Based Image Generation module stands as the epicentre of our image creation process, where the magic of turning text into images happens. Leveraging the sophisticated analysis provided by Stable Diffusion, this module harnesses the power of the U-Net architecture to craft images that capture the essence of the textual prompts. It serves as the creative engine of the system, adeptly converting abstract textual ideas into vivid, expressive visual art. This process embodies the system's capacity to bridge the gap between the imagination and visual representation, resulting in images that are both visually striking and semantically aligned with the input descriptions.

4.2 Inpainting

4.2.1 Mask Creation

The initial step of the inpainting process involves taking an image as input and designating a specific area for modification by drawing a rectangle. This action generates a mask, distinguishing the area targeted for inpainting. The mask facilitates the model's focus on the white (selected) regions for painting, while the black (unselected) areas remain unaffected, setting the groundwork for precise and targeted image modifications.

4.2.2 Noise

Our model utilizes a sophisticated blend of noise types to enhance the inpainting process, including Gaussian Noise and Perlin Noise. Gaussian Noise, with its probability density function mirroring the normal distribution, is applied during the image's initial generation phase through Stable Diffusion. Conversely, Perlin Noise, celebrated for its ability to produce sequences that mimic natural phenomena more closely than traditional random functions, is employed for post-generation image manipulation. These noise types are integral to the model's inpainting strategy, which relies on filling or outpainting the noisy areas with contextually relevant information, thereby achieving a seamless integration of the new content into the original image.

4.2.3 Realistic Noise Gen

A novelty of the inpainting process is the inclusion of Perlin Noise to add external noise to the masked (white) areas, enhancing the image's realism. This step is critical in creating a more lifelike and convincing image by introducing noise that mimics natural variations and textures. This step has to be accurate as then noise should not exceed the white part cause the next step involving inpainting would require the noised area so any mistake made while this step might lead to missing details in output image.

4.2.4   The Process of Image Inpainting

Following noise addition, the image, now stored in cache, undergoes a series of transformations via the U-Net module. This powerful component is adept at converting abstract textual inputs into detailed and expressive visual outputs. For inpainting, the model takes as inputs the mask and the noise-enhanced image, employing the noise as a guide to repaint the image. This process ensures that the final image not only looks realistic but also aligns closely with the original intent and context of the text prompt.

4.2.5   Adjusting the Aspect Ratio

To adjust an image's aspect ratio without compromising its integrity, our approach incorporates masking to delineate the image's borders according to the new desired ratio. This is followed by the strategic introduction of noise around these new borders. The process distinguishes between the central image (the black part) and the surrounding noise (the white part), where the noise is applied. Utilizing cache files, the model generates images with adjusted aspect ratios, ensuring that the final output maintains a high level of detail and realism, consistent with the original image's quality.

5   Results

5.1   Quantitative Measure

5.1.1   Clip Score

CLIP's effectiveness has been rigorously tested across a broad spectrum of benchmarks, covering a wide array of computer vision tasks ranging from OCR and texture recognition to detailed classification work. The comprehensive analysis presented in the study showcases CLIP's capabilities over numerous datasets, such as Food101, CIFAR10, CIFAR100, Birdsnap, SUN397, and many others, extending to complex tasks like fine-grained classification, object counting, and even extending into domains like Hateful Memes and various video datasets like UCF101 and Kinetics700. Despite its broad applicability, CLIP encounters certain challenges, particularly in areas requiring granular classification distinctions and precise object counts.

Image Generated
(See Figs. 3 and 4)

5.1.2   Evaluation Comparison

For inpainting process the most needed part is to denoise the image generated on noise which is done by schedulers present in the model playing a major role in completing the whole image without missing any details.

Fig. 3. Generated Image. CLIP SCORE: 0.258

Fig. 4. Inpainted Image. CLIP SCORE: 0.285

Scheduler Model Evaluation Details	
Scheduler	Score
DDIMS	34.0435

Diffusion Model uses DDIMS as its base scheduler to generate images. In this method we have configured the scheduler and obtained CLIP Score for each scheduler.

Configured Schedulers

Schedulers	Score
EulerAncestralDiscreteScheduler	33.7204
HeunDiscreteScheduler	37.3416
LMSDiscreteScheduler	37.3133
DPMSolverMultistepScheduler	38.6042

The above mentioned are the scores achieved by each scheduler. The score can be improved by using manual seed, fine-tune the model. Moreover, our goal is to achieve the image generation in low latency which might lead to problem in resolution of the image generated as the model cannot be used with its full potential (Fig. 5).

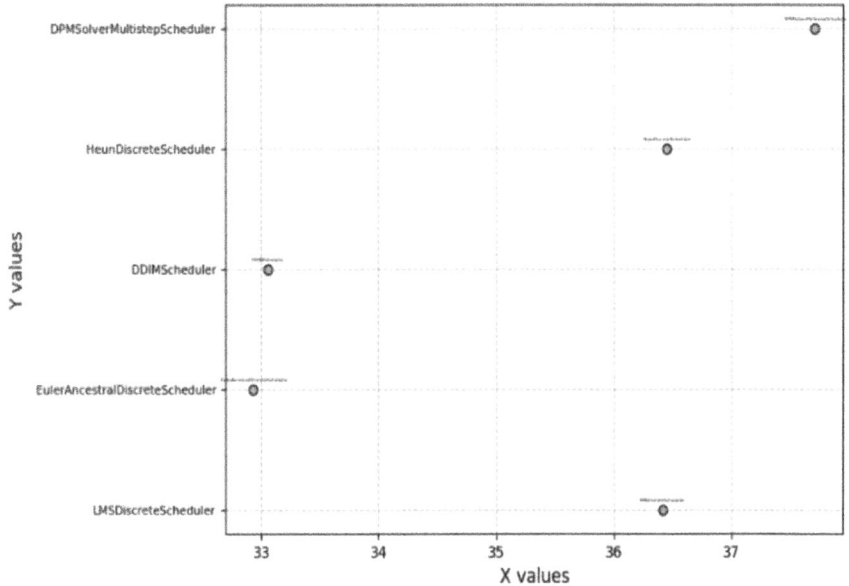

Fig. 5. Comparison chart between scheduler

The scheduler change in the model would provide a greater impact on the image generated which would give better results while running it locally which is our main goal (Fig. 6).

The above graph shows the comparison between other existing models where our model performs better while running them locally in a consumer grade GPU.

Cosine similarity is a metric is used to identify the similarity index between the prompt and the image generated

- + 1 indicates that the vectors are identical
- 0 meaning that they are orthogonal
- -1 suggesting that they are opposite (Figs. 7 and 8)

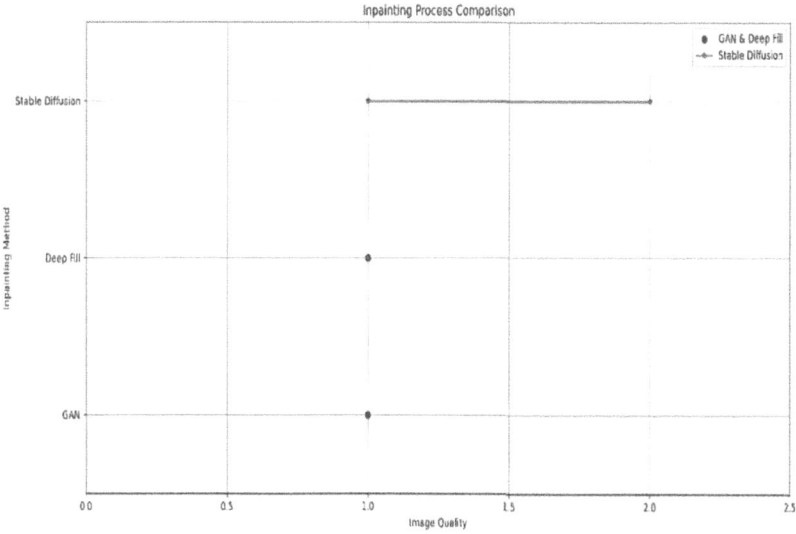

Fig. 6. Comparison with other models for running them locally

Fig. 7. Pikachu

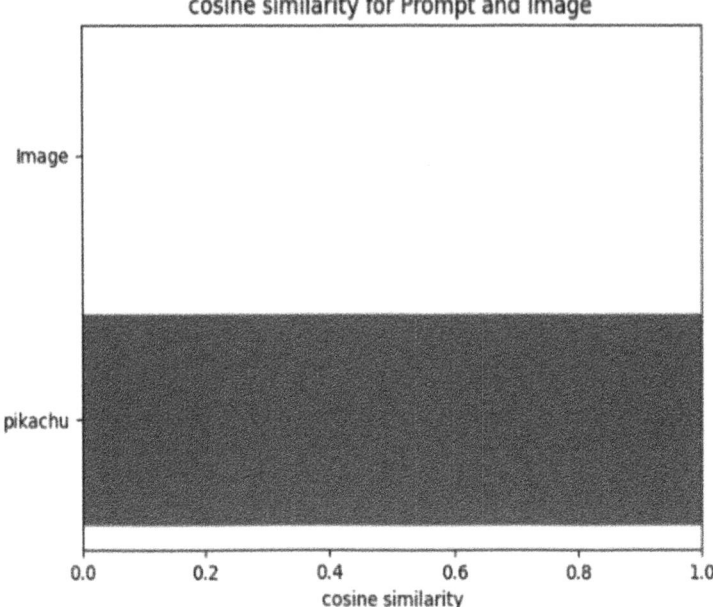

Fig. 8. Similarity Index

6 Conclusion

The proposed system can be described as a state-of-the-art method in Gen AI. The initial text processing step which acts as a base for the model in producing a quality output. Deep understanding and interpretation of textual prompts are made possible by the integration with a pre-trained model, and these insights are further improved by the U-Net-Based Image Generation module which has the potential to convert textual prompts into visuals. The inpainting procedure, which begins with Mask Creation and is refined by the purposeful application of Gaussian and Perlin Noise, demonstrates an advanced capacity to alter images successfully and convincingly. The system's additional approach to adjusting the aspect ratio increases its ability to produce images that maintain similarity to the original one, and not leaving any details in generated one which, creates a bridge between text and visual representation. The model is evaluated with performance metrices to show the efficiency of our work.CLIP Score: The CLIP score serves as a measure of the model's ability to align the generated images with textual prompts. Higher CLIP scores indicate better alignment and understanding of the input text. Cosine Similarity Values: Cosine similarity values are used to quantify the similarity between the generated images and the provided textual prompts. Higher cosine similarity values suggest a closer match between the generated images and the desired output. These metrics and graphical representations provide valuable insights into the performance and efficacy of the proposed system in generating quality images from textual inputs.

7 Future Work

In future work, the object inpainted can be moved, and the background can be changed by the same method by blacking out the objects. This change can open possibilities for further enhancement and refinement of the inpainting process, allowing for even greater flexibility and control over the generated images. Additionally, exploring techniques to improve the efficiency and speed of the model, particularly for low-latency image generation, would be valuable for real-world applications and study purpose. Continual research and development in this area will have a advancement in the field of generative AI systems.

References

1. Cao, H., et al.: A survey on generative diffusion models
2. Fawzy, N.M.: A survey on various image inpainting various image inpainting techniques
3. Iizuka, S., Simo-Serra, E., Ishikawa, H.: A survey of image inpainting Techniques
4. Karras, T., Aila, T., Laine, S., Lehtinen, J.: High-resolution image synthesis and semantic manipulation with conditional GANs
5. Johnson, J., Gupta, A., Fei-Fei, L.: Image generation from scene graphs
6. Jain, J., Zhou, Y., Yu, N., Shi, H.: Keys to better image inpainting: structure and texture go hand in hand
7. Yu, J., Lin, Z., Yang, J., Shen, X., Lu, X., Huang, T.S.: Generative image inpainting with contextual attention
8. Zhang, C., Zhang, C., Zhang, M., Kweon, I.S.: Text-to-image diffusion models in generative AI: a survey
9. Muller, V.C., Bostrom, N.: Future progress in artificial intelligence: a survey of expert opinion
10. Ntavelis, E., Romero, A., Kastanis, I., Van Gool, L., Timofte, R.: SESAME: semantic editing of scenes by adding, manipulating or erasing objects
11. Li, Y., Fang, C., Yang, J., Wang, Z., Lu, X., Yang, M.: Generative image inpainting with structural similarity loss. IEEE Trans. Image Process.
12. He, K., Zhang, X., Ren, S., Sun, J.: Deep residual learning for image recognition. In: IEEE Conference on Computer Vision and Pattern Recognition
13. Gupta, A., Vedaldi, A., Zisserman, A.: Synthesizing realistic images. In: European Conference on Computer Vision
14. Huang, H., Dong, C., You, S., Zhou, D.: Contextual attention for image inpainting. In: IEEE International Conference on Computer Vision
15. Zhang, R., Isola, P., Efros, A.: Split-brain autoencoders: unsupervised learning by cross-channel prediction. In: IEEE Conference on Computer Vision and Pattern Recognition
16. Wang, Z., Bovik, A.C., Sheikh, H.R., Simoncelli, E.P.: Image quality assessment: from error visibility to structural similarity. IEEE Trans. Image Process.
17. Ma, L., Jia, X., Sun, Q., Schiele, B., Tuytelaars, T., Van Gool, L.: Pose guided person image generation. In: IEEE Conference on Computer Vision and Pattern Recognition
18. Yang, Y., Yu, Y., Yuille, A., Darrell, T.: Deep layer aggregation. In: IEEE International Conference on Computer Vision
19. Wu, J., Zhang, C., Xue, T., Freeman, B., Tenenbaum, J.: Learning a probabilistic latent space of object shapes via 3D generative-adversarial modeling. Adv. Neural Inform. Process. Syst.
20. Pathak, D., Kr¨ahenb¨uhl, P., Donahue, J., Darrell, T., Efros, A.: Context encoders: feature learning by inpainting. In: IEEE Conference on Computer Vision and Pattern Recognition

Data Brain: Streamlining Data Science with AI-Assisted Conversations

Santhakumar S. Krithik, V. Aditya, Kurapati Praneeth Sai Reddy, and R. Gayathri[(✉)]

Department of Computer Science and Engineering (Emerging Technologies), SRM Institute of Science and Technology, Vadapalani, Chennai, India
`{sk8017,av7544,ks7812,gayathrr11}@srmist.edu.in`

Abstract. The "Data Brain" initiative introduces an innovative AI Data Assistant, revolutionising the data science landscape by streamlining the analysis process and making it more accessible. By integrating a user-friendly web interface via stream-lit, alongside robust data manipulation through pandas and visual insights via matplotlib, "Data Brain" caters to both novice and experienced data enthusiasts. OpenAI's advanced AI models and custom-built algorithms are employed to offer a conversational interface that guides users through exploratory data analysis, problem-solving, and algorithm selection, based on the analysis of uploaded CSV datasets. Furthermore, "Data Brain" enhances query understanding and response accuracy through Sentence Transformers and Pinecone's vector search technology, ensuring user queries are addressed with precision. This seamless integration of AI and machine learning technologies not only democratises data science by simplifying complex analysis tasks but also fosters an environment where users can effortlessly extract actionable insights from their data. "Data Brain" stands as a testament to the potential of AI in enhancing data accessibility, promoting an intuitive and interactive experience that empowers users to unlock the full value of their data. Through "Data Brain," the field of data science is made approachable, allowing users to navigate and analyse data with an unprecedented level of support and efficiency.

Keywords: Exploratory Data Analysis · Query understanding · Response Accuracy · Unprecedented Efficiency · Robust Data Understanding

1 Introduction

In the rapidly evolving landscape of data science and artificial intelligence (AI), the demand for tools that bridge the gap between complex data analysis and user-friendly interfaces has never been higher. Our project seeks to address this need by developing an innovative web-based application that leverages the power of AI to facilitate intuitive data exploration and analysis. Utilising Python for its versatile back-end capabilities, Streamlit for crafting a responsive web interface, and advanced AI models through LangChain integration, this application stands at the forefront of making data science accessible to a broader audience.

P. D. Sivakumar et al. (Eds.): IRCCTSD 2024, CCIS 2362, pp. 212–221, 2025.
https://doi.org/10.1007/978-3-031-82386-2_16

The core of our innovation lies in the seamless integration of natural language processing (NLP) techniques and machine learning (ML) models, enabling users to interact with their data through conversational AI. With features such as dynamic data visualisation, interactive exploration tools, and AI-powered analysis, our platform is designed to transform the way users engage with datasets, regardless of their technical expertise.

It not only represents a significant step forward in democratising data science but also serves as a testament to the potential of integrating cutting-edge AI technologies to enhance decision-making and insights generation in various domains. As we continue to refine and expand our application, we remain committed to unlocking the full potential of AI to empower users to discover, analyse, and visualise data in groundbreaking ways.

2 Literature Review

[1] Pandya and Holia (2023) describes a methodology for automating customer service that involves creating custom open-source GPT chatbots with LangChain, which improves organizational customer service automation efforts. [2] Radford et al. (2018) highlights the increases in language comprehension achieved using generative pre- training, which is a core strategy for many modern NLP models. [3] Johnson, Douze, and Jégou (2021) presents approaches for performing similarity searches on a billion-scale dataset using GPUs, considerably improving the efficiency and scalability of these searches. [4] Vaswani et al. (2017) presents the Transformer model, which emphasizes the importance of attention mechanisms in achieving cutting-edge performance in a variety of NLP tasks. [5] Kim et al. (2023) investigates EmbedDistill, a unique technique to knowledge distillation that enhances information retrieval using geometric methodologies. [6] Su et al. (2023) presents a versatile text embedding model that has been fine-tuned with instructions to perform effectively across a wide range of tasks. [7] Følstad and Skjuve (2019) analyzes user experience and motivation when utilizing chatbots for customer assistance, identifying characteristics that improve chatbot performance. [8] Bonifacio et al. (2022) introduces Inpars, an approach for generating unsupervised datasets for information retrieval that aids in the construction of robust retrieval systems. [9] Perer and Shneiderman (2008) explains how statistics and visualization can be integrated in exploratory data analysis to obtain clarity. [10] Ghosh et al. (2018) presents a comprehensive overview of methods for exploratory analysis of tabular industrial datasets, emphasizing numerous techniques and tools that aid in efficient data analysis.

[11] Topsakal and Akinci (2023) presents an overview of LangChain for developing large language model (LLM) applications, with a focus on quick development methodologies. [12] Pokhrel et al. (2024) presents a framework that combines OpenAI, LangChain, and Streamlit to create customizable chatbots for document summarizing and question answering. [13] Ye et al. (2023) does a complete capacity analysis of the GPT-3 and GPT-3.5 series models, assessing their performance across a variety of tasks. [14] Kuzlu et al. (2022) present a Streamlit-based AI trust platform developed for next-generation wireless networks, emphasizing its applications and advantages. [15] Kumar and Raschka (2021) introduce Pinecone, a straightforward and efficient framework for large language model inference that improves the scalability and usability of LLMs. [16]

Data et al. (2016) investigate exploratory data analysis in the context of electronic health records and demonstrate strategies for secondary analysis of such data. [17] Behrens (1997) presents the ideas and techniques for exploratory data analysis, providing fundamental insights into the methodology. [18] Provost and Fawcett (2013) investigate the relationship between data science, big data, and data-driven decision- making, focusing on the implications for diverse businesses. [19] Waller and Fawcett (2013) describe how data science, predictive analytics, and big data are changing supply chain design and management, emphasizing revolutionary trends and technology.

3 Methodology

3.1 Development Environment and Tool Integration

The project leverages Python for its backend capabilities, utilising libraries such as Pandas for data management and SentenceTransformers for natural language processing tasks. Streamlit is employed for developing an interactive web interface, enhancing user engagement with intuitive functionalities.

3.2 LangChain Implementation

LangChain is a pivotal component, seamlessly integrating with OpenAI models to empower the system with advanced natural language processing (NLP) capabilities. Serving as a bridge between the backend Python environment and user interactions, LangChain harnesses the power of sophisticated linguistic analysis, enabling the system to process intricate natural language queries effectively. By facilitating the integration of OpenAI models, LangChain significantly enhances the analytical engine's ability to extract precise and relevant insights from diverse datasets. This robust linguistic framework not only contributes to the system's capacity for nuanced AI-powered analysis but also plays a crucial role in the development of the ChatBox feature, providing users with real-time, contextually relevant AI-generated advice. LangChain thus stands as a cornerstone in elevating the project's user experience, enabling a more sophisticated and interactive dialogue between users and the advanced analytical capabilities of the application.

3.3 System Architecture and Data Processing

Our system architecture (refer to Fig 1.) combines Streamlit's frontend capabilities with a Python-based backend, ensuring a seamless user experience and efficient data processing. Streamlit serves as the core framework for developing the interactive web interface, providing an intuitive and dynamic platform for users to engage with the advanced data analysis tools seamlessly. As the frontend backbone, Streamlit facilitates the creation of visually appealing and user-friendly applications with minimal code, enabling rapid development and iteration of features. Its integration ensures a cohesive and responsive user experience, allowing for the elegant display of data visualizations and exploration outcomes. Leveraging Streamlit's capabilities enhances user engagement by providing an interactive and accessible interface for both technical and non-technical users. The framework's simplicity and versatility contribute to the overall user-centric design, ensuring that the application remains user-friendly while delivering robust functionalities for data exploration and analysis in a visually compelling manner. This framework supports agile data ingestion, cleaning, and preprocessing, underpinning the analytical engine with a reliable data management foundation. The architecture's dual nature facilitates both the elegance of user interface design and the robustness of backend operations.

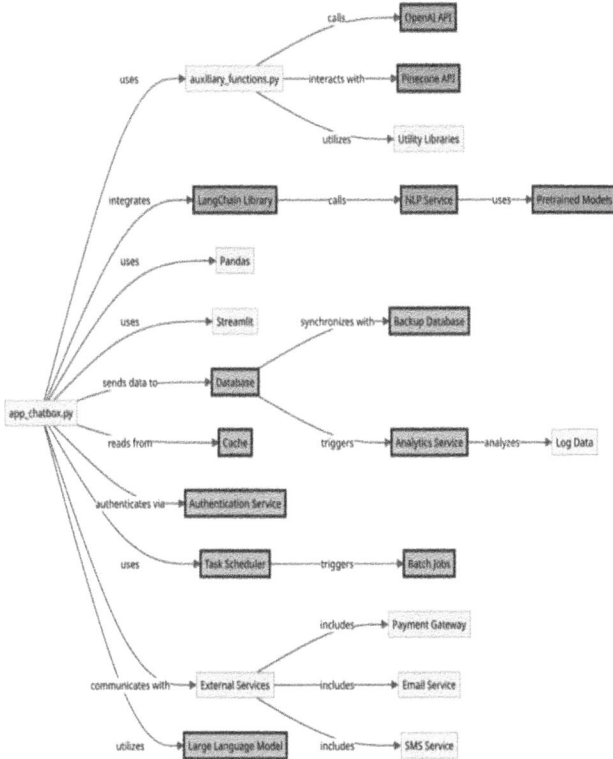

Fig. 1. An architecture diagram representing the working of the data brain which highlights how each module works and specifies how each component is connected to each other.

3.4 AI-Powered Analysis and User Experience

Incorporating GPT models allows the system to process natural language queries effectively, offering precise and relevant insights. Generative Pre-trained Transformer models are essential to the system's ability to process and respond to user queries through natural language processing. Making use of the capabilities of GPT models makes it easier to extract accurate and contextually relevant information from natural language inputs. By incorporating these models, the system can generate and comprehend text that is human-like, which enhances its comprehension and empowers it to provide intelligent answers to user queries. As shown in Fig. 2, this will improve the usage frequency of our model compared to the latest models. The usage of GPT models significantly improves the project's AI-powered analysis by allowing it to analyze and comprehend a wide range of natural language queries fast and provide users with relevant and personalized information. This integration of GPT models ensures that the system remains adaptable to evolving language patterns and user expectations, contributing to a more intelligent and user-friendly interaction model within the overarching framework of the application. The use of Pinecone's vector database enhances the system's ability to perform similarity searches, closely aligning user queries with data insights. The ChatBox feature, powered by ChatOpenAI, provides real-time, AI-generated advice, enriching the user interaction model with immediate, contextually relevant feedback.

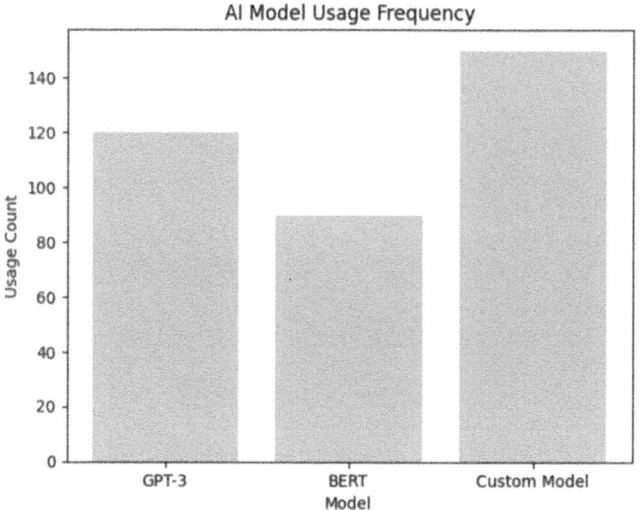

Fig. 2. A bar graph representing the model usage frequency of the various transformer models - including GPT-3, BERT and our custom model

3.5 Data Exploration, Visualization, and Interaction

The application democratises data science by enabling users to perform exploratory data analysis (EDA) through natural language queries. ED represents a transformative approach to democratizing data science, allowing users to engage in a sophisticated exploration of datasets through natural language queries. EDA serves as the cornerstone for user interaction, providing an innovative means for a wider audience to unravel complex datasets effortlessly. The application's EDA functionalities empower users to uncover patterns, trends, and anomalies, fostering a deeper understanding of the underlying data. Through the integration of natural language queries, users can dynamically inquire about specific aspects of the data, generating insightful visualizations that enhance the overall interactive experience. This innovative feature makes data exploration a more collaborative and intuitive process by making advanced data analysis tools accessible to users with varying levels of technical expertise. It also simplifies the traditionally complex process of data exploration. This invention fosters a new level of participation and insight by making sophisticated data analysis accessible to a wider audience. Visualizations generated from user queries further enhance this interactive experience, providing clear and actionable insights into the data exploration process.

3.6 Technical Challenges, Testing, and Security

To address the challenges of large dataset management and AI model fine-tuning, we've adopted strategic data handling and adaptive training techniques. A rigorous testing framework, including unit, integration, and User Acceptance Testing (UAT), ensures reliability and functionality. Our dedication to security is evident through the implementation of SSL/TLS, robust data encryption, and compliance with GDPR standards, ensuring user data protection and trust

3.7 Scalability, Performance Optimizations, and UI/UX Design

The application is designed for scalability, utilising cloud services, load balancing, and distributed computing to maintain performance amid growing user demand. Algorithmic optimizations and the use of caching and parallel processing significantly enhance system responsiveness. The UI/UX design follows foundational principles, ensuring responsive and intuitive navigation across devices, thus providing a seamless and inclusive user experience for accessing advanced data analysis tools.

4 Result

A thorough investigation was carried out with a focus on numerous important areas in order to assess the efficacy and usefulness of the created chat application powered by AI and designed with data science support. These elements include the user experience, functionality, AI response accuracy, and general project support value for data science initiatives.

4.1 User Interface and Experience

Streamlit was used in the development of the application's interface, which offers a clear and simple user interface. The chat box functionality and data analysis tools are neatly divided by the dual-tab arrangement, which also makes navigating simple. Users can interact with the application more easily and effectively when interactive elements like text inputs, selection boxes, and file uploaders are included.

4.2   Features and Data Analysis Tools

The data analysis component of the application provides all the standard EDA features, including dataset previewing, duplicate and missing value identification, and correlation analysis. These qualities make it easier to grasp the uploaded data at an early stage, which is important for defining data science tasks. Users benefit greatly in the early phases of their projects from these tools' smooth incorporation into the online interface.

4.3   Accuracy and Relevance of AI Responses

An assessment was conducted on the AI chatbot's capacity to interpret, hone, and react to user inquiries using context-aware insights. By employing advanced natural language processing (NLP) methods such as vector databases and semantic embeddings, the chatbot exhibits remarkable ability to produce perceptive and pertinent responses.

However, the complexity of the query and the clarity of the context offered determine how well the chatbot performs, suggesting that there is still opportunity for improvement when tackling unclear or extremely technical requests.As shown in Fig. 3, the performance of the model was aligned with the expected goals or the performance of the model as expected . The metrics measured included the response time, accuracy, user satisfaction, and engagement amongst the users.

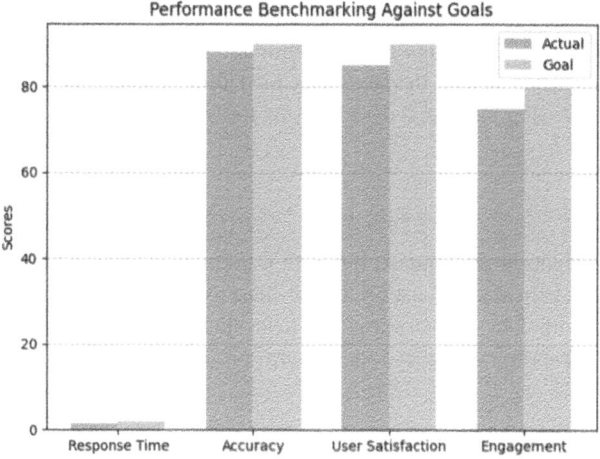

Fig. 3. A bar graph representing the benchmarking results of the custom model, comparing our predicted performance and its actual performance

4.4   Use in Project Support and Learning

The program is quite good at rephrasing business difficulties as data science issues and making suitable machine learning model recommendations. This feature helps with both the conceptual stage of data science projects and their instructional purpose by introducing users to different machine learning methods and how they can be applied to different kinds of data science challenges. As shown in Fig. 4, Over the course of time, the response time of the chatbot has seemed to significantly increase and decrease depending on the business difficulty which were addressed as data science issues over a wide range of datasets and problems, for which suitable model recommendations were required, thus leading to small fluctuations in response time of the proposed model, but nevertheless and irrespective of the response time the model has seemed to outperform the expectations and give proper insights to datasets and make suitable predictions and recommendations.

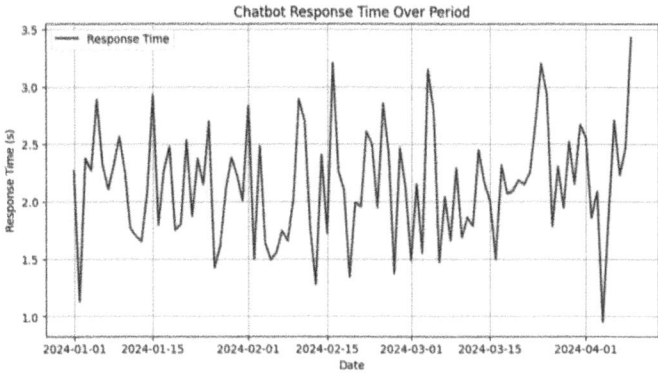

Fig. 4. A time series graph showing the response time over the period of the proposed data brain which has shown significant improvement over the course of time.

5 Conclusion

'Data Brain' presents an AI Assistant for Data Science, leveraging advanced technologies such as Streamlit, Pandas, Pinecone, Sentence Transformers, and OpenAI's API to facilitate interactive data analysis and machine learning model development within a user-friendly web application. By incorporating auxiliary functions for query refinement and conversational response generation alongside a main application interface for dataset upload, exploratory data analysis (EDA), and algorithm selection, the assistant streamlines the transition from business problems to data science solutions. A conversational ChatBox further enhances user engagement by providing dynamic support and guidance through context-aware responses. This integration of AI and machine learning technologies not only simplifies the data science workflow but also democratizes access to sophisticated data analysis and model development processes, making it an invaluable tool for data scientists.

6 Future Work

Extending features to include direct links to a greater range of data sources, such as databases, APIs, and cloud storage, as well as support for several data formats, would allow for more smooth data import and preprocessing in a variety of data science applications. Advanced data preprocessing, feature engineering, and automatic model selection and tuning methods, along with interactive data visualization tools, will enable users to do sophisticated analysis and generate insights. Collaborative capabilities for project sharing and team-based development, together with optimized model deployment tools, will provide efficient transitions from development to production, as well as model scalability and performance monitoring in real-world applications. Including a complete set of instructional resources, tutorials, and interactive guides customized to users of various ability levels will democratize access to data science information, encouraging skill development and fostering a community of practice. Strengthening security measures and privacy protections to protect sensitive data, ensuring compliance with global data protection regulations, and providing extensive customization and extensibility options will enable users to tailor the assistant to their specific needs and preferences, thereby improving user experience and application versatility. With these improvements, 'Data Brain' can keep its status as a top data science tool while benefiting a broader variety of users, from beginners to seasoned professionals.

References

1. Pandya, K., Holia, M.: Automating customer service using langchain: building custom open-source GPT chatbot for organizations. arXiv preprint arXiv:2310.05421 (2023)
2. Radford, A., Narasimhan, K., Salimans, T., Sutskever, I.: Improving language understanding by generative pre-training (2018)
3. Johnson, J., Douze, M., Jégou, H.: Billion-scale similarity search with GPUs. IEEE Trans. Big Data 7(3), 535–547 (2021)
4. Vaswani, A., et al.: Attention is all you need. CORR, abs/1706.03762 (2017)
5. Kim, S., Rawat, A.S., Zaheer, M., Jayasumana, S., Sadhanala, V., Jitkrittum, W., et al.: EmbedDistill: a geometric knowledge distillation for information retrieval arXiv:2301.12005 (2023)
6. Su, H., et al.: One embedder, any task: instruction-fine tuned text embeddings. In: Findings of the Association for Computational Linguistics: ACL 2023, pp. 1102–1121, Association for Computational Linguistics, Toronto, Canada (2023)
7. Følstad, A., Skjuve, M.: Chatbots for customer service: user experience and motivation. In: Proceedings of the 1st International Conference on Conversational User Interfaces, Dublin, Ireland, 22–23 August 2019
8. Bonifacio, L., Abonizio, H., Fadaee, M., Nogueira, R.: Inpars: unsupervised dataset generation for information retrieval. In: The 45th International ACM SIGIR Conference on Research and Development in Information Retrieval (2022)
9. Perer, A., Shneiderman, B.: Integrating statistics and visualization: case studies of gaining clarity during exploratory data analysis. In : Proceedings of 26th Annual SIGCHI Conference on Human Factors in Computing Systems. ACM Press, New York, NY, USA (2008)
10. Ghosh, A., Nashaat, M., Miller, J., Quader, S., Marston, C.: A comprehensive review of tools for exploratory analysis of tabular industrial datasets. Vis. Inform. 2(4), 235–253 (2018). https://doi.org/10.1016/J.VISINF.2018.12.004

11. Topsakal, O., Akinci, T.C.: Creating large language model applications utilizing langchain: a primer on developing LLM apps fast. In: The Proceedings of the 5 th International Conference on Applied Engineering and Natural Sciences (2023)
12. Pokhrel, S., Ganesan, S., Akther, T., Karunarathne, L.: Building customized chatbots for document summarization and question answering using large language models using a framework with OpenAI, lang chain, and streamlit. J. Inform. Technol. Digital World **6**(1), 70–86 (2024)
13. Ye, J., et al.: A comprehensive capability analysis of GPT-3 and GPT-3.5 Series Models
14. Kuzlu, M., Catak, F.O., Sarp, S., Cali, U., Gueler, O.: A streamlit-based artificial intelligence trust platform for next-generation wireless networks. In: 2022 IEEE Future Networks World Forum (FNWF). Montreal, QC, Canada (2022)
15. Kumar, A., Raschka, S.: Pinecone: a simple and efficient framework for large language model inference. arXiv preprint arXiv:2103.10811 (2021)
16. Data, MIT Critical, Komorowski, M., Marshall, D.C., Salciccioli, J.D., Crutain, Y.: Exploratory data analysis. Secondary analysis of electronic health records (2016)
17. Behrens, J.T.: Principles and procedures of exploratory data analysis. Psychol. Meth. **2.2** (1997)
18. Provost, F., Fawcett, T.: Data science and its relationship to big data and data-driven decision making. Big Data (2013)
19. Waller, M.A., Fawcett, S.E.: Data science, predictive analytics, and big data: a revolution that will transform supply chain design and management. J. Bus. Logist. **34.2** (2013)

Semantic Segmentation of Satellite Imagery Using Optimized U-Net Model

J. Kavipriya$^{(\boxtimes)}$ and G. Vadivu

Department of Data Science and Business Systems, School of Computing, SRM Institute of Science and Technology, Kattankulathur, Tamilnadu 603203, India
{Kj1101,vadivug}@srmist.edu.in

Abstract. The classification of land cover is crucial due to increasing demands and population expansion. Segmentation of the terrain is utilized in environmental monitoring to accurately identify and delineate areas of agriculture, buildings, and water bodies. Satellite imagery is quite valuable for the purpose of segmentation. Deep learning models offer automated and precise segmentation of satellite imagery. The U-Net architecture is widely employed as an encoder-decoder structure for the purpose of segmenting medical and remote sensing images. This study investigates the process of dividing satellite images into segments using an optimized U-Net model. Our model utilizes improved equal weights to calculate loss functions. In addition, we have implemented a dropout layer for each convolutional block in order to decrease the computational time required for feature extraction in the encoder layer of the optimized U-Net model. Our study examines the Dubai dataset, which has six distinct classes. The optimized U-Net model we propose achieves an accuracy of 81.5% and Intersection over Union (IoU) values of 61.7% for segmenting the dataset. These metrics are acquired with a reduced number of epochs, hence decreasing the computational expense.

Keywords: Satellite imagery · Segmentation · modified U-Net · Artificial Intelligence · Deep Learning

1 Introduction

Ensuring sustainable growth is necessary to adequately address the requirements of future generations. Each nation worldwide prioritizes sustainable development in order to achieve a harmonious equilibrium between technology advancement and human need. In order to ensure sustainability, prioritizing environmental preservation is crucial due to the rapid advancement of technology, which has resulted in the contamination of natural resources. Preservation of biodiversity in the environment assumes a significant importance. The examination of land cover and land usage facilitates the monitoring of the environmental consequences within a specific geographical region over a given timeframe. The monitoring of regional development and declination can be facilitated through the utilization of Land Use and Land Cover (LULC) diagrams. For instance, it is possible to analyze the expansion of urban areas as well as the decline in forested

P. D. Sivakumar et al. (Eds.): IRCCTSD 2024, CCIS 2362, pp. 222–232, 2025.
https://doi.org/10.1007/978-3-031-82386-2_17

areas and vegetation. The LULC maps provide comprehensive information regarding alterations in agriculture phenology growth rate, deforestation patterns, meteorological conditions, occurrences of flooding and forest fires, carbon emissions, and estimations of drought.

Multiple sources, such as satellite data, aircraft photography, Geographic Information System (GIS), drone imaging, and field surveys, can be utilized to acquire these Land Use and Land Cover (LULC) maps. Satellites such as LANDSAT and Sentinel, Moderate Resolution Imaging Spectroradiometer (MODIS) are frequently employed for the purpose of generating land use and land cover (LULC) maps. The process of creating land use and land cover (LULC) maps involves several key steps, including the acquisition of remote sensing pictures, image preprocessing, classification of the processed images, and final evaluations of metrics to ensure accuracy. Image preprocessing encompasses many techniques such as noise reduction, image sharpening, and scaling, among others. Techniques for classifying images include supervised, unsupervised, Reinforcement Learning and semi-supervised classification. Various metrics, including as accuracy, precision, recall, and F1 score, are employed to evaluate the performance of derived LULC maps.

Image classification is crucial for identifying land coverage and land utilization. Machine learning techniques can be utilized to facilitate the process of classification. Logistic regression, Support Vector Machines (SVM), Decision tree, Random Forest (RF), K-Nearest Neighbors (KNN), Naive Bayes, and neural networks are among the frequently employed methods in the field of machine learning.

Neural Networks (NN) are a rapidly growing area that aims to replicate the human brain and extract properties similar to those seen in humans. Neural networks (NN) can be categorized into three main types: Feedforward Neural Networks (FNN), Recurrent Neural Networks (RNN), and Convolutional Neural Networks (CNN).The CNN algorithm operates by utilizing convolutions and layers. CNN goes through four processes. The initial stage involves the convolutional operation, which encompasses elements such as filtering, padding, striding, and the Rectified Linear Unit (ReLU) operation. Features are extracted in this layer. Subsequently The second phase is max pooling, which is employed to reduce dimensionality and complexity.

The subsequent stage involves the process of transforming the reduced image into a unidimensional vector. The fourth stage is the Fully connected procedure, wherein images are classed according to the features that have been extracted in an earlier layer. In addition, activation functions such as softmax are employed in this layer. The classified images are subjected to back propagation and training iterations in order to enhance the learning process of the given image. To enhance the classification of an image, it is important to do a comprehensive analysis of each individual pixel inside the image. The accuracy of satellite imagery is enhanced by segmenting the images depending on their pixels prior to the classification procedure. Semantic segmentation is a technique used to distinguish and classify different objects inside an image by analyzing the individual pixels. Two widely used deep learning models in the field are U-net [1] and DeepLab. The objective of this study is to employ the modified U-Net model for the purpose of segmenting satellite imagery in order to get accurate classification outcomes. Figure 1 Depicts the how satellite images are used to obtain segmentation map.

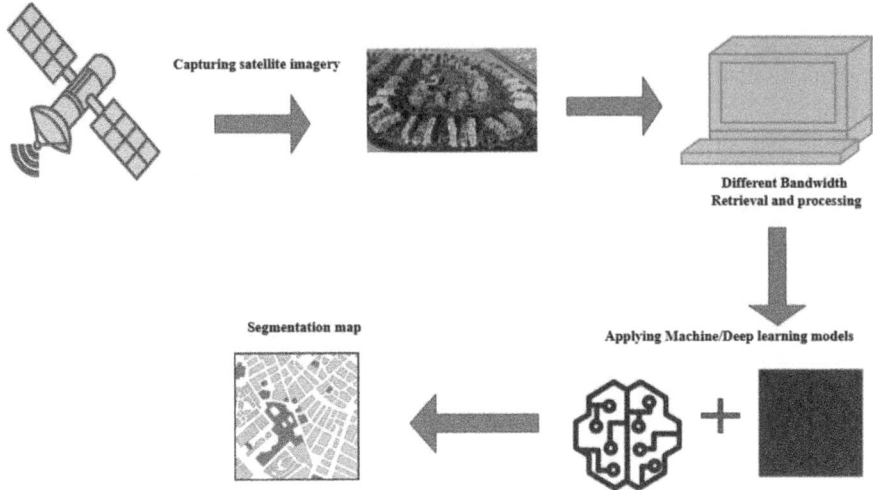

Fig. 1. Depicts work flow diagram for creating segmentation map

2 Related Works

2.1 Remote Sensing Applications

Featuring high-resolution images of South Korea, the paper cited as [1] conducts object segmentation on four classes: building, backdrop, water, and roads. The researchers have enhanced the U-Net model by introducing a novel pooling layer called the pyramid pooling layer. They then conducted a comparative analysis with other state-of-the-art models. The U-Net PSP model achieved an accuracy of 87.61% and a mean Intersection over Union of 79.52%. The research referenced [2] introduces the Shuffling Convolutional Neural Network (SCNNN) as a method for detecting small items in a benchmark dataset. The identification of small objects is achieved by the utilization of an Ensembled Deeper-CNN model that has been enhanced with Field of View Enhancement and SCNN. Shuffling parameters are employed to transform the feature maps that have been reduced in size into feature maps with a higher resolution. These feature maps are utilized for subsequent assessment. The study referenced [3] introduced L-Unet model that integrates traditional LSTM and U-Net architecture. L-UNet is derived from U-Net by substituting two convolutional layers with a ConvLSTM layer and a 2-dimensional convolutional layer. This model combines LSTM and Unet to extract both spatial and temporal data. The authors incorporate atrous pooling layer into the L-Unet design to mitigate spatial loss. Modified U-Net, when compared to other state-of-the-art models, demonstrates superior accuracy in comparison to DeepLab. The authors achieved a high level of accuracy in comparing the growth of structures over time using AL-UNet, in comparison to other models.

Paper [4] introduced a novel technique for decoding features, incorporating the use of Feature Pyramid Network (FPN), Multiscale Feature Fusion (MFF), and Segmentation Head (SH). FPN utilizes Context Aware Reassembly of Features (CARAFE) for the purpose of upsampling, as it integrates the attributes of context awareness. MFF utilizes sequential upsampling concatenation by employing four pooling layers with varying strides. Authors ultimately incorporated a segmentation head layer consisting of a 3X3 convolution layer and a 1X1 convolution layer. The proposed technique has the capability to accurately separate the background, standing water, waterway, and weed cluster in the given dataset.In the study [5], authors introduced a multitasking architecture that utilizes D3-Net for the purpose of calculating depth or height and performing semantic categorization. Authors considered equal weights, the GradNorm technique which adjusts the magnitudes of gradients based on the learnable parameters, and the Multi-Task Learning-Multiple Gradient Descent (MTL-MGDA) algorithm for optimization. The authors utilized multiple sources of data, including Digital Surface Map (DSM), Digital Elevation Model (DEM), Hyperspectral image, and Very High Resolution (VHR) RGB, as input data.

The research referenced [6] investigates the use of a completely conventional neural network for accurately segmenting roads in urban areas of England, Poland, and Germany. Google Earth optical pictures were used for ground truth validation. They have adjusted the tolerance and weight settings precisely by considering the number of road pixels in the ground truth images. The authors utilized a Fully Connected Conditional Random Field (FCCRF) for post-processing. The performance of the deep residual U-Net and Deep LabV3 models were compared for segmentation, resulting Deep LabV3 demonstrating superior performance. Paper [7] use deep convolutional neural networks (CNNs) to perform segmentation of highways, buildings, and backgrounds. The authors attempted to annotate the dataset using photos from Google Maps and OpenStreetMap. They have chosen to use the ISPRS Potsdam dataset for a sanity check since it contains well-annotated ground truth data. The authors' conclusion is that using a significant amount of training data with noisy pixels yields better results for segmenting a new validation dataset compared to using a benchmark dataset with high-resolution images that do not have noise. An investigation of densely populated places such as Tokyo, Chicago, Paris, and Zurich has been conducted by integrating data from OpenStreetMap (OSM) and Google Maps. The study referenced [8] examines the U-Net model (Table 1),

which combines LiDAR data and photogrammetric point clouds to segment buildings, roads, water, vegetation, and other objects. The Difference of Normal (DoN) approach has been utilized to calculate the point clouds in three specific locations within the Zhujiang town in southern China, where tall structures are present. The Euclidean distance threshold-based clustering algorithm is employed to classify point clouds. They have achieved optimal outcomes for vegetation segmentation with an accuracy rate of 87%.

The wind turbines in four private wind farms were segmented in a reference paper [9] in order to determine the flaws in the wind mill using UAV photography. They acquire 330 UAV images and enhance the data by applying rotation, horizontal flipping, vertical flipping, scaling, and shearing techniques. It was determined that convergence occurred after 20 epochs. Additionally, they employed 10-fold cross-validation for training. The

Table 1. Gives the dataset, CNN models and evaluation metric incorporated for segmentation

S.No	Authors	Data Used	CNN models Used	Evaluation metrics
1	Kin et al	High resolution images of south korea (Training-72400,Testing-9600)	FCN,U-Net, PSPNET, FCNPPL, UNetPPL	Accuracy-87.61 MeanIOU-79.52%
2	Chen et al	Vaihingen and Potsdam data sets	RDM,RDM-ASPP, Naive-SCNN, Naive-SCNN-ASPP, Deeper-SCNN, EDeeperSCNN	Vaihingen Dataset-Precision-0.8623 Potsdam data sets-0.8578
3	Sun et al	SZTAKI Air Change dataset and Beichuan dataset	U-Net, L-UNet, AL-Unet	SZTAKI Dataset-Accuracy-90.10 Kappa-0.7874 Beichuan dataset-Accuracy-88.37 Kappa-0.8499
4	Henry et al	Agriculture vision challenge dataset	U-Net, UperNet, DeeplabV+	Accuracy-81.3
5	Carvalho et al	IEEE Data Fusion Contest DFC 2018,ISPRS Vaihingen	D3-NetMTL-MGDA,GradNorm,Equal Weights	Accuracy -58.26 Kappa Coefficent-0.56
6	Henry et al	High resolution TerraSAR-X imagery	FCN,U-Net,DeepLabV3+	Precision-71.69 Recall-75.17 IoU-45.46
7	Kaiser et al	ISPRS potsdam dataset, Google maps and OSM Maps	Deep CNN (modified VGG16)	F1score-79.7 Precision-81.9 Recall-77.6
8	ZHANG et al	LiDAR and Aerial imagery photgrammetic point clouds	U-Net	F1score-92 Precision-91 Recall-93
9	Wang et al	DJI Mavic Pro UAV images of 2000X1500 image size, count -330	Improved U-Net(ResNet+ PSANet+ ECA-Net and U-Net)	IoU-63.06 Precision-77.98 Recall-72.34 DC-74.36
10	Donapati et al	NVIDIA Jetson nano enabled GPU platform takes photo of seeds every 5 secs in the lab environment	U-Net, CNN model	Accuracy-91 IoU-84 Precision-80

model validation process involved the utilization of Dice coefficient and IoU indicators. Their proposed model surpasses the performance of the classic U-Net approach. The publication referenced [10] provides an explanation of seed detection and germination analysis utilizing a fusion of the U-Net and CNN models. The authors utilize the U-net model to partition the images and determine the germination status using the pro posed CNN classification technique. The proposed model achieves a pixel accuracy of 91%, which is superior to state-of-the-art models such as Lenet, Inception, and Resnet50. The Table 1 provides a summary of the CNN models used for the segmentation task and their corresponding metrics used for validation.

3 Experiments and Results

3.1 Dataset Used

The dataset utilized in the study was obtained by satellites operated by the Mohammed Bin Rashid Space Centre (MBRSC). The dataset comprises six distinct classes: Building, Land, Road, Vegetation, Water, and Unlabeled. This dataset has six bigger tiles. Each tile is divided into 9 smaller tiles for processing. This dataset has a total of 72 images that encompass the city of Dubai, which serves as the capital of the United Arab Emirates. Additionally, it encompasses the masked images that have been manipulated using the one-hot encoding technique. This dataset is sourced from Kaggle and is freely accessible.

3.2 Optimized U-Net Model

The U-Net architecture enables pixel-level categorization in an image, resulting in the semantic segmentation of objects within the image. The U-Net architecture primarily consists of an encoder-decoder structure [11], which performs both upsampling and downsampling operations. During the downsampling process, the spatial dimensions of the images are decreased and characteristics are extracted. During the upsampling (expansion) procedure, layers are concatenated and features that were skipped are included. This phase facilitates the preservation of crucial traits that were overlooked during the downsampling procedure. The suggested methodology for optimizing the U-net model involves the use of convolutional layers (conv 2D layers), max pooling layers, drop out layers, and convolutional transpose layers (conv 2D transpose layers).The activation function utilized for each convolutional layer is rectified linear unit (ReLU). The purpose of using this activation function is to mitigate the issue of the vanishing gradient problem that occurs in deep neural networks [12].The most frequently utilized kernel initializer in conjunction with the ReLU activation function is he_normal, which employs a Gaussian distribution to manipulate weights. The downsampling convolutional layers have zero padding and zero strides.

$$F_{ReLu}(x) = \{0 \, if x < 0 \, x \, if x \geq 0 \tag{1}$$

Each Convolutional block is equipped with a dropout layer with a dropout rate of 20 percentage. During the process of upsampling, convolutional transpose layers utilize the concatenate function to combine the spatial features that may have been overlooked during downsampling. The upsampling procedure involves using strides of (2×2) and zero padding. The U-Net design consists of a total of 11 convolutional blocks, encompassing both upsampling and downsampling operations. Finally, the output layer utilizes the softmax activation function, which is specifically designed for multi-class classification tasks. Figure 2 explains the architecture of the Optimized U-net architecture for segmentation.

The total loss in this improved U-Net model is computed by adding the dice loss [14] and focus loss [13]. The dice loss is computed using the dice coefficient, which measures the similarity between the ground truth mask and the anticipated mask image. Due to the equal assignment of weights in our modified model, class imbalance is compromised. The focal loss function is also employed to address the issue of class imbalance [15].

The focal loss function assigns weights based on the modulating factor, resulting in a down weighting of less complicated pixels and an increased attention to complex pixels. In order to mitigate any potential bias caused by the class imbalance problem. When calculating the dice loss and focus loss, the weights for all the classes in the dataset are initialized to be equal. Next, the total loss is computed.

$$Total\ Loss = Dice\ Loss + Focal\ Loss \qquad (2)$$

The Adam optimizer [16] is utilized for model compilation. The system operates by adjusting the learning rate of the input parameters in an adaptive manner. Accuracy is a frequently employed metric in convolutional neural networks. It is determined by dividing the total number of correctly predicted images by the total number of images.This statistic is employed to assess the performance of the model. The higher the accuracy, the greater the possibility of making right predictions [17]. Accuracy is integrated into our model. The calculation of accuracy validation is determined using the following formula.

$$Accuracy = Predicted\ image/Total\ images \qquad (3)$$

The IoU or Jaccard coefficient is utilized to quantify the degree of similarity or dissimilarity between the provided datasets [18]. This is also referred to as the Jaccard Index. The Jaccard index is employed in our model to assess the similarity between the training and validation images. The Jaccard Index is computed using the following formula

$$JC(Tr, Val) = |Tr \cap Val| \div (|Tr| + |Val| - (|Tr \cap Val|)) \qquad (4)$$

4 Results and Discussion

The proposed modified U-Net model undergoes an iterative process, running through varying numbers of epochs. We have conducted experiments with up to 30 iterations, commencing with 4. Convergence begins after 12 epochs. Running after 12 epochs does not enhance the accuracy and IoU readings. In addition, we have experimented with varying dropout percentages ranging from 10% to 30% in each convolutional layer in order to get higher accuracy within a reduced number of epochs. The Table 2 presents a comparison of various epochs along with their respective accuracy and IoU values. From the table 74.82% accuracy is obtained for 4 epochs with 47.9% IoU. In next iteration accuracy got increased by 5% and IoU also increased by 5%. But in the next iteration that is 12th epoch accuracy growth rate got reduced by 3% but IoU growth rate increased by 9.

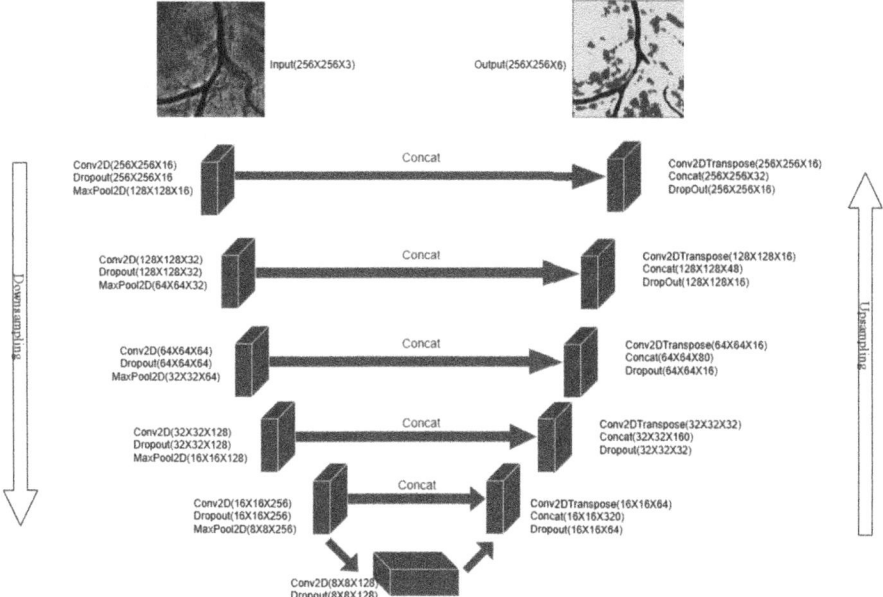

Fig. 2. Optimized U-Net architecture for segmentation

After 12 epochs there is no major change in the results since the convergence happens afterwards. Even in 30 epochs 85 percentage accuracy is achieved with 66.23%. This comparison make us to conclude the 12 epochs is enough to get optimized results. Also Jaccard coefficient growth rate also doesn't has that much impact after 12 epochs in segmenting the land cover areas when compared to the state of art models. Our model works better with vegetation land segmentation when compared to study[1] results. Most of models have segmented only two classes [3, 10] but our dataset has six classes. Our model performs well with tall structure buildings when compared to study [5]. Figure 3a and 3b gives the loss and IoU value comparisons with respect to epochs during training and validation.

Table 2. Comparison of metrics with different epochs of our optimized U-Net model

Epochs	Accuracy	Jaccard coefficient
4	74.82	47.9
8	79.48	53.72
12	81.52	62.70
30	85.86	66.23

Though our model performs well in terms of accuracy and IoU with more number of classes loss percentage is not reduced when compared to state of art models like [1]. There is no significant changes in the loss reduction even after 30th epochs. Increasing the drop out rate also doesn't impact on the loss reduction. As well as decreasing drop outs also doesn't have that much impact on the loss reduction. Also optimizing the weight values for loss function gives minor changes in the loss value. However loss value need to be concentrated by further fine tuning of our model.

Figure 3c provides the segmentation map which is obtained while running optimized U-Net model. If we noticed original masked image doesn't identify the buildings in the original image. But our U-Net segmented image gives details about the buildings in the original image which gives extra information.

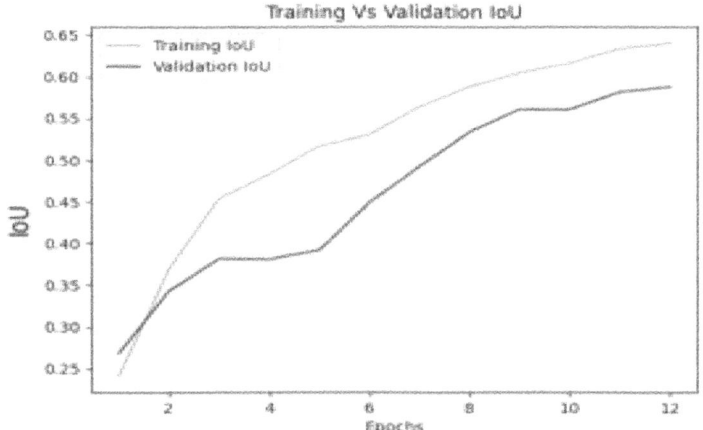

Fig. 3a. Epoch Vs IoU for 12 epochs

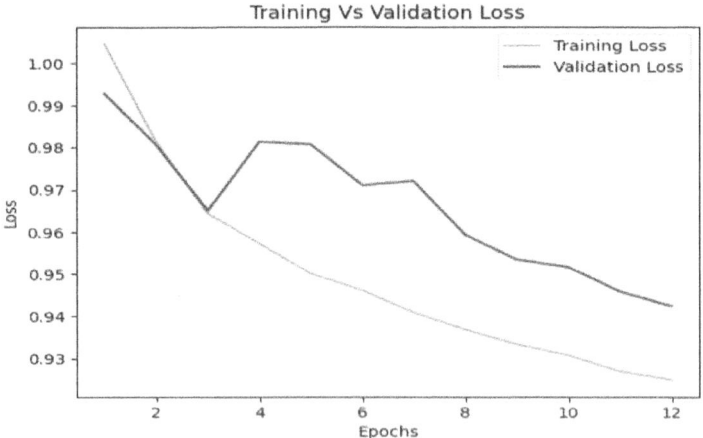

Fig. 3b. Epoch Vs Loss for 12 epochs

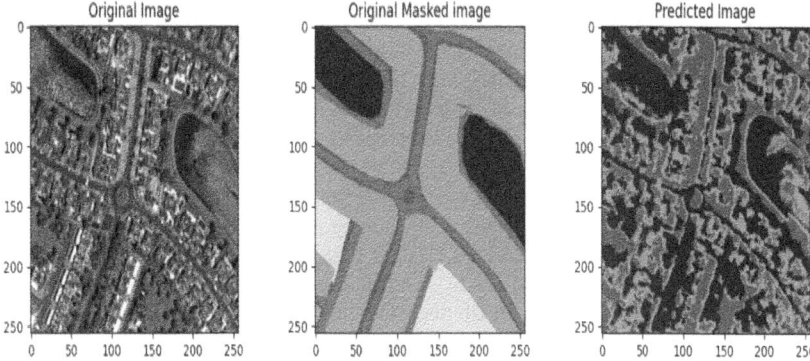

Fig. 3c. Segmentation map of optimized U-Net model

5   Conclusion

Our proposed research involves the development of a optimized version of the U-Net model, which is utilized for the purpose of segmenting the satellite images in order to obtain better landcover classification. With 12 epochs, the model that we have proposed achieves an accuracy of 81.5%. Additionally, in order to lessen the amount of computing that is required, we have incorporated drop out layers into our model. The value of the Jaccard Index is determined to be 62.70 for 12 epochs. A good segmentation of the vegetation, water, and structures is something that our model is able to identify with more accuracy. In terms of efficiency, our model performs very well when applied to the Dubai dataset. In spite of the fact that it works admirably for the input dataset that was used, we still need to investigate it using different benchmark datasets. In addition, this dataset has a smaller number of images because Dubai accounts for only 5% of the United Arab Emirates. As a result, our efforts in the future ought to be focused more on satellite images, which cover a greater area of the landscape.

References

1. Chen, K., Fu, K., Yan, M., Gao, X., Sun, X., Wei, X.: Semantic segmentation of aerial images with shuffling convolutional neural networks. IEEE Geosci. Remote Sens. Lett. **15**(2), 173–177 (2018). https://doi.org/10.1109/LGRS.2017.2778181
2. Kim, J.H., et al.: Objects segmentation from high-resolution aerial images using U-Net with pyramid pooling layers. IEEE Geosci. Remote Sens. Lett. **16**(1), 115–119 (2019). https://doi.org/10.1109/LGRS.2018.2868880
3. Sun, S., Mu, L., Wang, L., Liu, P.: L-UNet: an LSTM network for remote sensing image change detection. IEEE Geosci. Remote Sens. Lett. **19**, 1–5 (2022). https://doi.org/10.1109/LGRS.2020.3041530
4. Liu, C., Du, S., Lu, H., Li, D., Cao, Z.: Multispectral semantic land cover segmentation from aerial imagery with deep encoder–decoder network. IEEE Geosci. Remote Sens. Lett. **19**, 1–5 (2022). https://doi.org/10.1109/LGRS.2020.3037976
5. Carvalho, M., Le Saux, B., Trouve-Peloux, P., Champagnat, F., Almansa, A.: Multitask Learning of Height and Semantics From Aerial Images. IEEE Geosci. Remote Sens. Lett. **17**(8), 1391–1395 (2020). https://doi.org/10.1109/LGRS.2019.2947783

6. Henry, C., Azimi, S.M., Merkle, N.: Road segmentation in SAR satellite images with deep fully convolutional neural networks. IEEE Geosci. Remote Sens. Lett. **15**(12), 1867–1871 (2018). https://doi.org/10.1109/LGRS.2018.2864342

7. Kaiser, P., Wegner, J.D., Lucchi, A., Jaggi, M., Hofmann, T., Schindler, K.: Learning aerial image segmentation from online maps. IEEE Trans. Geosci. Remote Sens. **55**(11), 6054–6068 (2017). https://doi.org/10.1109/TGRS.2017.2719738

8. Zhang, P., et al.: 3D urban buildings extraction based on airborne LiDAR and photogrammetric point cloud fusion according to U-Net deep learning model segmentation. IEEE Access **10**, 20889–20897 (2022). https://doi.org/10.1109/ACCESS.2022.3152744

9. Wang, L., Yang, J., Huang, C., Luo, X.: An improved U-Net model for segmenting wind turbines from UAV-taken images. IEEE Sens. Lett. **6**(7), 1–4 (2022). https://doi.org/10.1109/LSENS.2022.3184521

10. Donapati, R.R., Cheruku, R., Kodali, P.: Real-time seed detection and germination analysis in precision agriculture: a fusion model with U-Net and CNN on jetson nano. IEEE Trans. AgriFood Electron. **1**(2), 145–155 (2023). https://doi.org/10.1109/TAFE.2023.3332495

11. Wiratama, W., Lee, J., Sim, D.: Change detection on multi-spectral images based on feature-level U-Net. IEEE Access **8**, 12279–12289 (2020). https://doi.org/10.1109/ACCESS.2020.2964798

12. Feng, H.-S., Yang, C.-H.: PolyLU: a simple and robust polynomial-based linear unit activation function for deep learning. IEEE Access **11**, 101347–101358 (2023). https://doi.org/10.1109/ACCESS.2023.3315308

13. Ming, Q., Xiao, X.: Towards accurate medical image segmentation with gradient-optimized dice loss. IEEE Signal Process. Lett. **31**, 191–195 (2024). https://doi.org/10.1109/LSP.2023.3329437

14. Zhang, J., et al.: Water body detection in high-resolution sar images with cascaded fully-convolutional network and variable focal loss. IEEE Trans. Geosci. Remote Sens. **59**(1), 316–332 (2021). https://doi.org/10.1109/TGRS.2020.2999405

15. Lin, T.-Y., Goyal, P., Girshick, R., He, K., Dollar, P.: Focal loss for dense object detection. in: 2017 IEEE International Conference on Computer Vision (ICCV), pp. 2999–3007. IEEE, Venice (2017). https://doi.org/10.1109/ICCV.2017.324

16. Khan, A.H., Cao, X., Li, S., Katsikis, V.N., Liao, L.: BAS-ADAM: an ADAM based approach to improve the performance of beetle antennae search optimizer. IEEECAA J. Autom. Sin. **7**(2), 461–471 (2020). https://doi.org/10.1109/JAS.2020.1003048

17. Li, M., Zang, S.: Mapping localized patterns of classification accuracies through incorporating image segmentation. IEEE Geosci. Remote Sens. Lett. **12**(7), 1571–1575 (2015). https://doi.org/10.1109/LGRS.2015.2413419

18. Samanthula, B.K., Jiang, W.: Secure multiset intersection cardinality and its application to jaccard coefficient. IEEE Trans. Dependable Secure Comput. **13**(5), 591–604 (2016). https://doi.org/10.1109/TDSC.2015.2415482

Video Caption Generation Using Deep Learning

N. Kopperundevi⬤, Adithya Biju, Kristef Chacko$^{(\boxtimes)}$, and Mayukh Saleel

School of Computer Science and Engineering, Vellore Institute of Technology, Vellore 632014, India

kopperundevi.n@vit.ac.in, {adithya.biju2020, kristefvarkey.chacko2020,mayukh.saleel2020}@vitstudent.ac.in

Abstract. A deep learning system addressing accessibility issues for blind users is the motivation of this proposed work. By integrating LSTM networks for temporal dependencies, it generates comprehensive video captions. Unlike traditional methods relying on visual cues, this system provides context-aware descriptions, enhancing understanding and enjoyment for blind users. LSTM networks capture temporal relationships, ensuring smooth narrative flow. Combining these approaches yields detailed audio descriptions, facilitating better access to online video content. This advancement marks a significant step towards inclusivity in digital media consumption, benefiting blind individuals and improving their online experience.

Keywords: Blind users · Accessibility barriers · Deep learning · Video captions · LSTM networks

1 Introduction

In today's digital landscape, the internet is replete with a vast array of visual content, from educational videos to entertainment media. However, this abundance of visual information presents a significant accessibility challenge for individuals with visual impairments. Despite advancements in technology, traditional methods of video captioning often fall short in providing meaningful access to the blind community due to their reliance on visual cues. Furthermore, existing audio descriptions typically lack the depth and detail necessary to convey the richness of visual media. In response to these challenges, the work aims to develop an innovative solution leveraging deep learning techniques to enhance accessibility for the visually impaired. Specifically, the use of Long Short-Term Memory (LSTM) networks is proposed to generate comprehensive video captions that cater to the unique needs of blind individuals. By accurately describing visual content, providing detailed audio descriptions, and conveying contextual information effectively, the system seeks to bridge the accessibility gap and empower blind users to fully engage with online visual content.

© The Author(s), under exclusive license to Springer Nature Switzerland AG 2025
P. D. Sivakumar et al. (Eds.): IRCCTSD 2024, CCIS 2362, pp. 233–243, 2025.
https://doi.org/10.1007/978-3-031-82386-2_18

2 Objective

The objective is to overcome the limitations of traditional video captioning algorithms through the application of sophisticated deep learning methods, specifically focusing on Long Short-Term Memory (LSTM) networks. Traditional video captioning solutions often rely heavily on visual cues, rendering them inaccessible to people with visual impairments. These methods may provide insufficient or vague descriptions of the visual information, thus preventing visually impaired people from understanding and engaging with the material. To address these drawbacks, the goal is to utilize LSTM networks. LSTM networks excel at capturing temporal dependencies in sequential data, making them ideal for creating captions that maintain coherence and continuity across the visual narrative. By analyzing the sequence of video frames, LSTM networks may generate captions that correctly represent the progression of events and actions in the video, improving overall comprehension for visually impaired viewers. LSTM models also provide significant advantages in capturing long-range dependencies and relationships within data sequences. By prioritizing key visual features and relationships, LSTM models can generate captions that provide detailed and contextual descriptions of the visual content. This allows visually challenged people to gain a better understanding of the topic and provides easier access to online video content. Using LSTM networks models, captions are aimed to be generated that are not just contextual and detailed, but also accurate. These captions are intended to significantly improve accessibility for visually impaired people by giving detailed descriptions of visual content. Finally, the goal is to contribute to a more inclusive digital environment in which people of all abilities may equally access and engage with online video content, creating greater diversity and inclusivity in the digital world.

3 Literature Review

In the 2023 paper titled "Enhanced transformer model for video caption generation", Soumya Varma and J. Dinesh Peter provides a literature survey on automatic video captioning, highlighting advancements from basic methods to advanced transformers. It proposes an enhanced transformer architecture with CNN for feature extraction, showing slight performance improvements compared to state-of-the-art methods, and suggests incorporating curriculum learning for further enhancement.

In the 2021 paper titled "Exploring Video Captioning Techniques: A Comprehensive Survey on Deep Learning Methods", Saiful Islam, Aurpan Dash and et.al show that deep learning revolutionizes video captioning, with ResNet, VGG, and LSTM popular, but GRU and Transformer emerging. MSVD and MSR-VTT datasets dominate, emphasizing need for more diverse samples. Despite progress, video captioning research has untapped potential in deep learning and dataset development.

In the 2019 paper titled "Phrase-based image caption generator with hierarchical LSTM network" in the Neurocomputing Volume 333, Ying Haua Tan and Chee Seng Chan presents a novel phrase-based hierarchical LSTM network, phi-LSTM, for image captioning. Compared to sequential models, phi-LSTM generates more accurate and novel captions by capturing temporal hierarchies. The proposed model achieves competitive results on popular datasets, showcasing its effectiveness in enhancing image captioning.

In this 2024 paper "Automatic image caption generation using deep learning" Akash Verma, Mohit Kumar and Divakar Yadav proposes an encoder-decoder model for image captioning, achieving significant performance improvements compared to state-of-the-art methods, validated through popular metrics and live sample images.

In the 2023 paper "Automatic image caption generation using deep learning", Adel Jalal Yousif and Mohammed H. Al-Jammas explores video captioning, which involves generating descriptive text for videos. It discusses how deep learning, using large datasets and neural networks, has advanced this field. The study examines deep learning techniques for video captioning, emphasizing important elements, and current developments.

In their 2013 paper titled "Framing Image Description as a Ranking Task: Data, Models and Evaluation Metrics" published in the Journal of Artificial Intelligence Research, M. Hodosh, P. Young, and J. Hockenmaier propose a novel approach to image description generation by formulating it as a ranking task. They introduce a large dataset paired with image descriptions and explore various models, including a baseline system and sophisticated neural network architectures, to generate descriptions. Through rigorous evaluation using diverse metrics, they provide insights into the effectiveness of different approaches and highlight the importance of designing appropriate evaluation methodologies for image description systems.

4 Data Set and Features

4.1 FLICKR8K Dataset

It consists of 8,000 unique images where each image is paired with five different captions, providing a variety of descriptions for the same visual content. The captions are written in English and enumerate the key elements, actions, and activities shown in the pictures.

- Relatively small size: While some datasets contain millions of images, FLICKR8K has 8,000 images, which is considered manageable. This smaller scale enables researchers with limited computer resources to train and experiment with picture captioning models more effectively. It also allows for faster iterations in model building and experimentation.
- Rich annotations: Each image in the FLICKR8K dataset has five distinct captions, resulting in a diverse variety of written descriptions for the same visual content. This complexity of annotations increases the dataset's usability by providing a variety of perspectives and interpretations of the photos. It allows machine learning models to learn various methods of characterising scenes, objects, and activities, resulting in more robust and complete picture understanding.
- Freely available: The FLICKR8K dataset is freely available, which implies that researchers and developers can use it without fee or limitation. This accessibility promotes collaboration, encourages creativity, and allows for reproducibility in research efforts focused on image captioning and similar tasks.

- Training image captioning models: The FLICKR8K dataset is primarily used to train deep learning models capable of automatically generating image captions. The dataset provides paired image-caption data, making it a great resource for training and fine-tuning neural networks and other machine learning algorithms for image recognition and natural language synthesis applications.
- Evaluating image captioning performance: The FLICKR8K dataset includes evaluation metrics such as the BLEU (Bilingual Evaluation Understudy) score, which assesses the similarity between generated and reference captions. Researchers and developers utilise these measures to evaluate the effectiveness of picture captioning models trained on the dataset, allowing them to identify strengths and weaknesses and fine-tune their approaches accordingly.
- Benchmarking progress in image captioning research: FLICKR8K provides a standardised benchmark dataset, allowing researchers to compare the performance of various image captioning methods in a consistent manner. This allows for the tracking of developments in the area over time and fosters healthy competition and collaboration among academics working on picture interpretation and natural language processing.
- Limited size: Despite its numerous advantages, the FLICKR8K dataset has several limitations. Its limited size may limit the complexity of models that can be trained on it, thus restricting the performance achievable using cutting-edge approaches..
- Caption diversity: the dataset provides numerous labels per image, these explanations may not cover all conceivable interpretations of the scene, resulting in annotation bias or incompleteness.

Overall, the FLICKR8K dataset remains a cornerstone in the field of picture captioning, offering researchers and developers a significant resource for training, assessing, and comparing image understanding and natural language generation algorithms. Its accessibility, extensive annotations, and role in stimulating innovation make it an essential instrument for advancing the state of the art in machine perception and understanding (Figs. 1, 2 and Table 1).

Table 1. Data used in model

	LSTM/CNN
Image Size	224×224
Data Format	RGB
No. of Images	8091

41999070_83808
9137e.jpg

42637986_135a9
786a6.jpg

42637987_86663
5edf6.jpg

44129946_9eeb3
85d77.jpg

53043785_c468d
6f931.jpg

54501196_a9ac9
d66f2.jpg

54723805_bcf7af
3f16.jpg

55135290_9bed5
c4ca3.jpg

Fig. 1. Examples of images in Dataset

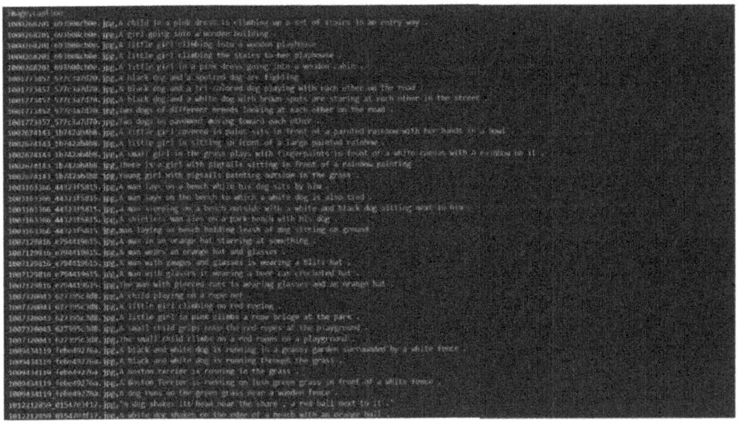

Fig. 2. Examples of captions for images in Dataset

5 Methodology

5.1 Data Collection

Gather a diverse dataset of video content spanning various genres, including educational, entertainment, news, and documentaries. Ensure the dataset includes videos with a wide range of visual complexity and content types to represent real-world scenarios.

5.2 Data Preprocessing

Extract visual features from the video frames using pre-trained convolutional neural networks (CNNs) such as ResNet or VGG. Preprocess the audio descriptions, including text normalization, tokenization, and removing irrelevant information. Align the extracted visual features with the corresponding audio descriptions to create training pairs for the models (Fig. 3).

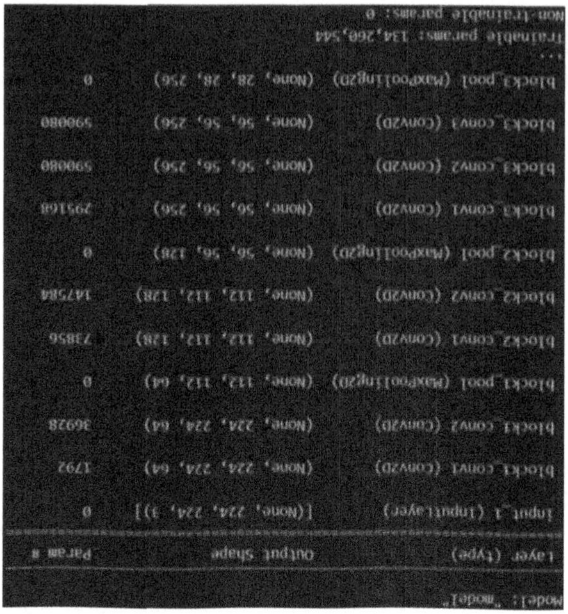

Fig. 3. VGG16 model used to extract features

5.3 Model Development

Implement LSTM networks to capture temporal dependencies in the video sequences and generate coherent captions over time. Develop LSTM-based models to leverage attention mechanisms for prioritizing relevant visual features and generating informative descriptions. Fine-tune the pre-trained models on the collected dataset to adapt them to the task of video captioning for the blind.

The methodology employed in this work begins with utilizing the VGG16 model, a convolutional neural network (CNN) that has been trained beforehand to extract features from pictures in the Flickr8k dataset. These features encapsulate essential visual information present in the images, forming a rich representation of their content. Subsequently, a Long Short-Term Memory (LSTM) network is utilized in conjunction with the extracted features to construct a comprehensive image captioning system. The LSTM network, known for its ability to capture temporal dependencies within sequential data, is tasked with generating descriptive captions for the images based on the extracted visual features. The model gains the ability to link visual characteristics to relevant textual descriptions during the training process, optimizing its caption generation capabilities. The Flickr8k dataset, consisting of 8,000 unique images, each paired with five distinct captions, provides a diverse and comprehensive training corpus for the model. The BLEU score metric, which measures the similarity of generated and reference captions and is a commonly used measure in natural language processing tasks, is used to assess the model's performance. By leveraging these methodologies, the work aims to develop an accurate and effective image captioning system capable of producing descriptive and contextually relevant captions for visually impaired individuals (Fig. 4).

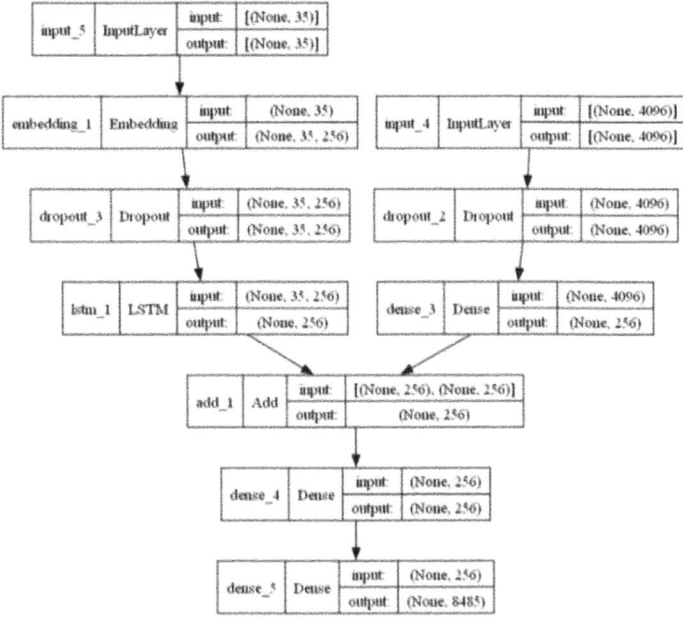

Fig. 4. LSTM/CNN Model

5.4 Geometric Average Precision Scores

The following formula is used to combine the Precision Scores. This can be calculated with various weight values and for various values of N. Usually, $N = 4$ is employed, and uniform weights $w_n = N/4$ are employed.

$$Geometri\ Average\ Precision(N) = exp\left(\sum_{n=1}^{N} \omega_n \log p_n\right)$$

$$= \prod_{n=1}^{N} p_n^{\omega_n}$$

$$= (p_1)\frac{1}{4}.(p_2)\frac{1}{4}.(p_3)\frac{1}{4}.(p_4)\frac{1}{4}$$

5.5 Brevity Penalty

The Brevity Penalty addresses overly concise output in text prediction by penalizing excessively short sentences, preventing inflated scores that prioritize brevity over meaningful content. If you look at the calculation of Precision, the result may have been a projected sentence with just one word, such "the" or "late." A perfect score of $1/1 = 1$ would have been obtained for this 1-g Precision. Given that it motivates the model to produce fewer words in order to achieve a high score, this is blatantly misleading. The Brevity Penalty penalizes sentences that are excessively brief in order to counteract this.

$$Brevity\ Penalty = \begin{cases} 1, if c > r \\ e^{(1-\frac{r}{c})}, if c \le r \end{cases}$$

- The number of words in the predicted sentence is denoted by c
- The number of words in the target sentence is denoted by r

5.6 Bleu Score

The Python Bleu score is a metric used to assess the quality of machine translation models. Even though it was initially just intended for translation models, it is currently employed in a variety of natural language processing applications. The BLEU score indicates how well a candidate sentence matches the list of reference sentences by comparing it to one or more reference sentences. It provides an output score ranging from 0 to 1. When a candidate sentence has a BLEU score of 1, it precisely matches one of the reference sentences. This score is a common metric of measurement for Image captioning models. Finally, to calculate the Bleu Score, the Brevity Penalty with the Geometric Average of the Precision Scores is multiplied.

$$Bleu(N) = Brevity\ Penalty.Geometric\ Average\ Precision\ Scores(N)$$

The Bleu Score can be calculated for various N numbers. Usually, N = 4 is employed.

- BLEU-1 employs the precision score for unigrams.
- BLEU-2 employs the geometric mean of unigrams and bigrams precision.
- BLEU-3 employs the geometric mean of unigram, bigram, and trigram.
- and so on.

6 System Architecture

The system architecture comprises several key components designed to facilitate the generation of descriptive captions for images. At its core is the VGG16 model, a pretrained convolutional neural network (CNN), utilized to extract high-level visual features from images within the Flickr8k dataset. These features serve as rich representations of the visual content and are subsequently fed into a Long Short-Term Memory (LSTM) network. The LSTM network is tasked with capturing temporal dependencies within the sequential data, enabling it to generate coherent and contextually relevant captions over time. Through an iterative training process, the model learns to associate the extracted visual features with corresponding textual descriptions, thereby optimizing its caption generation capabilities.

The system architecture also encompasses data pre-processing steps, including text normalization and tokenization of audio descriptions, to ensure alignment with the extracted visual features. Additionally, the architecture includes mechanisms for fine-tuning the LSTM network parameters to adapt to the task of image captioning.

Overall, the system architecture is designed to seamlessly integrate the VGG16 model, LSTM network, and data pre-processing components to facilitate the generation of accurate and informative captions for visually impaired individuals. Through this architecture, the work aims to develop an effective image captioning system that enhances accessibility and inclusivity in accessing visual content (Fig. 5).

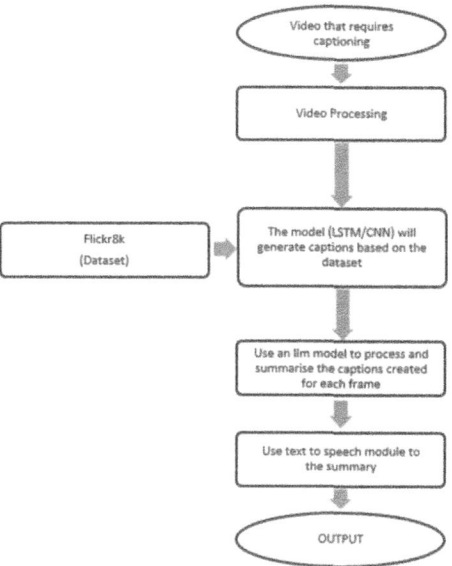

Fig. 5. System Architecture

7   Simulated Results

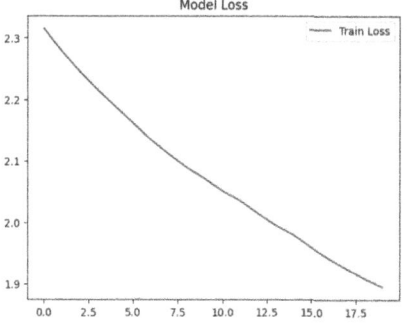

Fig. 6. Loss graph for 20 epochs

The model's performance shows a steady decline as longer n-grams is considered. The highest score of 0.54 for BLEU-1 (unigram) indicates the model is good at capturing individual words that match the reference text. However, the significant drop to 0.31 for

BLEU-2 (bigram) suggests the model struggles to generate sequences of two words that appear in the reference. This trend continues with BLEU-3 (trigram) and BLEU-4 (tetragram) scores dropping even further, highlighting the model's difficulty in capturing longer meaningful phrases and sentence structure (Fig. 6).

Hence, the model seems to be proficient at generating words that are found in the reference text, but struggles with accurately replicating the word order and forming coherent sequences, especially for longer phrases. This suggests potential areas for improvement, such as focusing on improving the model's ability to capture word dependencies and context within a sentence (Fig. 7).

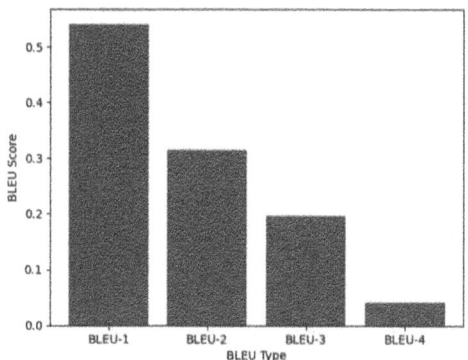

Fig. 7. BLEU Scores for Image Captioning

8 Conclusion

In conclusion, the proposed methodology presents a systematic approach to tackle the longstanding accessibility challenges encountered by the visually impaired when engaging with online visual content. Through the integration of advanced technologies such as LSTM networks-based models, the objective is to craft a sophisticated video captioning system uniquely attuned to the distinctive requirements of blind individuals. The anticipated insights from the comprehensive comparative study will not only shed light on the strengths and weaknesses inherent in each model architecture but will also serve as a crucial foundation for refining and optimizing the system, thereby advancing the cause of inclusive technology for the visually impaired in the digital landscape.

References

1. Varma, S., Dinesh Peter, J.: Enhanced transformer model for video caption generation (2023). https://doi.org/10.1111/exsy.13392
2. Islam, S., Dash, A., Seum, A., et al.: Exploring video captioning techniques: a comprehensive survey on deep learning methods (2021)

3. Tan, Y.H., Chan, C.S.: Phrase-based image caption generator with hierarchical LSTM network. Neurocomputing **333**, 86–100
4. Verma, A., Yadav, A.K., Kumar, M., Yadav, D.: Automatic image caption generation using deep learning, vol. 83, pp. 5309–5325, (2024)
5. Yousif, A.J., Al-Jammas, M.H.: Automatic image caption generation using deep learning, vol. 6, p. 100372 (2023)
6. Hodosh, M., Young, P., Hockenmaier, J.: Framing image description as a ranking task: data, models and evaluation metrics. J. Artific. Intell. Res. **47**, 853–899 (2013)
7. Kopperundevi, N., et al.: Sentiment analysis for online product reviews and recommendation using deep learning based optimization algorithm. Int. J. Recent Innov. Trends Comput. Commun. ISSN: 2321–8169 **11**(9), 3629–3640 (2023)

Leveraging LangChain Framework and Large Language Models for Conversational Chatbot Development

R. Ashish Tarun, B. Priyadarshini, M. Sneha, and K. Akila[✉]

Department of Computer Science and Engineering, SRM Institute of Science and Technology,
Vadapalani Campus, Chennai, India
{ra1058,bb0923,sm7061,akilak}@srmist.edu.in

Abstract. Conversational AI technologies have transformed human- machine interactions in educational settings, such as the SRM Institute of Science and Technology, where intelligent chatbots play pivotal roles in supporting students, faculty, and staff. This research presents a sophisticated chatbot system tailored for the SRM Institute, leveraging cutting-edge technologies like the LangChain framework and Large Language Models (LLMs), including OpenAI's GPT-3.5 Turbo, to enhance language understanding and generation capabilities. The methodology emphasizes meticulous data preparation, including data collection, preprocessing, and embedding creation, integrated with vector databases to establish a robust knowledge base. The chatbot's user interface, built with Flask, ensures an intuitive user experience with a visually appealing and responsive layout. Evaluation results demonstrate a 95% accuracy rate and high reliability in handling diverse user inquiries, highlighting the chatbot's capacity to manage growing volumes of queries while maintaining consistent accuracy. These outcomes underscore significant improvements in user engagement and satisfaction, marking a substantial advancement in educational chatbot technology.

Keywords: Conversational AI · Chatbot system · LangChain framework · Large Language Models(LLMs) · OpenAI's GPT-3.5 Turbo

1 Introduction

Early detection and treatment of diseases of the eyes like macular degeneration and diabetic retinopathy depend heavily on accurate diagnosis. Disparities in healthcare, medication, and therapy programs, and a lack of medical system integration all contribute to the underfunding of ocular disorders. Thanks to its sophisticated tools for segmenting and classifying eye fundus images, artificial intelligence (AI) has completely changed the field of ophthalmology. Large datasets are analyzed by machine learning algorithms, which makes it possible to automatically identify and categorize minute pathological changes. By training on a variety of datasets, AI models for diagnosing eye fundus diseases identify patterns and characteristics suggestive of particular diseases, opening the door to reliable and precise automated diagnostics. Refraction error, cataracts, glaucoma,

P. D. Sivakumar et al. (Eds.): IRCCTSD 2024, CCIS 2362, pp. 244–255, 2025.
https://doi.org/10.1007/978-3-031-82386-2_19

corneal opacities, macular degeneration, diabetic retinopathy, and presbyopia are among the conditions that, as defined by the World Health Organization, impact roughly 1.2 billion individuals globally. In recent times, conversational AI technologies have become valuable tools for automating interactions between humans and machines through natural language conversations. However, existing chatbot systems frequently encounter challenges in adequately addressing user queries and delivering contextually relevant responses within educational environments.

The challenges faced by existing chatbot systems in educational settings encompass limited language understanding capabilities, inefficient query processing pipelines, and suboptimal user experience. These obstacles hinder the smooth integration of chatbots into academic institutions and restrict their capacity to offer timely and accurate support to students, faculty, and staff.

This research paper presents a sophisticated chatbot system designed specifically for the SRM Institute of Science and Technology.

The main contributions of our work include:

Implementation of advanced language models and semantic indexing mechanisms for enhanced language understanding and response generation.

Development of an efficient query processing pipeline integrating LangChain framework and OpenAI technologies.

Design of a user-friendly interface prioritizing visual appeal, usability, and error handling mechanisms.

Comprehensive evaluation of the chatbot system's performance and accuracy across varying question volumes.

With a comprehensive overview provided, Sect. 2 delves into the exploration of related works, Sect. 3 elaborates on the methodology employed in crafting the chatbot system, Sect. 4 presents the experimental results and Sect. 5 concludes the paper with a summary of findings and implications.

2 Related Works

Attigeri, Agrawal, and Kolekar focus on personalized information delivery in education, particularly for university decision-making, in their study [1]. They explore chatbot technology utilizing advanced NLP models, including neural networks, TF-IDF vectorization, sequential modeling, and pattern matching. Their comparative analysis favors neural network-based models, emphasizing the critical need to prevent overfitting through sequential modeling techniques.

The paper by S. Banu and S. D. Patil presents the design and implementation of an intelligent web app chatbot leveraging Natural Language Processing (NLP) and Microsoft Azure's services, particularly Language Understanding Intelligent Service (LUIS) [2]. The chatbot is integrated into a web application, allowing users to interact through conversational interfaces. LUIS is employed to predict user intents and entities based on trained models, ensuring accurate and fast responses. Enhanced authentication secures the chatbot from unauthorized access.

Authored by A. Mondal, M. Dey, D. Das, S. Nagpal, and K. Garda, this work introduces the development of a chatbot in the educational domain, focusing on question-answering capabilities [3]. The process involved data collection and preprocessing to create a structured dataset of questions and answers. Various features were extracted from this dataset, including word count, question type, and linguistic elements. The chatbot was built using an ensemble learning approach, specifically random forest, which improved response classification. Evaluation using precision, recall, and F-measure demonstrated effective performance.

The paper by Sanjay Chakraborty et al. initiates with a thorough literature review exploring chatbots' potential in healthcare, particularly in combating infectious diseases [4]. They enhance human interaction with databases through a preliminary training model and study report. Using natural language processing, the study analyzes human behaviors and chatbot characteristics. Their proposed model employs a deep feedforward multilayer perceptron for interaction and prediction.

N. N. Khin and K. M. Soe introduce the design of a University Chatbot, employing AI algorithms to deliver precise and efficient responses to user inquiries [5]. The system is implemented using Artificial Intelligence Markup Language (AIML) and Pandorabots, utilizing AIML tags to generate responses. Additionally, the system segments user input through a combination of word processing and pattern matching techniques.

B. R. Ranoliya, N. Raghuwanshi, and S. Singh focus on crafting a chatbot specifically tailored to address university FAQs, employing a blend of Artificial Intelligence Markup Language (AIML) and Latent Semantic Analysis (LSA) techniques [6]. Their research explores the integration of pattern matching methodologies to enable the chatbot to efficiently process user queries, aiming to provide relevant responses to user inquiries.

The research paper authored by N. Bhartiya, N. Jangid, S. Jannu, P. Shukla, and R. Chapaneri presents the design and implementation of a University Counselling Auto-Reply Bot customized for the field of Engineering, utilizing Natural Language Processing (NLP) techniques [7]. The dataset is trained using a Feedforward neural model, and the Chat Application is deployed on Facebook Messenger as the primary interaction platform.

M. -T. Nguyen, M. Tran-Tien, A. P. Viet, H. -T. Vu, and V. -H. Nguyen present an intelligent chatbot designed specifically for university admission processes, constructed on the Rasa platform [8]. Their work delves into the exploration of different natural language understanding components to determine the optimal pipeline. Experimental results indicate a preference for the pipeline utilizing DIET (Dual Intent and Entity Transformer) with features extracted from pre-trained language models. Furthermore, the chatbot is deployed on Facebook to provide support for the admission process.

3 Methodology

Our proposed work focuses on improving the efficiency and effectiveness of conversational AI systems, particularly in educational settings like the SRM Institute. We meticulously gather and preprocess data from various sources within the institute to establish a strong knowledge base capable of addressing diverse user queries. Utilizing advanced techniques such as data segmentation, embeddings, and similarity search ensures a precise comprehension of user intentions and the delivery of contextually appropriate responses. Additionally, the integration of OpenAI's GPT-3.5 Turbo enhances the chatbot's language understanding capabilities, facilitating the real-time generation of coherent responses. Furthermore, the implementation of a conversational retrieval chain enhances user queries by considering past context and incorporating relevant history for deeper comprehension (Fig. 1).

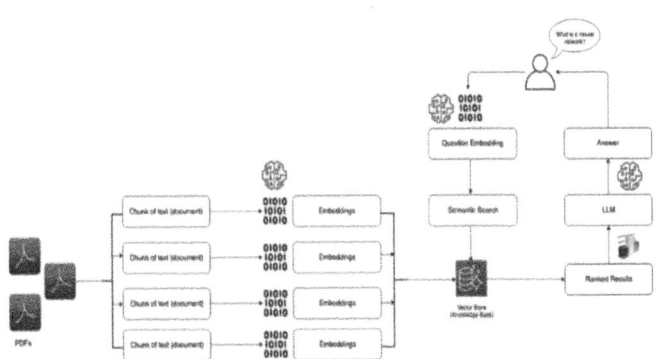

Fig. 1. Working of Chatbot using LangChain Framework

3.1 Data Preparation and Pre-processing

(See Fig. 2)

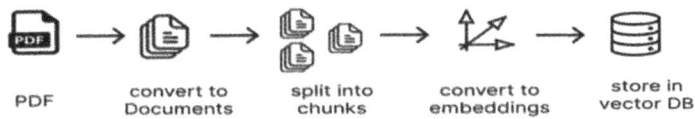

Fig. 2. Workflow of Data Processing and Loading into Database

3.1.1 Data Collection

In the data collection phase, we gather information from various sources within the SRM Institute of Science and Technology, such as course details, campus facilities, and frequently asked questions (FAQs).

3.1.2 Data Loading

The data was loaded into the Large Language Model (LLM) using the LangChain Data Loader, with a text file serving as the data source. The loading process was made efficient by utilizing the TextLoader module.

3.1.3 Data Segmentation

The loaded data is segmented using LangChain's TextSplitter. This involves employing a recursive character text splitter to break down the data into meaningful chunks.

3.1.4 Data Embeddings

The text chunks are converted into numerical vectors using the OpenAI embedding model "Text-Embedding-3-Large." These vectors enable the chatbot to comprehend the user's intent more effectively.

3.1.5 Vector Database

The embeddings generated from the text chunks are stored in a database called Chroma. This database allows for efficient storage and retrieval of the embeddings, enhancing the chatbot's ability to understand and respond to user queries.

3.2 User Interface

3.2.1 Web Interface Design

The web interface design of our chatbot incorporates a visually appealing and user-friendly layout, ensuring smooth interaction between users and the chatbot. The interface is built using HTML and CSS. It includes a header section showcasing the SRM Institute of Science and Technology logo along with the chatbot's name.The chatbot comprises a chat box where messages are displayed dynamically, enabling users to engage in conversation with the chatbot. Additionally, a form group with a text area allows users to input their messages, where a "Send" button triggers the submission of the message for processing.

3.2.2 Output Presentation

The output presentation of our chatbot is designed to provide users with clear and concise responses in a visually appealing format. Chats between the user and the chatbot are displayed within a chat box, which is scrollable to accommodate a potentially large conversation history. Each message is enclosed within a rounded container, with user messages distinguished from bot responses through background color variations. This clear visual distinction aids in easily identifying the origin of each message within the conversation.

3.2.3 Flask Web Application

The chatbot is built using Flask, a lightweight and flexible Python web framework. Flask allows us to create routes for handling HTTP requests, facilitating communication between the client-side interface and the server-side logic of the chatbot. The main Python script initializes the Flask application, defines routes for user queries, and manages the interaction between the user interface and the chatbot logic. The Flask application runs in debug mode for easier development and troubleshooting. With Flask, our chatbot has a reliable and scalable web infrastructure, ensuring consistent performance and accessibility for users.

3.3 Processing User Queries

3.3.1 Conversion of User Query

User queries are broken down into smaller segments or pieces of information. These segments are then converted into numerical vectors, which represent them in a format that can be processed by the system. This conversion allows for easier handling and analysis of the user's input, enabling more efficient communication between the user and the system.

3.3.2 Similarity Search

Similarity search involves using cosine similarity as a metric to gauge the similarity between user queries and stored data vectors. By comparing the angles between the vectors in a multi-dimensional space, cosine similarity quantifies the resemblance between the query and the stored data. The system then identifies the data vector with the highest cosine similarity to the user query and retrieves the corresponding response. This method allows for efficient retrieval of relevant information or responses based on the degree of similarity between the user's input and the stored data vectors.

3.3.3 OpenAI Integration

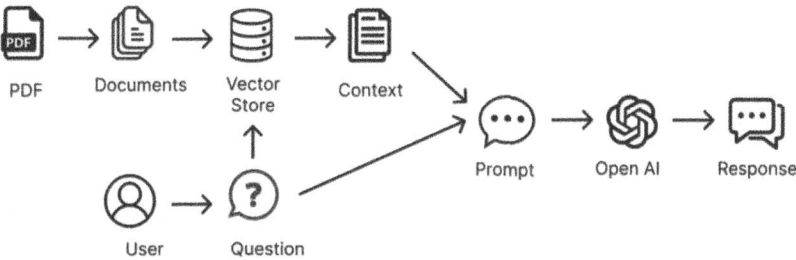

Fig. 3. Workflow of Chatbot with Open AI Integration

To generate coherent responses, the chatbot is Integrated with OpenAI to enhance language understanding capabilities. By leveraging GPT-3.5 Turbo, the chatbot can produce contextually relevant responses to user queries. This integration improves the ability of the chatbot to understand and interpret natural language input more effectively. The integration of GPT-3.5 Turbo model enhances the responses that are contextually appropriate and linguistically coherent (Fig. 3).

3.3.4 Conversational Retrieval Chain

The Conversational Retrieval Chain process enhances user queries by considering past context, which aids in better understanding user needs. It incorporates pertinent historical data to provide a richer context, enabling the system to grasp the nuances of user queries more effectively. By utilizing a retriever mechanism, the system selects documents based on embeddings, ensuring the retrieval of relevant information aligned with the user's query. The responses generated through this process are seamlessly integrated into the chat history, contributing to a continuous and coherent conversation experience.

3.4 Response Generation

Response generation is facilitated by leveraging GPT-3.5 Turbo from OpenAI, which prompts the model with rephrased queries and relevant context. This approach ensures that the generated responses are aligned with the conversation history, maintaining coherence and relevance throughout the interaction. The system seamlessly integrates these responses into the chat interface, providing users with natural and contextually appropriate replies. This process enhances the conversational flow and overall user experience by delivering timely and accurate responses that are reflective of the ongoing dialogue.

3.5 User Experience

User experience is a top priority, with a focus on efficiency and usability to ensure seamless interactions. We enhance performance through model persistence, which improves the system's responsiveness and reliability, resulting in a more satisfying experience for users.

4 Results and Analysis

Figures 4 and 5 display the chatbot system's response to user queries. Each screenshot illustrates the chat interface initiated by users alongside the corresponding answers provided by the chatbot.

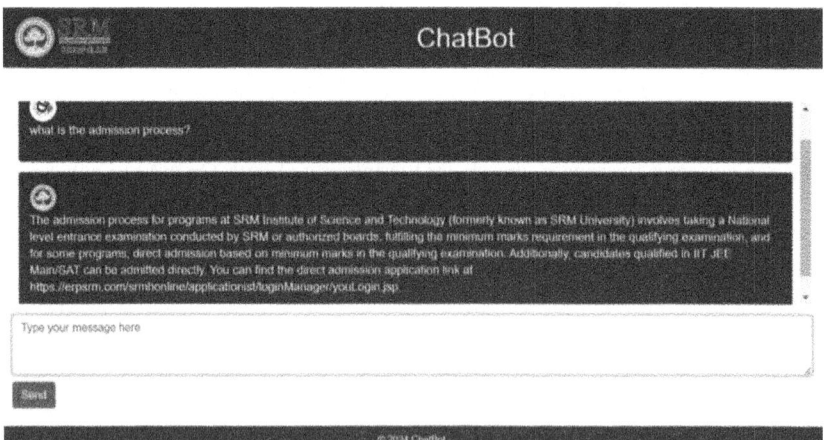

Fig. 4. Chatbot Response 1

The chatbot consistently delivers accurate and relevant responses, offering users contextually appropriate information.

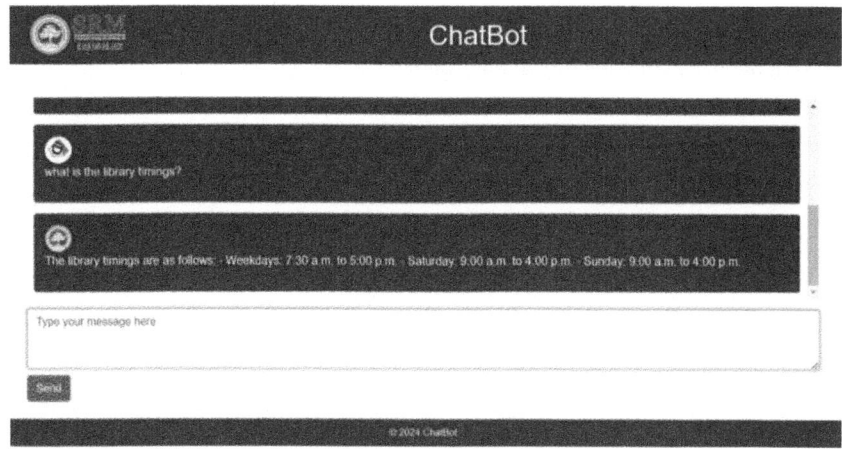

Fig. 5. Chatbot Response 2

These outcomes highlight the effectiveness of our chatbot architecture, which integrates advanced language models seamlessly to provide accurate and insightful responses to user queries. Additionally, the user interface design promotes a seamless and intuitive interaction experience.

Figure 6 depicts the performance of our chatbot system in relation to the number of questions tested and the corresponding number of accurate answers provided. With an increasing number of questions tested, there is a consistent increase in the number of correct answers generated by the chatbot. This graph showcases the chatbot's ability to effectively handle a growing volume of user inquiries while sustaining a high level of accuracy in its responses.

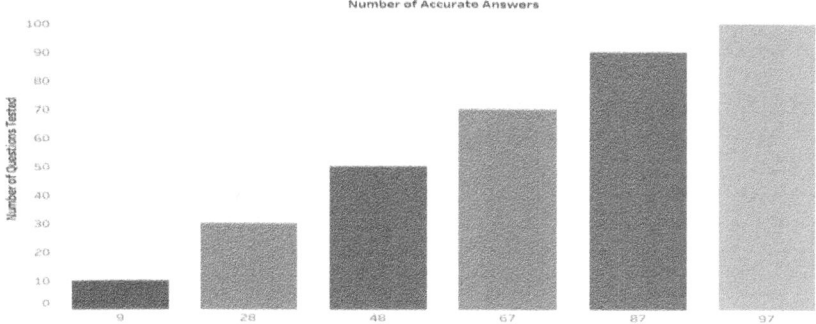

Fig. 6. Analysis of Accuracy variation with Varied Question Count

The correlation between the number of questions tested and the accuracy of answers underscores the robustness and reliability of our chatbot architecture. These results validate the effectiveness of our approach in creating a proficient and dependable chatbot system customized for educational settings. The system demonstrates a consistent ability to provide accurate and relevant information to users, reaffirming its suitability for addressing the needs of educational environments.

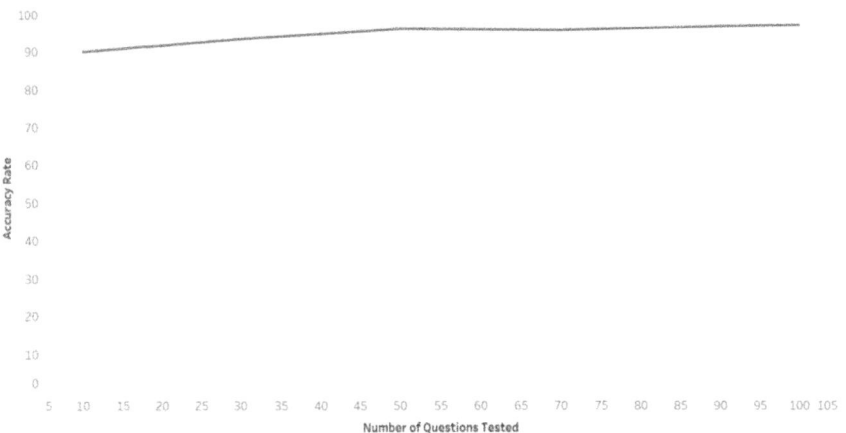

Fig. 7. Analysis of Accuracy Rate with Varied Question Count

Figure 7 indicates a direct relationship between the number of questions tested and the chatbot's accuracy rate. As the number of questions increases, the accuracy rate steadily rises, demonstrating the chatbot's capability to manage a growing volume of user inquiries while maintaining high accuracy in its responses. This graph highlights the robustness of our chatbot architecture in delivering accurate and relevant information within educational environments.

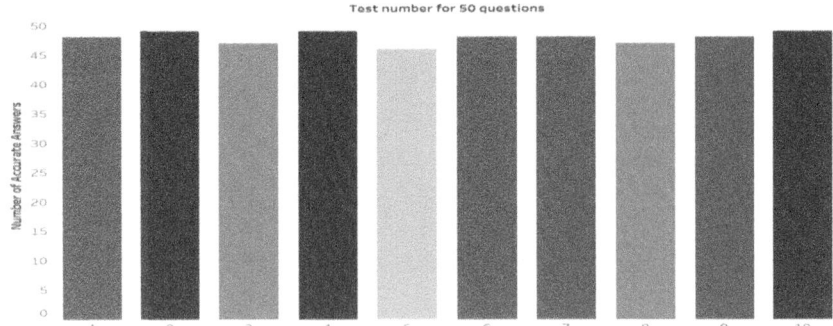

Fig. 8. Analysis of Accuracy variation with Uniform Question Count

Figure 8 illustrates the trend in the number of accurate answers across ten sets of test cases, each consisting of fifty questions. This visualization demonstrates the consistent performance of our chatbot system as the number of questions tested increases. The graph shows the chatbot's ability to maintain a high level of accuracy across different volumes of questions.

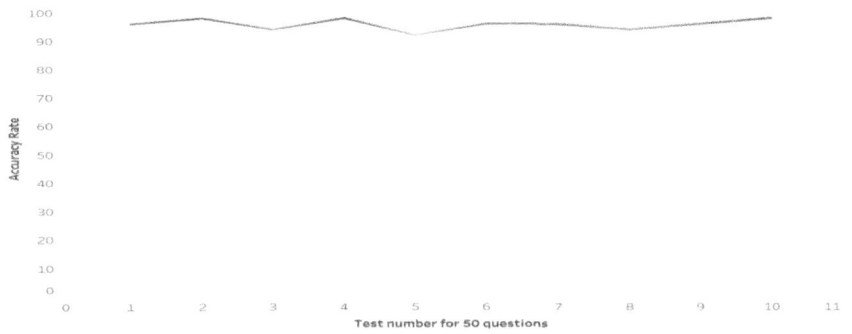

Fig. 9.   Analysis of Accuracy Rate with Uniform Question Count

Figure 9 illustrates the accuracy rate of our chatbot system across different test cases, each consisting of fifty questions. The y-axis represents the accuracy rate, indicating the percentage of correctly answered questions out of the total tested. As we move through the test cases on the x-axis, variations in the accuracy rate are observable. However, the majority of test cases demonstrate a high accuracy rate, with most exceeding 95%. These results confirm the reliability and effectiveness of our chatbot system in consistently providing accurate responses to user queries, thereby enhancing user satisfaction and engagement within educational environments.

5   Future Research Directions

In future endeavors, we plan to explore multi-modal interaction modalities to elevate the capabilities of our chatbot. By integrating various modes of interaction such as voice and visual inputs, we aim to enhance user engagement and provide a more intuitive experience. Additionally, we seek to enhance response diversity to offer users richer and more varied answers, thereby improving the overall conversational experience. Another aspect of our future work involves expanding domain-specific knowledge bases to augment the chatbot's expertise across a broader range of topics. This expansion will enable the chatbot to provide more comprehensive and accurate responses tailored to specific domains or industries, further enhancing its utility and effectiveness in various contexts.

6   Conclusion

In this research paper, we introduced a sophisticated chatbot system tailored specifically for the SRM Institute of Science and Technology, addressing the challenges commonly faced by chatbots in educational settings. Our approach integrates advanced language models, semantic indexing, and a streamlined query processing pipeline built on the LangChain framework and OpenAI technologies, resulting in significant enhancements in language understanding and response generation. The chatbot's performance was rigorously evaluated, and the results demonstrated its ability to consistently provide accurate and contextually relevant responses, boasting a high accuracy rate, with most surpassing 95%. Analysis showed a positive correlation between the number of user queries and the chatbot's accuracy, maintaining a high accuracy rate even with increasing query volumes. The user interface, designed for visual appeal and ease of use, further contributed to a seamless user experience. Overall, our chatbot system proves to be an effective and reliable tool for educational environments, capable of offering timely and precise support to students, faculty, and staff, thereby enhancing engagement and satisfaction within the academic community.

References

1. Attigeri, G., Agrawal, A., Kolekar, S.V.: Advanced NLP models for technical university information chatbots: development and comparative analysis. IEEE Access **12**, 29633–29647 (2024). https://doi.org/10.1109/ACCESS.2024.3368382
2. Banu, S., Patil, S.D.: An intelligent web App Chatbot. In: 2020 International Conference on Smart Technologies in Computing, Electrical and Electronics (ICSTCEE), pp. 309–315. Bengaluru, India (2020). https://doi.org/10.1109/ICSTCEE49637.2020.9276948
3. Mondal, A., Dey, M., Das, D., Nagpal, S., Garda, K.: Chatbot: an automated conversation system for the educational domain. In: 2018 International Joint Symposium on Artificial Intelligence and Natural Language Processing (iSAI-NLP), pp. 1–5. Pattaya, Thailand (2018). https://doi.org/10.1109/iSAI-NLP.2018.8692927
4. Chakraborty, S., et al.: An AI-based medical chatbot model for infectious disease prediction. IEEE Access **10**, 128469–128483 (2022). https://doi.org/10.1109/ACCESS.2022.3227208
5. Khin, N.N., Soe, K.M.: University Chatbot using artificial intelligence markup language. In: 2020 IEEE Conference on Computer Applications (ICCA), pp. 1–5. Yangon, Myanmar (2020). https://doi.org/10.1109/ICCA49400.2020.9022814
6. Ranoliya, B.R., Raghuwanshi, N., Singh, S.: Chatbot for university related FAQs. In: 2017 International Conference on Advances in Computing, Communications and Informatics (ICACCI), pp. 1525–1530. Udupi, India (2017). https://doi.org/10.1109/ICACCI.2017.812 6057
7. Bhartiya, N., Jangid, N., Jannu, S., Shukla, P., Chapaneri, R.: Artificial neural network based university chatbot system. In: 2019 IEEE Bombay Section Signature Conference (IBSSC), pp. 1–6. Mumbai, India (2019). https://doi.org/10.1109/IBSSC47189.2019.8973095
8. Nguyen, M.-T., Tran-Tien, M., Viet, A.P., Vu, H.-T., Nguyen, V.-H.: Building a Chatbot for supporting the admission of universities. In: 2021 13th International Conference on Knowledge and Systems Engineering (KSE), pp. 1–6. Bangkok, Thailand (2021). https://doi.org/10.1109/KSE53942.2021.9648677
9. Agus Santoso, H., et al.: Dinus Intelligent Assistance (DINA) Chatbot for university admission services. In: Proceedings - 2018 International Seminor on Application Technology Information and Communication Creative Technology Human Life, iSemantic 2018, pp. 417–423 (2018). https://doi.org/10.1109/ISEMANTIC.2018.8549797
10. Mohamed Firdhous, M.F., Elbreiki, W., Abdullahi, I., Sudantha, B.H., Budiarto, R.: WormGPT: a large language model chatbot for criminals. In: 2023 24th International Arab Conference on Information Technology (ACIT), pp. 1–6. Ajman, United Arab Emirates (2023). https://doi.org/10.1109/ACIT58888.2023.10453752

11. Dean, M., Bond, R.R., McTear, M.F., Mulvenna, M.D.: ChatPapers: an AI Chatbot for inter-acting with academic research. In: 2023 31st Irish Conference on Artificial Intelligence and Cognitive Science (AICS), pp. 1–7. Letterkenny, Ireland (2023). https://doi.org/10.1109/AIC S60730.2023.10470521

12. Khadija, M.A., Aziz, A., Nurharjadmo, W.: Automating information retrieval from faculty guidelines: designing a PDF-Driven Chatbot powered by OpenAI ChatGPT. In: 2023 Interna-tional Conference on Computer, Control, Informatics and its Applications (IC3INA), pp.394–399. Bandung,Indonesia (2023).https://doi.org/10.1109/IC3INA60834.2023.10285808

13. Akilesh, S., Sheik Abdullah, A., Abinaya, R., Dhanushkodi, S., Sekar, R.: A Novel AI-based chatbot application for personalized medical diagnosis and review using large language mod-els. In: 2023 International Conference on Research Methodologies in Knowledge Manage-ment, Artificial Intelligence and Telecommunication Engineering (RMKMATE), pp. 1–5. Chennai, India (2023). https://doi.org/10.1109/RMKMATE59243.2023.10368616

14. Patel, N.P., Parikh, D.R., Patel, D.A., Patel, R.R.: AI and Web-based human-like interactive university Chatbot (UNIBOT). In: 2019 3rd International conference on Electronics, Com-munication and Aerospace Technology (ICECA), , pp. 148–150. Coimbatore, India (2019). https://doi.org/10.1109/ICECA.2019.8822176

15. Villanueva, D.P.P., Aguilar-Alonso, I.: A Chatbot as a support system for educational insti-tutions. In: 2021 62nd International Scientific Conference on Information Technology and Management Science of Riga Technical University (ITMS), pp. 1–6. Riga, Latvia (2021). https://doi.org/10.1109/ITMS52826.2021.9615271

16. Kim, J.K., Chua, M., Rickard, M., Lorenzo, A.: ChatGPT and large language model (LLM) chatbots: the current state of acceptability and a proposal for guidelines on utilization in academic medicine. J. Pediatr. Urol. S1477513123002243 (2023)

17. Naveed, H., et al.: A comprehensive overview of large language models, arXiv, Aug. 2023. http://arxiv.org/abs/2307.06435

18. You, Y., et al.: TI-Prompt: towards a prompt tuning method for few-shot threat intelligence twitter classification. In: 2022 IEEE 46th Annual Computers Software and Applications Conference (COMPSAC), pp. 272–279 (2022)

19. Kamnis, S.: Generative pre-trained transformers (GPT) for surface engineering. Surf. Coat. Technol. **466**, 129680 (2023)

20. Lanyo, K., Wausi, A.: A comparative study of supervised and unsupervised classifiers utilizing extractive text summarization techniques to support automated customer query question-answering. In: 2018 5th International Conference on Soft Computing Machine Intelligence (ISCMI), pp. 88–92 (2018)

Designing of In-Memory Computing SRAM Energy-Efficient for Artificial Intelligence

Preethi Karmakar[✉], E. John Alex, and K. Niranjan Reddy

Department of Electronics and Communication Engineering, CMR Institute of Technology,
Hyderabad, India
karmakarpreethi289@gmail.com

Abstract. Utilizing the potential of Static Random-Access Memory (SRAM), In-Memory Computing has surfaced as available approach to tackle the computational and energy-efficiency issues. An overview of memory computing's energy-efficient SRAM for AI applications is given in this paper. When compared to traditional SRAM, memory technologies provide benefits in the form of reduced leakage power, increased durability, and enhanced reliability, which supports long-term and reliable AI hardware implementation. To sum up, memory computing with energy-efficiency the SRAM offers available solution to the computational and energy issues in AI hardware design. SRAM-build on In- Memory Computing has the potential to completely change the field of the AI computing by utilizing cutting-edge architectural improvements and newly developed memory technologies to create more effective and scalable AI systems for a wide range of application areas. SRAM array based on 10 T SRAM is constructed in 180-nm SCL technique to examine the advocated IMC macro architecture's functionality and performance. In this paper we have used supply voltage of 1.8 V, Array size of 136×32, energy efficiency obtained 1 V and achieves an area efficiency of 65.2%.

Keywords: SRAM · In-Memory Computing (IMC) · Artificial intelligence (AI) · energy-efficient

1 Introduction

Artificial intelligence (AI) to increase real-time application scalability, efficiency, security, and privacy [3]. The Artificial intelligence edge devices are powered by Convolutional Neural Networks (CNNs). Von Neumann Architecture (VNA) as seen in Fig. 1, which includes separate storage (memory) and Central Processing Unit (CPU), is often the foundation of the most sophisticated AI edge devices [14].

1.1 IMC Architecture

The capacity to carry out computation and data storage inside the same memory array is known as in-memory computing. An integrated computing unit and a storage unit are by

P. D. Sivakumar et al. (Eds.): IRCCTSD 2024, CCIS 2362, pp. 256–266, 2025.
https://doi.org/10.1007/978-3-031-82386-2_20

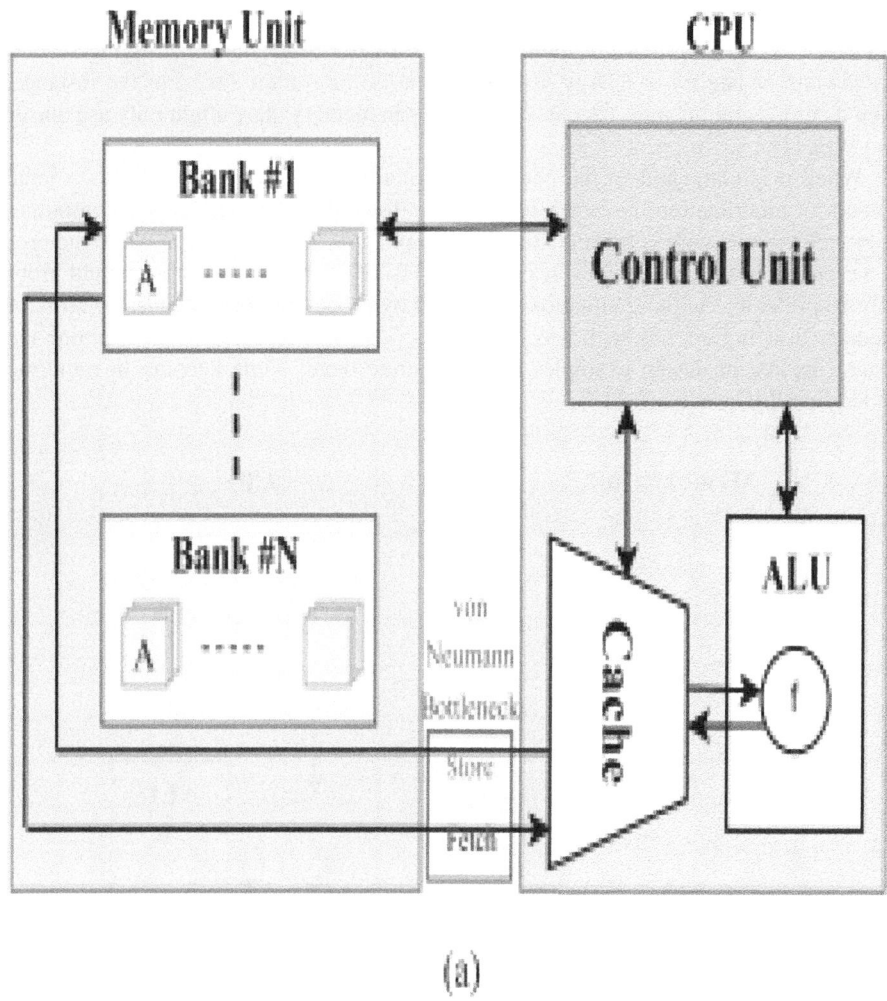

Fig. 1. Von-Neumann Architecture

components of the in-memory computation (IMC), as seen in Fig. 2, this used less energy for the computing [1]. IMC is the technique that can be used to increase AI's throughput and energy efficiency. Using chances to address computing energy and productivity, in-memory computing (IMC) leverages the dataflow in MVM and the structural alignment of a large 2D array of bit cells [2].

The trade-off between bandwidth, latency, and energy versus SNR that results from IMC is what has resulted to the successes and shortcomings seen in the most current prototypes [15]. Compared to fully optimized digital accelerators, these prototypes have demonstrated the potential for increase of 10 times in area-normalized throughput and energy at the same time. However, the integration and scale of In-Memory Computing in the heterogeneous architecture necessary for real-world computing systems have also been limited. Recently, other IMC strategies have been put out [2]. It helps to frame this discussion by relating these to the basic trade-offs that were discussed in the section "bandwidth or latency or energy v/s signal-to-noise ratio trade-off". For the instance, even if certain architectures execute calculations in memory, they might only use one or two WLs [1].

When it is compared to the conventional memory accessing, this avoids a large amount of amortization; however, it presents difficulties when integrating computing in bit-cell circuits with constraints requires adopting of a 10 T bit cells, while requires numerous memory-operation cycles [12]. Accessing of standard memory would probably be preferable in these situations, followed by processing just outside the array of memory utilizing circuits with less constraints. The subsequent sections examine the issues, use recent design examples to demonstrate them, while keeping in mind the underlying IMC trade-offs [13].

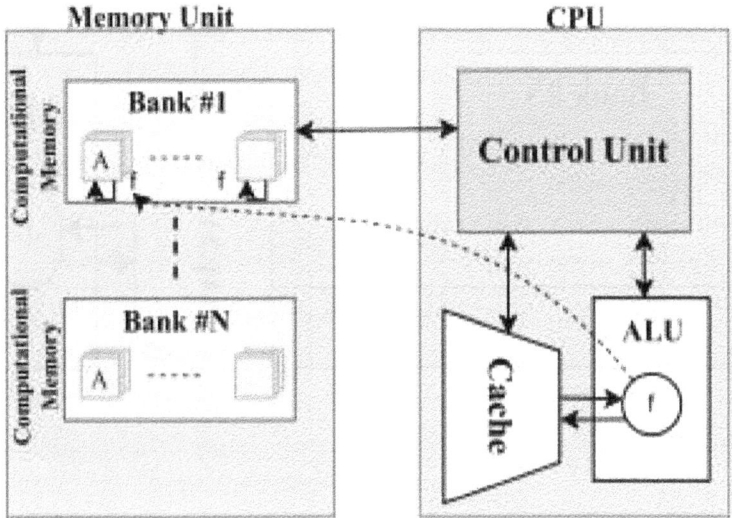

Fig. 2. In-Memory Computing Architecture

1.2 Proposed 10T SRAM

Below Fig. 3 illustrates the suggested IMC macro's architecture has used a 4 Kb (136 × 32) array size 10-T SRAM bit cells. 16 pairs of Sense Amplifiers (SA), a 2 × 1 column multiplexer (MUX) and sense amplifier, a GC1 decoder and a high virtual ground Decoder, also one write driver, 32 precharge circuits, one write decoder and one read decoder, 136 read driver, and one Vref generator make up the peripheral circuit and macro. An array size of 136 × 32 planned on a 10-T SRAM bit cells make up the core of the proposed IMC macro. 32-bit cells in the same row share a pair of RWL, WWL and RWLB, while 136-bit cells in the same column are connected by 2 read-bit lines and 1 write-bit line. The 136-bit cells share the high virtual ground decoder and GC1 decoder signals. The suggested 10T SRAM features single-ended writing through an M6 transistor and differential reads over a read decoupling path M7, M8, M9 and M10. To improve the write margin, M5 is added to the 10T SRAM core as a second transistor [4].

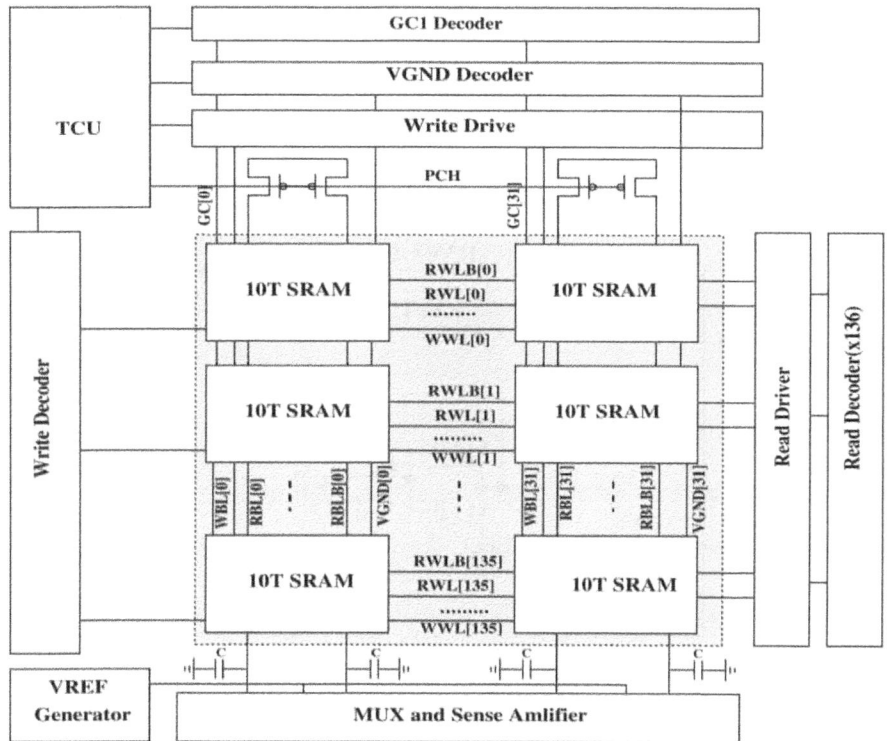

Fig. 3. IMC Architecture used on 10T SRAM

2 Operation

2.1 Memory Mode

In computer architecture and memory management, memory mode usually refers to the way memory access is configured or operates inside a system. It includes a range of options and setups that control how hardware and software components access, uses, and manage the memory. Memory mode in CPUs can refer to many operating modes, including protected mode and real mode. Real mode, which is frequently present in older x86 processors, offers a straightforward memory model with direct access to hardware resources but constrained memory addressing capabilities. In contrast, protected mode allows current operating systems to effectively manage memory by providing features like multitasking and memory protection.

2.2 Read Operation

The read operation is a basic procedure in computer memory systems that retrieves data from memory cells so that the CPU or other components can process it [11]. This process entails gaining access to particular memory regions and sending the information contained there to the party making the request. Figure 4 shows the read operation and the suggested 10T SRAM operating requirement for reading data. The read bit line and RBLB are precharge to VDD, GC1 is maintained at VDD, and high virtual ground is holding at ground during the read operation. The difference between read bit line and RBLB values is detected by the sensing amplifier [4].

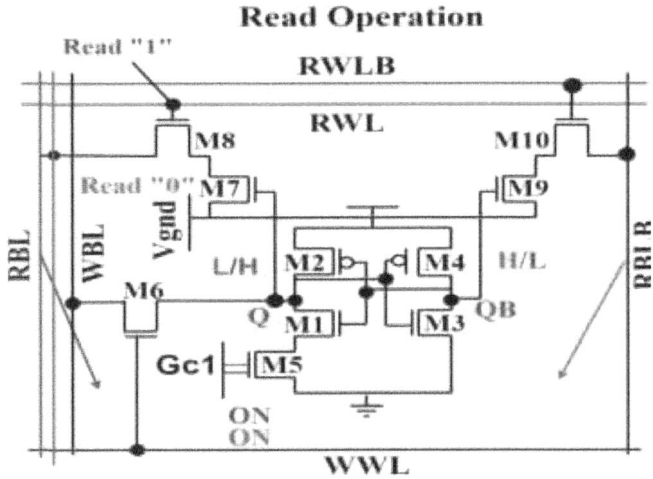

Fig. 4. .

2.3 Write Operation

The memory controller or related hardware starts a write operation by delivering a write command that includes the target memory address and the data that needs to be placed. After that, the memory subsystem writes the new data into the relevant memory cells that match the given address and activates those memory cells [11]. Figure 5 shows the write operation and different mechanisms may be involved in the write operation depending on the type of memory technology being employed. This operation of the suggested 10-T SRAM is single-ended. In a write operation, write data is put through WBL, VGND has been maintained at VDD, GC1 is maintained on ground to write "1" and at VDD to write "0," the RWL and RWLB are placed at ground. The data is written using transistor M6, and to reduce power consumption resulting from the write '1' operation's off state, another transistor, M5, is added. Furthermore, during write operations, the open VGND signal is employed to minimize leakage current [4].

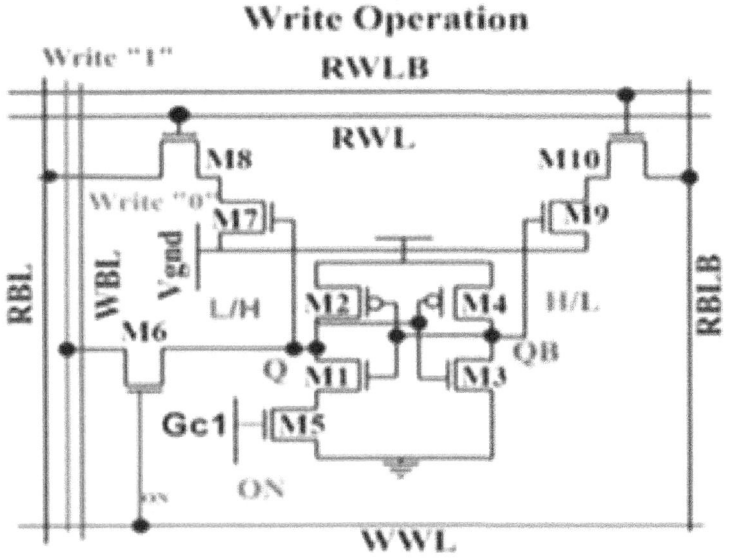

Fig. 5. Write Operation

2.4 Arithmetic Operation

Arithmetic calculations performed in memory using the In-Memory Computing architecture used on a 10-T SRAM. The carry ripple adder close to array is implemented with one XOR gate and two nor gates in order to facilitate addition. Every peripheral circuit in a column gets carry in from preceding columns of peripheral circuit and gives carryout to the following column's peripheral circuit [7]. Simulation of one-bit addition under various operational circumstances. The two's complement method can be used to execute subtraction operations. The shift and add algorithm can be used to carry out the multiplication operation [4].

2.5 In Memory Dot Operation

They are classified into logic operation and current accumulation.

In-Memory logic operation:

In-Memory logic operation on 2 distinct operands (rows 0 and row 1) summarize the behaviours of the SRAM bit cells circuit and the findings of the IMDP [8]. A precharge driver drags the read bit lines using 2 pmos transistors before the IMDP process starts, precharging both the read bit line and RBLB voltages to higher variable IMDPA instances with variable input and weight values are shown in Fig. 6a and b. No matter how much weight is reserved in the SRAM, when input is zero or low, no current discharges read bit lines. The proper nmos transistor, whose node at gate is connected to the storage node of SRAM at a high voltage, is used to discharge RBL or RBLB [9].

The RBL is briefly discharged in response to the pulse width of a positive pulse applied to the RWL, as seen in Fig. 6b. The binary IMDP result is reflected in the voltage differential that results between RBL and RBLB, which can be either positive or negative [4].

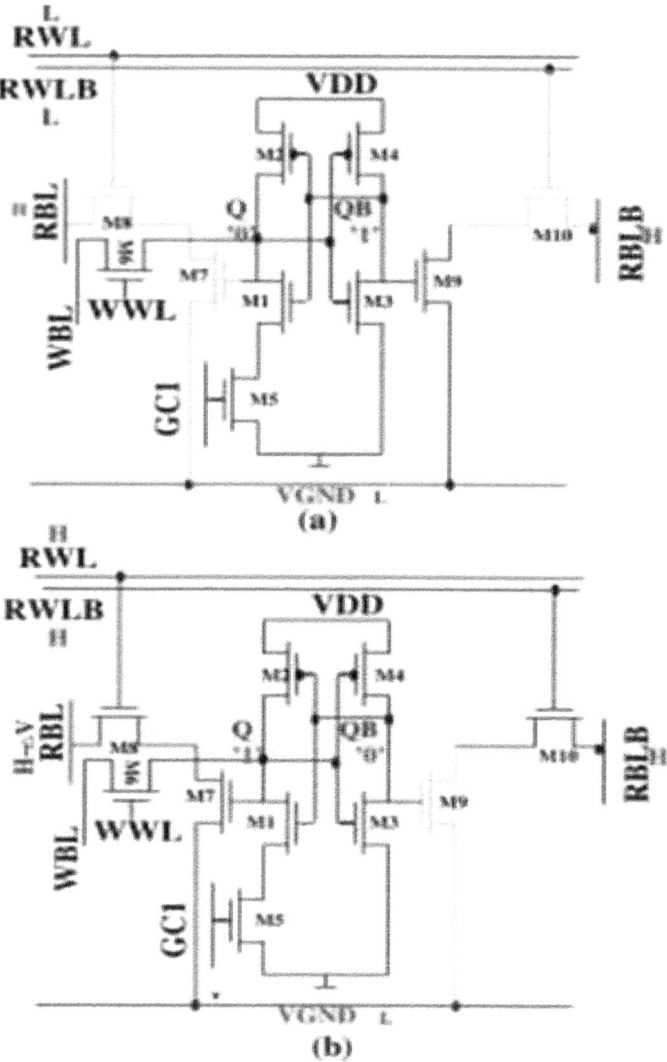

Fig. 6. In-Memory operation

In-Memory current accumulation

The given 10-T SRAM bit cells IMDP findings are aggregated to form a column-based neuron. A pair of RBLs and an RBLB establish a vertical connection between two 128 of bit cells. The read bit lines are utilized to aggregate the IMDP outcomes by utilizing the current discharge process. To increase accumulation linearity, a 1.9 V core of SRAM supply is used in this work. A dynamic of higher range of RBL is offered by the decoupling read path based 10 T SRAM, which improves linearity during accumulation. One cycle of accumulation [8].

3 Analysis

3.1 Cost Comparison

The terms read and write energy, access time, the price of 10-T proposed is contrasted to that of the 8 T and 12 T SRAM bit cells. The Read Access Time (RAT) of the 10-T SRAM measures how much time it needs to switch from the RBL of supply voltage from 48% to 12%. RAT is the same. Because of the precise read process. Write Access Time (WAT) is said as the time between the WWL edge at 48% VDD and at node 'Q' is 12% VDD. In comparison to the other SRAM, the 10- T bit cell has a wide range of write access time [1].

3.2 Read Energy and Write Energy

The 10 T SRAM bit cell's read energy and write energy in comparison to the 8 T SRAM and 12 T SRAM bit cells [6]. Comparing the bit cell to the 8 T SRAM bit cell, then read energy is reduced by 1.5%. Comparing to the 8 T SRAM bit cells, the suggested SRAM has 88% of reduced write energy [1].

3.3 Area

Using the layout editor of Cadence Virtuoso and the 180 nm SCLPDK designs standard, 10 T SRAM is created. The SRAM bit cell is 13% more than a 9 T SRAM since it has an extra transistor, but it is roughly similar in area to 12 T SRAM bit cell [1].

3.4 Leakage Power

The bit cells itself generates majority of unwanted l power that gets leaked or wasted in SRAM during hold operation the leakage power consumption is decreased by using in VGND. Furthermore, the suggested 10T SRAM bit cells only have one write operation. The access transistor M6 thus provides a single channel from WBL to ground. Therefore, in contrasted to 8 T and 12 T SRAM, the given SRAM cell's leakage power decreases by 85% and 47% [1] (Fig. 7).

4 Result

Figure 7 shows the layout of test chip 180 nm.This works has offered the In-Memory Computing architecture for conventional operation of memory, IMC and the IMDP operation using the 10 T SRAM with better write stability and energy -efficient. With comparison to a 8 T SRAM it is operating at 1.7 V supply voltage, the proposed SRAM bit cells gives approximately the same RSNM but with an average WSNM that is 40% higher, 89% less write energy, and 85.8% less power is leaked. Figure 8 shows photo of test chip 180 nm [9, 10].

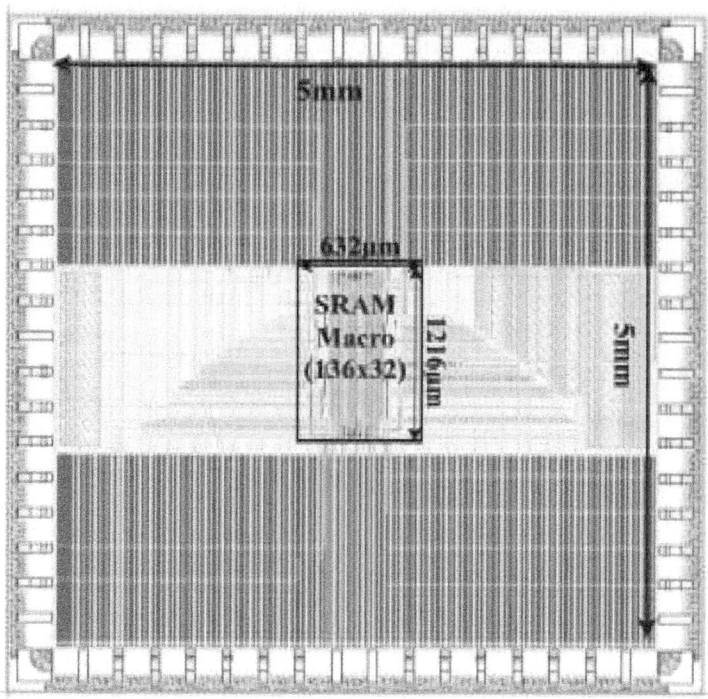

Fig. 7. Test chip 180 nm layout.

Table 1. Analysis of SRAM

Technique	180 nm	180 nm	180 nm
Voltage supply	1.2 V	1.2 V	1.2 V
SRAM	8 T	12 T	10 T
Maximum Frequency	40 MHz	200 MHz	60 MHz
Array size	128 × 16	256 × 64	136 × 32
Logic Energy (fJ/bit)	195	118.43	291.5
Normalized Logic Energy (fJ/bit)	201	211.21	291.5
Throughput (GOPS/kb)	7.45	9.1	15
Energy-Efficient (Normalized)	0.5 V	1.2 V	1 V
Accuracy MNIST CIFAR	97% 87%	98.84% 88.78%	97.02% 88.39%

Using the SCL 180 nm of technology, an IMC architecture based on a 4 Kilobyte SRAM array size is created. The suggested IMC design operates for IMDP, logic, and arithmetic functions at 60 MHz's. Using the MNIST dataset, the picture classification

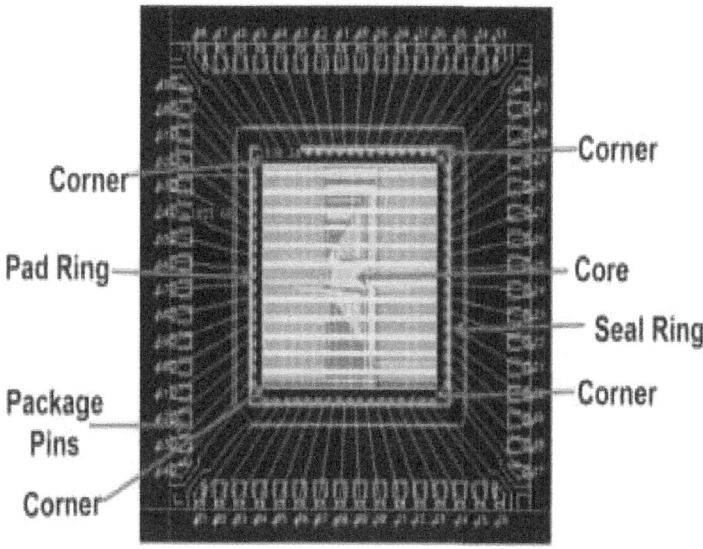

Fig. 8. Test chip 180 nm die photo

accuracy on MLP is 97.02%, 0.1% less than the software baseline. The Table 1 shows the analysis of SRAM by using 8 T, 12 T and 10 T.

5   Conclusion

A novel paradigm known as In-Memory Computing (IMC) is beginning to take on the role of the VNB in data-concentrated applications. In comparison to 9T SRAM, the 10T SRAM have a roughly identical read energy and a lower leakage power and less write energy of 89% and 83.8%, respectively. Additionally, to examine the functionality and the performance of the suggested In- Memory Computing architecture, 4Kb array size based on 10-T bit cell SRAM is constructed in 185-nm SCL technique. The energy efficiency of suggested In-Memory Computing design 4.5 TOPS for 1-bit add operation, and 5.3 TOPS for 1-bit logic operation. The suggested IMC macro architecture's 136 × 32 array achieves an area efficiency of 65.2%.

References

1. Verma, N., et al.: In-memory computing: advances and prospects. IEEE Solid-State Circuits Mag. **11**(3), 43–55 (2019). https://doi.org/10.1109/MSSC.2019.2922889
2. An 8T SRAM Array with Configurable Word Lines for In-Memory Computing Operation Jin Zhang, Zhiting Lin *, Xiulong Wu, Chunyu Peng, Wenjuan Lu, Qiang Zhao and Junning Chen
3. Distributed Active Learning Pengcheng Shen, Chunguang Li, (Senior Member, IEEE), Zhaoyang Zhang, (Member, IEEE) Zhejiang Provincial Key Laboratory of Information Processing, Communication and Networking and College of Information Science and Electronic Engineering, p. 310027Zhejiang University, Hangzhou, China

4. Rajput, A.K., Pattanaik, M., Kaushal, G.: An energy-efficient 10T SRAM in-memory computing macro for artificial intelligence edge processor

5. Kang, M., Gonugondla, S., Patil, A., Shanbhag, N.: A 481pJ/decision 3.4M decision/s multifunctional deep in-memory inference processor using standard 6T SRAMarray (2016). HTTPs://arxiv.org/abs/1610.07501

6. Zhang, J., Wang, Z., Verma, N.: A machine-learning classifier implemented in a standard 6T SRAM array. In: IEEE Symposium on VLSI Circuits (VLSI-Circuits), pp. C252–C253 (2016)

7. Simon, W.A., Qureshi, Y.M., Rios, M., Levisse, A., Zapater, M., Atienza, D.: BLADE: an in-cache computing architecture for edge devices. IEEE Trans. Comput. **69**(9), 1349–1363 (2020). https://doi.org/10.1109/TC.2020.2972528

8. Chen, J., Zhao, W., Wang, Y.H.: Analysis and optimization strategies towardreliable and high-speed 6T compute SRAM. IEEE Trans Circuits Syst. I. Regul. Pap. **68**(4), 1520–1531 (2021). https://doi.org/10.1109/TCSI.2021.3054972

9. Rajput, A.K., Pattanaik, M.: Implementation of Boolean and arithmetic functions with 8T SRAM cell for in- memory computation. In: 2020 International Conference for Emerging Technology, INCET, pp. 1–5. (2020). https://doi.org/10.1109/INCET49848.2020.9154137

10. Agrawal, A., Jaiswal, A., Lee, C., Roy, K.: X-SRAM: enabling in-memory Boolean computations in CMOS static random-access memories. IEEE Trans. Circuits Siti. Regular. Pap. **65**(12), 4219–4232 (2018). https://doi.org/10.1109/TCSI.2018

11. Agrawal, A., et al.: Xcel-RAM: accelerating binary neural networks in high-throughput SRAM compute arrays. IEEE Trans. Circuits Syst. I Regul. Pap. **66**, 3064–3076 (2019)

12. Qing, D., Supreet, J., Mehdi, S., Yejoong, K., Sylvester, D.: A 0.3V VDDmin 4+2T SRAM for searching and in-memory computing using 55 nm DDC technology. In: Proceedings of the 2017 Symposium on VLSI Circuits, 12–15 June 2017, pp. C160–C161. Kyoto, Japan (2017)

13. Jeloka, S., Akesh, N.B., Sylvester, D., Blaauw, D.: A 28 nm configurable memory (TCAM/BCAM/SRAM) using push-rule 6T bit cell enabling logic-in-memory. IEEE J. Solid-State Circuits **51**, 1009–1021 (2016)

14. Petrenko, S., Asadullin, A., Petrenko, A.: Evolution of the von Neumann architecture, Protect. Inf. Inside **2**(74), 18–28 (2017)

15. Chen, Y.-H., Krishna, T., Emer, J.S., Sze, V.: Eyeriss: an energy– efficient reconfigurable accelerator for deep convolutional neural networks. IEEE J. Solid-State Circuits **52**(1), 127–138 (2017)

Application of AI for Education

Integrating Students as Academic Partners in Software Engineering: A Group Software Development Case Study

Raj Ramachandran[1](✉), Helga Gunnarsdottir[2], Krishan Ranjan[2], Khoa Phung[2] (iD), and Emmanuel Ogunshile[2] (iD)

[1] Cardiff School of Technologies, Cardiff Metropolitan University, Cardiff, UK
RRamachandran@cardiffmet.ac.uk
[2] School of Computing and Creative Technologies, University of the West of England, Bristol, England
{helga.gunnarsdottir,khoa.phung,emmanuel.ogunshile}@uwe.ac.uk

Abstract. Exploring the role of a Student as an Academic Partner (SAP) in software engineering education, this study investigates the interactions between students and faculty and evaluates the perceptions of students on SAP's integration within a Group Software Development Project module at the University of the West of England (UWE). Utilizing a mixed-methods approach, findings reveal nuanced insights into collaborative dynamics and the potential benefits of student-faculty partnerships. This research underscores the transformative potential of SAP roles in enhancing educational practices and student engagement, with plans to extend this model to international contexts such as in India.

Keywords: Student as Academic Partner · Co-creation · Software Engineering

1 Introduction

In the evolving landscape of higher education, the concept of student engagement through partnership has emerged as a transformative approach to learning and teaching. This paper examines the integration of the Student as Academic Partner (SAP) model within a Group Software Development Project (GDSP) module at the University of the West of England (UWE). The study investigates the interactions between students and faculty within this framework, aiming to evaluate its impact on fostering a supportive and collaborative learning environment.

Recent pedagogical research underscores the value of involving students as partners in their education, suggesting that such collaborations can enhance learning outcomes, increase student satisfaction, and encourage deeper engagement [1, 2]. In the context of software engineering education, which often faces criticisms for its traditionally didactic and rigid curriculum structures, the SAP model offers a promising avenue for pedagogical innovation [3].

Supported by the University of the West of England

P. D. Sivakumar et al. (Eds.): IRCCTSD 2024, CCIS 2362, pp. 269–277, 2025.
https://doi.org/10.1007/978-3-031-82386-2_21

The significance of this study is twofold. Firstly, it contributes to the broader discourse on student engagement strategies in STEM education, providing empirical insights into the practical benefits and challenges of implementing student- faculty partnerships. Secondly, by exploring the SAP model's application in a technical discipline, the research addresses a gap in the literature which has primarily focused on more general or liberal arts contexts [4].

The structure of this paper is as follows: we begin with a review of the literature surrounding students as partners in higher education, establishing a theoretical framework for our investigation. We then describe the methodology employed to integrate the SAP model at UWE and its intended replication in a similar module in India. Subsequent sections present the findings of this integration, discuss their implications for software engineering education, and conclude with recommendations for educators and institutions interested in adopting this innovative approach.

2 Literature Review

There is a growing realisation in higher education that partnerships with students are highly beneficial. The past decade has seen an increase in literature relating to 'students as partners' [1–3], and since 2017, the 'International Journal for Students as Partners' has brought a wealth of research and case studies on this subject. This concept in higher education has gained substantial attention over the past decade, reflecting a shift towards more inclusive and participatory educational practices. Healey [1] define this partnership as a collaborative, reciprocal process where all participants actively contribute to pedagogical planning and decision-making. This model has been shown to improve learning outcomes, enhance student engagement, and foster a deeper connection to the curriculum [2].

Much of the literature focuses on students either as co-creators of the curriculum or as pedagogical consultants [2] within their own modules or programmes of study, however, with an increasing interest in looking at partnerships in a wider context, such as the library [5], co-creating a new capstone module [6], organising a conference [7] or as a community-engaged departmental research project [8]. In software engineering education, the application of the partnership model remains under explored, with most studies focusing on more traditional teaching environments. Matthews et al. [3] argue that while the model has been successful in various disciplines, STEM fields, including software engineering, present unique challenges due to their highly technical content and traditional lecture-based approaches. Cook-Sather et al. [4] suggest that these environments can benefit significantly from integrating students as co-creators, which can lead to innovative teaching and learning practices.

However, there is a noticeable gap in the literature regarding the specifics of implementing such partnerships in the context of software development projects within university settings. This gap is particularly evident in the limited discussion on the roles students can play beyond mere consultation, especially technical and applied disciplines. The current study addresses this gap by exploring the role of an SAP within a module such as GSDP, focusing on the impacts of this integration on student learning and project outcomes.

Furthermore, the literature points to several potential challenges in fostering effective student-staff partnerships. Cook-Sather et al. [4] identify resistance from both faculty and students to new roles and unfamiliar forms of collaboration as significant obstacles. Additionally, the question of how to measure the impact of these partnerships on educational outcomes remains largely unanswered. By examining these issues within the context of a software engineering module, the present study contributes new insights into the practical application of the partnership model in a discipline characterized by rapid technological change and project-based learning.

This paper further contributes to the range of exceptions by looking at a Student as an Academic Partner (SAP), where an international student from a different Master's (MSc) Programme (e.g., MSc. Real Estate Finance and Investment) was recruited into the module teaching team to support students. Other examples of paid partnerships exist such as the impact of students being employed as partners through the student union [9]. Although the initial recruitment of the SAP initially followed the existing Peer Assisted Learning (PAL) scheme - which fosters academic success through peer-led support and collaboration, enhancing students' understanding of course material, study skills, and confidence, it was soon evident that the SAP role went beyond what was expected of a PAL leader and the employment type was therefore adapted to suit the purpose of the SAP role.

The role does, however, still fit with the literature and various proposed frameworks in several ways. Bovill et al. [10] identified four frequently assumed and overlapping roles for students as co-creators which cover all the aspects of the SAP role:

- A co-researcher in terms of co-authoring this paper,
- pedagogical co-designer, in terms of designing and facilitating additional support sessions for students on the module,
- a consultancy role, coming from another MSc Programme unrelated to Computer Science offering a fresh perspective, and
- a representative, both as a student feeding back to the academic team, but also as a current student representing the module team when working with groups to resolve various issues.

Having engaged with the existing literature on the subject before the start of the partnership, the module team were aware of the difficulties that students can experience when starting to work in a student-staff partnership, such as trust issues, the uncertainty of roles and responsibilities [10] and therefore involved the SAP in all discussions, both regarding the definition of responsibilities and relating to the day-to-day module leadership. Further, similar to Kapadia [11]'s reflection of the gradual integration into the partnership with staff, the SAP had also been a stakeholder in a previous module run before their role as SAP, which was helpful in terms of prior knowledge of the module as well as the module team, which was probably one reason that the partnership did not run into any considerable tension, as otherwise often reported, e.g. communication issues and other frustrations as reported by Knaggs et al. [6] for example.

Whereas it is always difficult to determine the level of equality in a partnership, it was decided from the start of the partnership that the SAP would be a named co-author and therefore be able to contribute to the writing of the paper. However, as Owen et al. [12] noted 'the assumptions of power are deeply embedded within the structures and roles of higher education' and changing the mindset of students and staff is difficult with short-term partnerships; however, the module team made every effort to attempt an as an equal partnership on all levels as possible.

3 Methodology

3.1 Structure and Delivery of the Module

The module was a 30-credit core module in the MSc Information Technology (IT) programme. The students in this module came from a range of backgrounds – both in terms of geographical origin and linguistics as well as in terms of domain knowledge.

The module itself was broadly divided into two parts. The first part consisted of six weeks of programming in Java. Each week, the students had one lecture, one tutorial and one support session, each lasting two hours and led by a lecturer who delivered face-to-face.

The second part of the module consisted of a six-week group project, focusing on the Software Development Life Cycle (SDLC) and the application of concepts to a project chosen by the students. The students were allocated into groups of four or five students based on the results of a personality type indicator test, a module questionnaire, and an interview. The interview was conducted by a permanent member of staff and the SAP, although the interviews were planned and led by the SAP.

In this part of the module the lecture and the tutorial were merged into one four-hour flipped learning online lectorial, where the students were provided with a task in advance of the session and were expected to engage in pre-session activities, including a weekly team meeting. The lectorial sessions were primarily led by students and the lecturer facilitated the discussion, drawing upon experiences and theory. As the module took an evidence-based approach, students were expected to maintain individual logs through the module and record their team meetings as part of their evidence.

The module was traditionally seen as a highly successful module with over 95% pass rates. However, over the last few runs, the module's pass percentage was as low as 19%. The module was slightly redesigned to offer a better learning experience and a more authentic learning environment simulating work-like situations and took an evidence-based approach. Issues such as plagiarism, lack group working skills, time management, English language abilities, and academic writing skills were identified.

The SAP came into play at various times throughout the module. Before the start of the module, the whole module team had two one-hour introductory sessions with the students, introducing the module and the module team, including the SAP.

One of these sessions also had two former students speaking of their experiences. The SAP was introduced as a part of the module teaching team. The SAP was also involved in the group allocation process leading the support sessions and proactively responding to student e-mails after the session. In the latter part of the module, the students were also offered the opportunity to book optional support sessions led by the SAP.

3.2 Development of the Role

The profile for the SAP role was based on the above issues, where we specified that we were looking for an international student with group work experience which included project work. By introducing an international current student as a member of the teaching team to the cohort, the idea was to bring in their experiences to benefit both the module team and the students in the cohort as 'students spend many years experiencing and, in some cases, analyzing learning that might or might not be optimal and engaging. The partnership brings these forms of experience and expertise into dialogue in ways that inform and support more intentional action [4].

Exploring the existing channels at UWE for student partnerships, we started by exploring the PAL Scheme which was already well established for undergraduate programmes, and started the recruitment process through this scheme. The successful candidate had already been an external stakeholder in the module and therefore had a good understanding of the module.

It soon became clear however that the PAL role was somewhat limited and not entirely suitable for the support planned for the students, leading to the creation of the Student as Academic Partner role, where the SAP became a part of the module teaching team, consisting of one Senior Lecturer and one Lecturer, who were both co-leading the module, the SAP and a Digital Learning Manager.

3.3 Data Collection

The research element in the module was introduced to the cohort during the 'setting expectation sessions'. There were a total of 86 students, the vast majority of them being international students. An anonymous module feedback survey was gathered at the end of the module. Students were also invited to participate in the focus group. Participating in the research was optional and voluntary; therefore, it did not influence the assessment outcome. Students agreeing to participate in the focus group were provided with the participant information sheet and had to email the signed consent form to the module leader. Two students participated in the focus group and received 23 anonymous responses to the module feedback survey. The response rate was approximately 27%.The university's ethical approval process was followed and data were collected only after obtaining the ethical approval (Ref: FET.21.05.054). The focus group was conducted by a Senior Learning Technologist from a different faculty. The session was facilitated online. The meeting was recorded with the participant's permission and subsequently transcribed, anonymous and redacted where necessary. During the focus group, both students lost connectivity with the session; however, the discussion continued and each participant was asked the same questions when they returned to the session.

4   Findings and Analysis

The cohort perception of the SAP was interesting. The SAP was introduced to the cohort as a 'support session lead' to facilitate integration with the module teaching team and help change the cohort's perception of SAP.

When asked as part of the module feedback survey "I saw the support session lead as a part of the module teaching team", the overall result was 'neither agree nor disagree'. 8.7% of the students strongly disagreed with the view that SAP was fully a part of the module teaching team, 13.0% disagreed and 34.8% neither agreed nor disagreed.

Equally, 34.8% agreed and 8.7% strongly agreed with the view that SAP was completely a part of the module teaching team. We assume this split perception of the SAP role to be due to inconsistent delivery of the support sessions. Some of the support sessions were covered by the module leader to enable SAP to pursue activities that contribute to his chosen career path. The reflection of the module leader, SAP and the perception of the cohort indicate a gap that needs to be further explored and filled in. For example, the module leader's positive outlook on the SAP role was reflected in the following comment: "SAP was beneficial, and its benefits were seen beyond this module as described by some students. I think we managed it well with the cohort. Personally, I would not make the role too rigid as I think it would limit innovation. If I am offering a safe space to the students to make mistakes and learn from them, it is fair to extend the same to SAP."

The SAP was also quite optimistic and felt a part of the module teaching team. To a question on experience as an Academic Partner inclusion and empowerment, the comments from SAP were: "It was a whole new experience, something that I never experienced before. I have a thing for helping and mentoring students and I think that worked for me in this role. Yes, I felt included both in terms of the students and the faculty side. I received proper guidance and support from the faculty team and a good positive response from the students. All this accounted for me feeling empowered within this role."

The comment contradicted an observation the module leader made on the SAP's hesitation to make optimal use of Blackboard in the same way as other members of the teaching team. Despite measures to integrate students as Academic Partner, there is intangible evidence of the maintenance of an unspoken hierarchy which can be seen in the following reflection of SAP:

"I was aware somewhere in the back of my mind that I have access to the blackboard announcements, however, I think I felt that if the message is coming from Raj, it would be more impactful." The module leader perceives this to be a failure in achieving a truly equal partnership. In a situation where everyone has equal access to the Blackboard site and resources, where everyone can post announcements on their sessions, the reflection of SAP reiterates the existence of the 'authority' of the module leader, something that we were trying to get away from. It can be argued that the lack of effective usage of Blackboard might have contributed towards the cohort's perception of the SAP as a full member of the module teaching team.

The support sessions were largely led by the needs of the student with a skeleton structure to ensure basics were covered.

There seems to have been a disproportionate focus on some topics in the support session as seen in comments from one of the focus group participants: "There's only so much you can touch on it. Don't use it as a copout. Actually, try it and talk about some project management tips and that kind of stuff. I think that's all I'd say. They were great. There was just a little bit of a focus on referencing maybe more than it needed to be."

The organisation of the support session by SAP was almost autonomous with little initial guidance from the module leader and the learning technologist. Therefore, the comments from the students would need to be viewed in the same manner as though it was delivered by a member of staff as it is not uncommon for members of staff (tutors) to receive criticisms on aspects that they might think to be a well-designed or well-delivered session. Contrary to the literature [6], it can be construed that the partnership between the student and the academics was truly achieved at some level as seen in the comments from one of the module leaders: "For me and the module team, it is about having a fresh perspective, new thoughts, new ways of working and helped us to revisit aspects/ areas of module without compromising on the academic rigour and professional body requirements. Personally, I have learnt quite a few things from SAP."

This suggests a degree of receptiveness, openness, and an inclination to learn from the SAP. The module leader's presence along with the SAP during the sessions could have contributed to the power dynamics, not because of distrust within the partnership such as with Knaggs et al. [6], but in the cohort's perception of the presence of the lecturer in the room. Anecdotal evidence points to the power dynamics contributing to the student perception of the SAP. However, the presence of the module leader in the support session was to support the SAP as the candidate was new to the Higher Education teaching domain. There were also constraints within the university system that limited who could be timetabled. This must not be construed as mentoring or coaching SAP as the candidate had all the necessary skills required to lead the support session independently and autonomously. Although there is unanimous consensus among the module team that the SAP for the cohort was an 'ideal' candidate who has set the bar high for future SAPs, there still seem to be gaps and varying expectations that need closer attention. The role of SAP was not diluted and the expectation of the module team was on par with other colleagues who may be on similar contracts.

"We did manage a good balance of including the SAP in the module team, they were a part of the conversation, always copied in to emails and worked with us through some of the more difficult incidents with students.

I believe they were given responsibilities that were suitable for the number of hours they worked, and I think we were very conscious of not limiting their input in terms of it being easier if a staff member did the task. There were a couple of smaller issues that arose, such as timetabling, where it was clear that they did not have an as deep understanding of the university processes as the rest of the module team, however, if the duration of the role had been longer than one cohort, this would just have been part of the learning as with any new job."

Overall, the support sessions seemed to be useful regardless of prior United Kingdom (UK) academic learning experience which can be seen in the following comment: "In my Bachelor in Quantity Surveyor, I used the APA style for referencing. So, I found it is similar and needed to change the format of the referencing. From the start, I made some mistakes and used the APA style as I used to. Practice makes perfect. Now, I can use UWE Harvard referencing without any help.

Furthermore, critical writing was taught as it is very important to write a quality report with rationalisation and argument. I found this very challenging and tried my best. I have yet to ace the critical writing technique but I am improving."

5 Conclusion

Overall, the journey with SAP in Group Software Development was a rewarding experience for the permanent members of staff within the module team, SAP and the cohort alike. It is imperative to ensure careful selection of the candidate for this role as this may influence the student learning experience which in turn may affect the engagement, overall results and outcome of key indicators such as module feedback, student surveys etc. In the case of this module, the SAP has had a positive impact on the overall module experience. Plans for future work include replicating the process with a Home student (i.e. UK domiciled student) and replicating the concept of SAP in other learning contexts such as in India through collaborative efforts. The potential application of the SAP model extends beyond the UK context, offering valuable insights for educational reforms in countries like India, where traditional educational practices often dominate. In India's dynamic and diverse educational landscape, integrating the SAP model could address several persistent challenges such as student disengagement, passive learning, and the gap between theoretical knowledge and practical application. By adopting this model, Indian institutions could foster a more active learning environment that encourages student initiative and a deeper connection with the academic content.

However, implementing such a model in India would require careful consideration of cultural, institutional, and logistical factors. For example, the hierarchical nature of many educational institutions in India may challenge the egalitarian ethos of the SAP model. Therefore, pilot projects could be initiated in select universities to tailor the approach to local conditions and evaluate its effectiveness.

Throughout this journey, we recognise that the role needs to be further defined but since this is in its inception, there was a deliberate attempt to provide more autonomy to the SAP and allow them to explore a largely unsupervised environment. In turn, this research has provided massive learning opportunities for the module team, embraced good practices that emerge, and understood some of the shortcomings for the future runs of the module. Since the marking was incomplete upon completion of this paper, we are unable to comment on or link the effectiveness of SAP to overall module results. We are deliberately avoiding the link as the focus is on learning, ways of learning and providing real-world academic exposure to the student undertaking the role. Although the initial intention of SAP was to improve the pass percentage of the module, with time, there are other pedagogical benefits such as disrupting the traditional hierarchy and providing students with an opportunity to learn from someone with whom they can resonate.

Acknowledgements. We gratefully acknowledge Samantha Jimenez from the Universidad Autonoma de Baja California for her important role in providing practical projects for our Group Software Development Project module. Her contributions significantly enriched the students' learning experience by linking academic theories with real-world applications. We deeply appreciate her commitment and support.

References

1. Healey, M.: Students as partners in learning and teaching in higher education. In: Workshop Presented at University College Cork, vol. 12. 1, p. 15 (2014)
2. Mercer-Mapstone, L., et al.: A systematic literature review of students as partners in higher education. Int. J. Stud. Partners (2017)
3. Matthews, K.E., et al.: Enhancing outcomes and reducing inhibitors to the engagement of students and staff in learning and teaching partner- ships: Implications for academic development. Int. J. Acad. Develop. **24.3**, 246–259 (2019)

4. Cook-Sather, A., Bovill, C., Felten, P.: Engaging students as partners in learning and teaching: a guide for faculty. Wiley (2014)
5. Salisbury, F., Dollinger, M., Vanderlelie, J.: Students as partners in the academic library: Co-designing for transformation. New Rev. Acad. Librarianship **26.2–4**, 304–321 (2020)
6. Knaggs, A., et al.: Partnership status: it's complicated. Reflections on the "undiscussables" in a student-staff partnership. Int. J. Stud. Partners **5.1**, 131–137 (2021)
7. Pedersen, C.L., et al.: Organizing an undergraduate psychology conference: the successes and challenges of employing a student-led approach. Psychol. Learn. Teach. **12.1**, 83–91 (2013)
8. Hunt, L.Y.A., Hunt, L.J.: The Importance of a whole-of-department framework in learning partnerships. Int. J. Stud. Partners **1.2**, 92–99 (2017)
9. Freeman, R., et al.: Student academic partners: student employment for collaborative learning and teaching development. Innov. Educ. Teach. Int. **51.3**, 233–243 (2014)
10. Bovill, C., et al.: Addressing potential challenges in co-creating learning and teaching: Over-coming resistance, navigating institutional norms and ensuring inclusivity in student–staff partnerships. High. Educ. **71**, 195–208 (2016)
11. Kapadia, S.: Academic representation and students as partners: Bridging the gap. Int. J. Stud. Partners **5.2**, 169–173 (2021)
12. Owen, J., Wasiuk, C.: An agile approach to co-creation of the curriculum. Int. J. Stud. Partners **5.2**, 89–97 (2021)

Chattitude in Education: Chatbot Expertise for Educational System

K. S. Kaavya Shree[1], Padigi Reddy Sangeetha[1], Shravani Upadhyay[1], and S. Prasanna Devi[2(✉)]

[1] Department of Computer Science and Engineering (Emerging Technologies), SRM Institute of Science and Technology, Vadapalani, Chennai, India
{kk2735,ps9349,su3230}@srmist.edu.in
[2] Department of Computer Science and Engineering, SRM Institute of Science and Technology, Vadapalani, Chennai, India
hod.cse.vdp@srmist.edu.in

Abstract. Chattitude, a Chatbot-enabled educational system combines the capabilities of a Chatbot with the expertise of an intelligent Educational System. This integration allows for interactive and personalized information delivery through a conversational interface which helps in revolutionizing the educational landscape by serving intelligent, accessible, and personalized tools that enhance the user learning experience and administrative processes of Educational Institutions and Universities. Employing chatbots in the educational domain helps to streamline and enhance student information portals, ultimately contributing to a more effective and Student-centric educational environment. This intelligent chatbot works by leveraging Artificial Intelligence particularly, Natural Language Processing (NLP) techniques, to build the conversational interface. This also includes Machine Learning Algorithms, such as K-Means Clustering and Feedback generation Mechanism, to embark our chatbot by providing summary reports, academic details and all the other features, thus helping to enhance a unique user experience.

Keywords: Artificial Intelligence · Machine Learning · Natural Language Processing · Feedback System · Information Retrieval

1 Introduction

Rapid technological advancements and shifting pedagogical paradigms have generated significant advancements in the educational environment in the recent years. In this revolutionary environment, incorporating chatbot technology into educational systems provides a compelling means to use Artificial Intelligence to improve teaching and learning processes. With their ability to interpret Natural Language, chatbots present a unique chance to completely transform the way that educational information is accessible, presented, and tailored. Traditional educational institutions frequently struggle with issues including accessibility, scalability, and personalized support. The term "Chattitude", which refers to the combination of chatbots and context related to the educational institution, is a paradigm change to approach these issues. Chattitude aims to provide learners with customized help and advice in a variety of circumstances by integrating chatbots with educational knowledge and contextual awareness, therefore overcoming the constraints of traditional learning environment.

2 Existing System

The background study regarding the Chatbot assistance in the Educational sector, seeded the main ideology for this proposed system. There were some traditional existing systems for employing chatbots in the Educational sector. Those existing system used the below techniques,

- Rule – based System: They follow a set of predefined rules and patterns, which includes simple if-else-then statements to respond to specific queries. These system fails in understanding complex queries and has a maximal time-consuming process.
- Information Retrieval System: These systems are an enhanced version of Rule based system, that creates a knowledge base and often incorporates a Information Retrieval technique to produce a relevant answer to the user queries. In this system, updating the knowledge base is not possible, thus it fails to provide accurate results for user queries.

Table 1. Comparison of the Existing System with the Novelty of the Proposed System

Paper Title	Method	Limitations	Novelty of the Proposed System
An Automated Chatbot for Educational Institutions using Natural Language Processing	Student Academic Information retrieval using Natural Language Processing	Uses the simple Information Retrieval technique in NLP for getting student marks	Through the seamless integration of NLP and ML, Chattitude presents a strategy that improves student access to academic material through NLP driven interactions and powered responses
UniBud: A Virtual Academic Advisor	Utilizes the usability principle, to build voice-based interaction with students	Fails when the system mis-understands the voice statements, to provide inaccurate response	Chattitude provides an improved user experience, for the interaction and advice. The "Fuzzywuzzy" module is included to give appropriate responses even in situations where voice statements are misinterpreted
Interactive Advising with Bots: Improving Academic Excellence in Educational Establishments	"AdvisorBots" are Bot based framework reflecting on a virtual support system with unsupervised learning	Limitations when the underlying patterns have worse correlation	In comparison, our Chattitude provides individualized academic supportt, eliminating the constraints using ML algorithms and feedback mechanisms, and improves academic performance

(*continued*)

Table 1. (*continued*)

Paper Title	Method	Limitations	Novelty of the Proposed System
Chatbot for Educational System	To use local database and web database to provide response. Also involves techniques such as pattern matching, data processing	Lacks to support when the database is not updated with respect to time	To address database staleness with real-time changes, the Chattitude innovates by connecting local and web databases

Smart Educational framework [2], and the future trends, play a major key role, in cultivating advanced technology with the Educational Sector, to change the world of Education. This method insists on the usage of Artificial Intelligence, IoT, Smart Class room policy, and Data Science to improve the overall Institutional Educational framework. As a part of this futuristic ideology, employing chatbots for academic and administrative advisory, was one of the key insights to drive over this proposed ideology. There are some of the chatbot technologies that are existing, which are primarily focusing on the educational sector. Automated Chatbot using Natural Language Processing [1], proposed a system, which provides the student academic details. These details are extracted from the student database using the traditional Information Retrieval technique. Unibud Bots [3], were employed in the same task which is built to work based on usability principle, having voice-based interaction with the students. AdvisorBots [4] are a virtual support system that aims at employing unsupervised learning techniques [5, 6] with student data, to provide real – time interactions. The key features of all the discussed methodologies with the Novelty of the Proposed system is summarized in Table 1.

3 Dataset

In our educational system project, we integrate three diverse datasets—student academic records, university information, and placement outcomes—alongside extracurricular activities data to create a comprehensive platform supporting student success and holistic development. These datasets are briefly explained below.

3.1 Student Academic Database

In this project, we have created a student database, that includes all the academic student information. This data is collected using Google-forms, that facilitates to get details from the students, and comparing with the University's academic backup database. This database involves a vast amount of data giving answers for all the possible questions in varied fields related to the academic details of the students. The list of fields that could be taken as the key features and possess a major interest are Name, Registration Number, Department, Section, Year of Passing, each semester Course details, Grade for all the registered courses, Overall CGPA, Overall Attendance Percentage, Arrears/Backlogs details and Future career path details.

3.2 Chatbot Intent Dataset

This Chatbot Intent Dataset refers to the set of data [7], that includes information related to all the University, that can help the Chatbot to answer the queries related to the University's policy [8, 9], About the Courses offered, In-depth details about all the Academic departments and also about the surplus details about the University. This chatbot mainly helps in handling these four sub-modules of the Chatbot, such as,

- About SRM Institute of Science and Technology
- Faculty of Engineering and Technology
- Student Extra-curricular Dataset

In this project, we have created a separate student database that includes all the academic achievements, non-academic achievements of the student and professional career related information. This data is collected using Google-forms, that facilitates to get details from the students, and comparing with the University's academic backup database. This database involves a vast amount of data giving answers for all the possible questions in varied fields related to the extra-curricular details of the students. The list of fields that could be taken as the key features and possess a major interest are Name, Register Number, Event Participation details, Club work details, Event Organization details, Sports Achievements, Academic Achievements, Internship/PPO details, Future career aspiration, Future career domain and many more.

3.3 Student Extra-Curricular Dataset

In this project, we have created a separate student database that includes all the academic achievements, non-academic achievements of the student and professional career related information. This data is collected using Google-forms, that facilitates to get details from the students, and comparing with the University's academic backup database. This database involves a vast amount of data giving answers for all the possible questions in varied fields related to the extra-curricular details of the students. The list of fields that could be taken as the key features and possess a major interest are Name, Register Number, Event Participation details, Club work details, Event Organization details, Sports Achievements, Academic Achievements, Internship/PPO details, Future career aspiration, Future career domain and many more.

4 Proposed System

Employing Web-based Chatbots in the Educational domain helps to streamline and enhance student information portals, ultimately contributing to a more effective and Student – centric educational environment. The Chattitude, is built by incorporating various Artificial Intelligence techniques, that contributes to an advanced and a well supportive framework, which will be useful for the students and the Educational users. The Architectural diagram for our Educational Chatbot is given below.

The "Fig. 1" gives out the architectural flow of the proposed methodology. The list of modules, that are used in the Chattitude are briefed below.

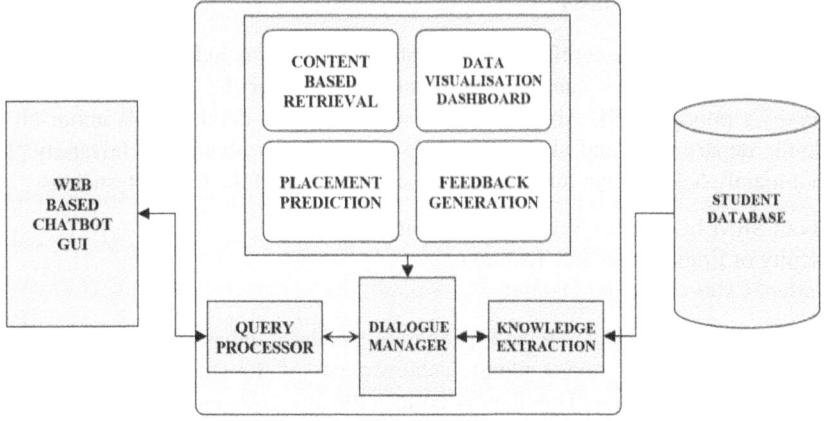

Fig. 1. Proposed Design

Query Processor

By incorporating cutting-edge technology into the current educational system, this significant initiative seeks to completely transform the educational environment. Our intelligent instructional chatbot relies heavily on the Query Processor [11, 12], which gives it the ability to understand and reply to the user inquiries. The proposed Query Processor makes use of Machine Learning and Natural Language Processing (NLP) technologies to comprehend the subtleties of educational questions. Through the use of Context-based retrieval algorithm in Natural Language Processing the system consistently refines its comprehension of user input, guaranteeing responsive and contextually appropriate solutions. The module is made to manage a wide variety of requests, such as those for general querying, academic help, Institutional details, Course Details offered by the Institution, Student academic statistics and curriculum-related information.

Data Visualization Dashboard

To portray complicated educational data in an understandable way, the Data Visualization Dashboard uses state-of-art-visualization approaches as in [13, 14]. This offers a interactive dashboard that facilitates effective exploration and analysis of the presented data. While presenting a summarized report for a Class, this also involves Machine Learning algorithm such as K-Means Clustering and Classification, to categorize students into different subsets, such that, it helps in identifying the particular student groups for the academic faculty. A comprehensive and technologically sophisticated strategy, the integration of Data Visualization Dashboard into the Chattitude in Education Project is positioned to improve educational practices via deliberate planning and well-informed decision making.

Placement Prediction

A comprehensive overview of Chatittude into the field of education, emphasizing the "Placement Prediction" module in particular as in [15]. The program evaluates students' qualifications and forecasts their eligibility for many placement categories using an eligibility-checking algorithm using Python. The module classifies students into several placement categories based on their academic achievement, coding abilities, aptitude, and interview evaluations. The check_eligibility function, at the center of the system, assesses important parameters including skill ratings, CGPA, and academic scores to establish a candidate's eligibility for Marquee, Super Dream, Dream, and Non-Engineering placements. This method is used by the placement_eligibility function to give students immediate feedback on their eligibility for campus placements.

Feedback Generation Mechanism
In the context of digital education, providing immediate, tailored feedback can greatly improve the student experience [16]. A complex input analysis system that makes use of natural language processing (NLP) to correctly interpret student responses is at the heart of the system. The ability of this system to comprehend a broad range of textual inputs, interpret semantic meaning, and determine contextual relevance is essential for precisely determining the student's needs and current knowledge level. The mechanism evaluates the student's knowledge and comprehension after input analysis. By using machine learning algorithms [17, 18], it determines accurate answers, misunderstandings, and knowledge gaps by comparing the examined input to a comprehensive knowledge database.

Web – Based Chatbot Graphical User Interface
The development of a graphical user interface (GUI) for a web-based chatbot designed exclusively for educational support offers [19–21] a special chance to improve the educational experiences of students. This web interface provides a simple and easy-to-use platform for the users to participate in interactive learning experiences. The design has a strong emphasis on simplicity and minimalism to make sure that students can use the chatbot efficiently and without needless complication. The web-based chatbot's graphical user interface (GUI) is designed to provide a simple and straightforward learning experience. It has a clear design with a large chat window and a text input field for inquiries. This accessibility is crucial in democratizing educational support and allowing students to continue their learning journey outside the traditional classroom environment. The Methodology overview for the proposed system is summarized in Table 2.

5   Methodology

The development of a Chatbot-enabled Educational System requires a carefully planned methodology to ensure the successful integration of artificial intelligence (AI) technologies into the Educational domain. This section outlines the systematic approach adopted to design, develop, and implement the chatbot system, which aims to provide personalized learning experiences and expert guidance to users.

As discussed earlier, the Chattitude has Five different key functional frameworks, that serve the purpose of its working. These framework modules are briefed below.

Table 2. Methodology Overview

Module	Overview
Natural Language Processing for User Query Processing	The system interprets user inquiries and determines their underlying intentions by using advanced Natural Language Processing algorithms. The system derives complex meanings from user inputs using extensive pre-processing techniques including tokenization, lemmatization, stemming, keyword extraction, and intent classification
Query Processing and Response generation	Using an already-built knowledge base that contains educational resources and institutional data, the system enables quick and easy query processing and answer creation. Through the use of sophisticated vectorization techniques such as Term Frequency-Inverse Document Frequency (TF-IDF) and pattern matching, the system creates contextual correlations between user inputs and the knowledge base. Furthermore, the incorporation of the "Fuzzywuzzy" module guarantees exact answers to even vague or unclear inquiries
Data mining and Data Visualization	The chatbot presents both student-specific and general academic data in an easy-to-understand manner by utilizing interactive dashboard frameworks like Plotly and PowerBI. The system employs data integration strategies to provide informative replies by carefully mining pertinent information from many data sources using data retrieval techniques
Feedback Generation Mechanism	Sophisticated Natural Language Generation (NLG) techniques are used to generate personalized feedback for students. The chatbot is highly skilled at creating customized feedback responses that promote improved student engagement and learning outcomes. This is achieved by dynamically obtaining relevant student data instances and utilizing machine learning algorithms that are guided by heuristic approaches and past interactions

5.1 Query Processor Module

This Query Processor Module is the first key module that facilitates the entire flow of the Chatbot working. This Query processor module helps in understanding and pre-processing all the user queries, to provide relevant and precise response. This module primarily works to serve the purpose of the following sub-tasks by the Educational Chatbot, leveraging to respond to queries relating to these below topics.

1. About SRM Institute of Science and Technology Module – This module provides users with comprehensive information about the SRM Institute of Science and Technology, including its vision and mission, history of the SRM Group of Institutions, SRM University Campus tour, and further details providing access to the University's official website especially about the Vadapalani Campus of SRMIST.
2. Faculty of Engineering Module – This module offers detailed information about the Faculty of Engineering at SRM Institute of Science and Technology, Vadapalani Campus particularly focusing on the information relating to all the Academic Departments of the Campus. Users can explore the different departments within the Faculty of Engineering, and learn about the different programs available in each department.
3. Academic Assistance Module – This module serves as a comprehensive academic support system for students, offering personalized assistance by using their registration number as the key feature. Students can access their semester marks, overall attendance, CGPA, and academic performance summary reports, facilitating self-assessment and goal setting. The chatbot also provide insights into placement statistics, internship opportunities, and career guidance services available to students, helping them make informed decisions about their academic and professional career development.

5.2 Data Visualization Dashboard

This Data Visualization dashboard helps to view all the Visualizations of the Student academic, non-academic and also their overall performance, such that, they can have a better experience in understanding their prospects. This module mainly helps to visualize the analytical data such as Attendance percentage, student grade comparison and so on.

1. Summary Report Generation – This part allows Educators to input their Year of study and Department with section to access their Overall Class Academic summary report. This Academic summary report is automatically generated using NLP and ML Techniques, integrated as a Web-based summary report. In this report, ML Algorithms such as Clustering and Classification are used to group and classify students based on their various aspects. The module may also include graphical representations, to visualize the student's academic progress over time and compare their performance with class averages or benchmarks.

5.3 Placement Prediction Module

This module retrieving the student's academic information and all the necessary fields required, to assess the student for their eligibility for placement opportunities based on predefined placement criteria and conditions of the University's guidelines. The chatbot evaluates the student's academic performance, extracurricular activities, soft skills, and other relevant factors to predict their likelihood of securing a job placement along with the details of the placement category they are likely to get placed. Students receive feedback on their placement prospects, along with recommendations for improving their employ- ability and enhancing their chances of success in the career path for their professional development.

5.4 Feedback Generation Module

This module emphasizes the importance of feedback in supporting students' academic and personal growth. Using advanced analytics and machine learning algorithms, the chatbot generates personalized and precise feedback for each student. Feedback is tai- lored to the individual student's strengths, weaknesses, learning style, and goals, pro- viding actionable insights for improvement. The chatbot identifies specific areas where the student can make enhancements, such as academic performance, study habits, time management skills, communication skills, and interpersonal skills.

5.5 Web-Based Chatbot GUI

Streamlit provides an intuitive and user-friendly framework for building web-based chatbot GUIs. With its simple yet powerful syntax, developers can create interactive chat interfaces with ease. Leveraging Streamlit's functionality, you can design a chatbot interface that not only allows users to converse with the bot but also enables them to input commands, receive responses, and visualize data seamlessly. By integrating Streamlit with your chatbot backend, you can offer a dynamic user experience, complete with text inputs, buttons, and even multimedia elements, thus, the GUI of Chattitude stays strong and versatile to provide Educational support and guidance.

6 Experiment and Results

The results of the experiment using "Chattitude in Education: Chatbot Expertise for Educational Systems" are presented in this paper. The Data visualization, feedback cre- ation, query processing, machine learning, and natural language processing are just a few of the features that research's sophisticated system is powered by. This experiment's findings demonstrate the applicability and flexibility of Chatbot emphasizing how it may streamline academic assistance, career guidance, and information retrieval in a class- room setting. Natural Language Processing (NLP) context-based retrieval, Python, and machine learning methods are used to implement the suggested modules. It has func- tions like data processing and query processing. Features including data input, placement prediction, visualization, and query processing are available. Users may ask questions to get access to different modules. The experimental results are given in the following Figures.

The system is organized into different modules, each serving a specific purpose, such as under the "Fig. 2" About SRM Institute of Science and Technology module we can view information about the vision and mission of the institute, history of SRM

Chatbot: Hii there! 👋Welcome to Chattitude, your virtual companion on the exciting journey of exploration!🎇
Chatbot: Let's embark on this educational adventure of SRM Institute of Science and Technology.
This is exclusively Vadapalani Campus, where every question is a step towards knowledge and understanding.
Chatbot: Ready to dive into the world of learning? Let's get started! 🚀
Chatbot: Here are some of the menu options listed below, that I can support you to move further in this adventure!
1. About SRM Institute of Science and Technology
2. Faculty of Engineering @ SRM Vadapalani
3. Academic Assistance of SRM Vadapalani
4. Type in 'Exit' to end our conversation.
User: About SRM
Chatbot:
Let's start exploring about SRMIST Vadapalani!
Give me an option.
a) Mission and Vision of SRMIST
b) History of SRM Group of Institutions
c) SRM Vadapalani Campus Tour.
d) SRM Vadapalani Website
User: Vadapalani Campus
Chatbot:
SRM University, City Campus located in the prime heart of the city, Vadapalani, is well connected by road, rail and air.
The Campus hosts well planned,
1. Academic block
2. Computer Center
3. Lecture hall
4. Library
5. Laboratories
6. Conference Hall
7. Canteen
8. Wifi-enabled Campus
9. Class rooms are equipped with centralized Air conditionin and LCD projectors.
To know more, please visit our official website, link - www.srmistvdp.edu.in

Fig. 2. About SRMIST

Group of Institutions, Campus tour and access to the official website to know more. This implementation gives us the overview of the user input query -Vadapalani Campus where we can see the Chatbot responding about the infrastructure and campus facilities available. "Fig. 3' Faculty of Engineering module gives us the information about the various courses offered by each department like Computer Science, Electrical and Mechanical. The third module "Fig. 4" and "Fig. 5" academic assistance takes the Registration Number field as the primary key and gives us information about semester marks, overall attendance and CGPA, summary report and placement prediction statistics for each student. We have given a sample registration number to check academic summary report. "Fig. 6" Student academic information shows us the semester grades for semester 7 for a user-input registration number given. Then based on the eligibility criteria and placement conditions we check the placement prediction "Fig. 7" of the student. The last module emphasizes the feedback mechanism where it gives a very personalized, precise and highly predictive feedback "Fig. 8" about each student and areas where they can make improvement.

User: Faculty of Engineering
Chatbot: With the far-reaching approach, research- oriented and relevance-based practices,
SRM Vadapalani is an institution dedicated to building an environment that fosters intellectual freedom.
The Faculty of Engineering at SRM Vadapalani, has various departments in it, which offers a variety of courses.
The list of departments that are offered in SRMIST Vadapalani are listed below.
1. Department of Computer Science Engineering
2. Department of Electronics and Communication Engineering
3. Department of Mechanical Engineering
4. Department of Career Development Centre
5. Department of Chemistry
6. Department of English & Foreign Languages
7. Department of Mathematics
8. Department of Physics
To know more about the department, Please enter the name of the department.
User: computer science
Chatbot:
About the Department
Computer Science is a rapidly evolving discipline today and at SRM, we go to great lengths to ensure that,
our faculty and students can devote themselves to take maximal advantage of modern computer science and engineering -
to solve a wide range of complex scientific, technological and social problems.
Department Website Link - www.srmistvdp.edu.in/department-of-computer-science-and-engineering
Programs Offered by the Department
Under Graduate Programs
B.Tech - Computer Science and Engineering
B.Tech - Computer Science and Engineering with specialization in Artificial Intelligence and Machine Learning
B.Tech - Computer Science & Business System (in collaboration with TCS)
B.Tech - Computer Science and Engineering with specialization in Big Data Analytics
B.Tech - Computer Science and Engineering with specialization in Cyber Security
Post Graduate Programs
M.tech - Computer Science and Engineering
Doctorate Programs
Ph.D in Computer Science and Engineering
User: [   ]

Fig. 3. Faculty of Engineering and Courses offered

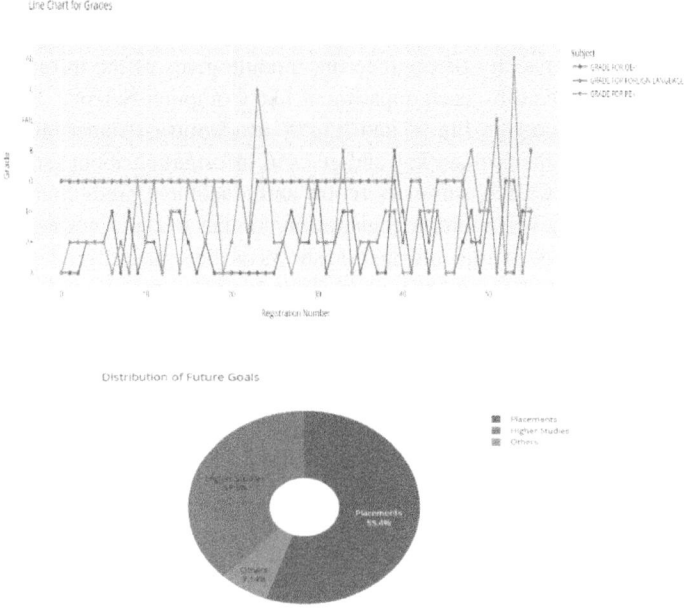

Fig. 4. Grades of students in courses

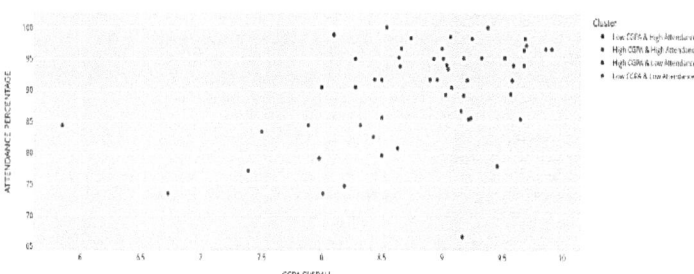

Fig. 5. Academic Summary Report

```
Please enter any option from the main menu
User:academic
This section is specialized for retrieving the student information of CSE-AIML (2020 - 2024)
Please enter the student register number
RA2011026040006
Student Name:  Kaavya Shree K S
Register Number: RA2011026040006
Department: B.Tech CSE
Section: AIML
Year of Passing: 2024.0
Mail ID: kk2735@srmist.edu.in

Give an option number to continue!

1. Semester Marks
2. Overall Attendance and CGPA
3. CSE-AIML Summary Report
4. Placement Prediction
sem
Enter the Semester Number (1 - 7)
7
Student Grade for Semester 7 is:
SEM NO                         7
REGISTER NUMBER     RA2011026040006
STUDENT NAME        Kaavya Shree K S
18CSE359T                     O
18CSE480T                    A+
18MBO412T                    A+
18CSP107L                     O
SGPA SEM 7:                  9.5
Name: 5, dtype: object
```

Fig. 6. Student Academic Information

```
User: academic
This section is specialized for retrieving the student information of CSE-AIML (2020 - 2024)
Please enter the student register number
RA2011026040006
Student Name:  Kaavya Shree K S
Register Number: RA2011026040006
Department: B.Tech CSE
Section: AIML
Year of Passing: 2024.0
Mail ID: kk2735@srmist.edu.in

Give an option number to continue!

1. Semester Marks
2. Overall Attendance and CGPA
3. CSE-AIML Summary Report
4. Placement Prediction
Placement Prediction
This module gives the Student's Placement Eligibility details.

10th Percentage: 94.6
12th Percentage: 76.0
Overall CGPA: 9.25
Kaavya Shree K S is Eligible for Placement
Further more details about Placements!
Kaavya Shree K S (RA2011026040006): Your are Eligible to the below offers in SRM Campus Placement
 - Super Dream Placement: Congratulations! You meet the criteria for Super Dream Placement. Keep up the good work!
   Dream Placement: Congratulations! You are eligible for Dream Placement. Keep up the good work!
 - Non-Engineering, IT, Core Placement: You qualify for Non-Engineering, IT, Core Placement. Keep improving your coding skills!
Want to return back to main menu? (Yes/No)
```

Fig. 7. Placement Statistics

```
User: exit
This is the Last module of Chattitude. This section provides a Personalised Feedback for students!
Please Enter your Student Register NumberRA2011026040006
Chatbot: Feedback for Kaavya Shree K S(RA2011026040006):
Excellent performance in your 10th, 12th grade and Undergraduate degree, in which you have secured Distinction!.
It was really a awesome and a appreciatable result.
I personally congratulate you for your active participation in college clubs named White Hat Hackers
and leading College Club Activities by serving as 'Event head '.Hope You will lead high places.
I also appreciate your preliminary work experience as you have completed 1 internships,
in the domain of Machine Learning
. I hope you will find a suitable firm for your
Placements that best suites your aspirations, in your dream career domain of Data Science.Congrats and Wishing you Best of luck!
Chatbot: Thanks for choosing me! Hoping you would've had a good experience! Untill next time, Goodbye!
```

Fig. 8. Personalized Feedback

7 Conclusion

7.1 Conclusion

The exploration of Chattitude, the fusion of Chatbot technology with educational expertise, presents a promising frontier in reshaping education for the digital age. Through our research, we've illuminated the transformative potential of Chattitude in revolutionizing the learning landscape. By harnessing its capabilities, we've witnessed the power it holds in enhancing personalized learning journeys, offering real-time assistance, and optimizing both academic and administrative support functions within educational institutions. Moreover, Chattitude boasts a user-friendly interface, ensuring accessibility and ease of use for all stakeholders involved. However, this journey has just begun. To ensure Chattitude's seamless integration and global adoption within educational systems, it's imperative to engage stakeholders at every level, fostering interdisciplinary collaboration to refine and expand its capabilities. Continuous iteration based on insights gained from real-world application will be essential in creating a dynamic, inclusive, and productive learning environment that transcends traditional boundaries. Our ultimate goal is to cultivate an educational ecosystem that caters to the diverse needs and backgrounds of learners of all ages, thus unlocking the full potential of Chattitude to go beyond mere augmentation and truly revolutionize the educational experience.

In conclusion, Chattitude represents a groundbreaking fusion of Chatbot technology and educational expertise, offering transformative potential in the digital era. By prioritizing personalized learning experiences, real-time assistance, and streamlined support functions, Chattitude promises to revolutionize education on a global scale. However, realizing this vision requires concerted efforts to engage stakeholders, foster collaboration, and continuously iterate upon our findings. Through interdisciplinary collaboration and a commitment to continuous improvement, we can create a more dynamic, inclusive, and productive learning environment that transcends traditional boundaries and empowers learners of all ages and backgrounds to succeed.

7.2 Future Work

Looking ahead, the future work in Chattitude promises to be both exciting and challenging as we strive to realize its full potential in reshaping the educational landscape. One avenue of exploration involves further refining Chattitude's capabilities to enhance its effectiveness in delivering personalized learning experiences. This entails leveraging advances in artificial intelligence and machine learning to develop more sophisticated algorithms capable of dynamically adapting content and support mechanisms to meet the evolving needs of individual learners. Additionally, there is a pressing need to expand the scope of Chattitude beyond traditional academic subjects to encompass a broader range of learning domains, including vocational skills, emotional intelligence, and lifelong learning competencies. Moreover, as Chattitude becomes more deeply integrated into educational systems worldwide, attention must be given to issues of equity, inclusion, and accessibility to ensure that all learners have equal access to its benefits. Furthermore, ongoing research and development efforts should focus on evaluating the long-term impact of Chattitude on learning outcomes, student engagement, and institutional performance, providing valuable insights for future iterations and refinements. By embracing these challenges and opportunities, we can continue to push the boundaries of educational technology and create a more dynamic, adaptive, and learner-centered educational experience for generations to come.

References

1. Senthil Kumar, M., Dharshani, J., Divyabharathi, K., Sneha, S.: An automated Chatbot for educational institution using natural language processing. Int. J. Creat. Res. Thoughts **10**(5) (2022). ISSN: 2320-2882
2. Diaz-Parra, O., Fuentes-Penna, A., Barrera-Camara, R.A., Treho-Macotela, F.R.: Smart education and future trends. Int. J. Combin. Optim. Probl. Inform. 13th ed., vol. 1. ISSN: 2007–1558 (2021)
3. Alkhoori, A., Kuhail, M.A., Alkhoori, A.: UniBud: a virtual academic adviser (2020)
4. Nwankwo, W.: Interactive advising with bots: improving academic excellence in educational establishments. Am. J. Oper. Manage. Inform. Syst. (2018)
5. Lakshmi, V., Majid, I.: Chatbots in education system (2022)
6. Zhang, Y., Qin, G., Cheng, L., Marimuthu, L., Santhosh Kumar, B.: Interactive smart educational system using AI for students in the higher education platform (2021)
7. Munawar, S., Toor, S.K., Aslam, M., Hamid, M.: Move to smart learning environment: exploratory research of challenges in computer laboratory and design intelligent virtual laboratory for elearning technology. EURASIA J. Math. Sci. Technol. Educ. ISSN:1305-8223 (2018)
8. Hiremath, G., Bhosale, P., Hajare, A., Nanaware, R., Wagh, K.S.: Chatbot for education system. Int. J. Adv. Res. Ideas Innov. Technol. ISSN: 2454-132X **4** (2020)
9. Griol, D., García-Herrero, J., Molina, J.M.: The EducAgent platform: intelligent conversational agents for e-learning applications
10. Ait-Mlouk, A., Jiang, L.: KBot: a knowledge graph-based ChatBot for natural language understanding over linked data (2020)
11. Santos, G.A.: De Andrade, G.G.: A conversation-driven approach for Chatbot management (2022)
12. Patel, N.P., Parikh, D.R., Patel, D.A., Patel, R.R.: AI and web-based human-like interactive university Chatbot (UNIBOT) (2019)
13. JAICOB: A data science Chatbot daniel Carlander-Reuterfelt, Álvaro Carrera (2020)
14. Rivas, P., Holzmayer, K., Hernandez, C., Grippaldi, C.: Excitement and concerns about machine learning-based Chatbots and Talkbots: a survey (2018)

15. Vianden, J.C.B., Barlow, P.J.: Strengthen the bond: relationships between academic advising quality and undergraduate student loyalty. NACADA J. **35**(2), 15–27 (2015)
16. Elliott, K.M., Healy, M.A.: Key factors influencing student satisfaction related to recruitment and retention. J. Mark. High. Educ. **10**(4), 1–11 (2001)
17. Google. Dialogflow, natural language processing platform (2020). https://dialogflow.com
18. Jackson, P.: Introduction to Expert Systems, 3 edn. Addison Wesley, p. 2 (1998), ISBN 978-0-201-87686-4
19. Keston, L., Goodridge, W.: AdviseMe: an intelligent web-based application for academic advising. Int. J. Adv. Comput. Sci. Appl. **6**(8) (2015)
20. Daramola, O., Emebo, O., Afolabi, I., Ayo, C.:Implementation of an intelligent course advisory expert system. Int. J. Adv. Res. Artific. Intell. **3**(5) (2014)
21. Engin, G.C.B., et al.: Rule-based expert systems for supporting university students. Procedia Comput. Science **31**, 22–31 (2014)

PDFQA: A PDF Question Answering System

Harihara Subramanian[1], Shahawar Alim[1], Jhevaan Reddy[1], and B. Prabha[2(✉)]

[1] Department of Computer Science and Engineering, SRM Institute of Science and Technology, Chennai, India
{hm3727,ss0558,j19274}@srmist.edu.in
[2] Department of Computer Science and Engineering (Emerging Technologies), SRM Institute of Science and Technology, Chennai, India
jemi.prabha@gmail.com

Abstract. In this paper, we present PDFQA, a PDF Question Answering System designed to extract information from PDF documents and provide accurate answers to user queries. The system leverages advancements in Natural Language Processing (NLP), utilizing frameworks such as Streamlit, PyPDF2, spaCy, and the Hugging Face Transformers library. PDFQA enables users to upload PDF files, pose questions related to the content of those files, and receive answers based on the extracted text. The system employs a combination of text extraction, semantic similarity analysis, and BERT-based question answering to deliver efficient and accurate responses. Through extensive testing and evaluation, we demonstrate the effectiveness and usability of PDFQA in navigating and extracting insights from PDF documents.

Keywords: PDF Question Answering · Natural Language Processing · Streamlit · Transformers

1 Introduction

PyPDF2, spaCy, Bert PDF documents serve as a ubiquitous medium for storing and sharing information across various domains [1]. However, extracting relevant information from PDFs, especially when dealing with large volumes of text, can be a daunting task. Traditional methods of manual extraction are not only time consuming but also inefficient, underscoring the pressing need for automated solutions. PDFQA addresses this challenge head-on by providing a user-friendly interface tailored for querying PDF documents and retrieving specific information through natural language questions. Revolutionizing the landscape of PDF analysis and information retrieval. In summary, PDFQA represents a significant advancement in the field by bridging the gap between user needs and technological capabilities, ultimately empowering users to unlock the wealth of knowledge embedded within PDF documents effortlessly.

P. D. Sivakumar et al. (Eds.): IRCCTSD 2024, CCIS 2362, pp. 293–304, 2025.
https://doi.org/10.1007/978-3-031-82386-2_23

2 Literature Survey

The paper introduces a novel approach to knowledge grounded conversation models. Existing models often rely on single documents, ignoring the broader context of the conversation. In contrast, the proposed model retrieves multiple relevant documents based on both the conversation topic and local context, enhancing the generation of knowledge-grounded responses. The model incorporates topic words from the conversation and tokens preceding the response, selecting representations for comparison between conversation and document contexts. A new data-weighting scheme encourages knowledge-grounded response generation during training, resulting in more knowledgeable, diverse, and relevant responses. The paper addresses the limitations of current models and demonstrates improvements in response quality through both automatic and human evaluation on a large-scale dataset [2]. The paper also discusses various aspects of knowledge grounded dialog systems, response generation methods, and frameworks for effective conversation. Additionally, it highlights the use of the By harnessing the power of natural language processing (NLP) technologies, PDFQA streamlines the extraction process, enabling users to swiftly navigate through PDF content with ease. Its sophisticated backend architecture, coupled with a user-centric design, ensures seamless access to relevant data. This not only enhances efficiency but also reduces reliance on cumbersome manual methods, thereby RAG-token model for open-domain QA tasks, showcasing its versatility and effectiveness in different contexts. Finally, the paper emphasizes the importance of focusing on knowledge selection mechanisms within conversation models, providing a comprehensive overview of the state of-the-art techniques in this domain.

The paper introduces an innovative approach [3] to information retrieval using Generative AI, specifically targeting faculty guidelines. While e-books offer benefits like searchability and linking to additional resources, concerns about extended reading comfort persist. To address this, the paper proposes a PDF-Driven Chatbot leveraging Large Language Models (LLMs), specifically OpenAI's ChatGPT (GPT-3.5 Turbo), and Pinecone for response generation. This chatbot automates information retrieval from PDF documents, providing coherent responses aligned with the context of the guidelines. The literature review included in the paper delves into automatic question-answering systems and chatbots, emphasizing the relevance of previous studies in this area. By incorporating insights from existing literature and leveraging advanced AI technologies, the research contributes to the development of intelligent chatbot solutions tailored to specific domains, such as faculty guidelines.

The article delves into the development of chatbots [4] customized for academic functions within middle school settings, recognizing the imperative for refined teaching and learning processes. Despite the rising integration of chatbots in commercial and entertainment realms, their utilization in educational contexts remains relatively limited. This study endeavors to address this gap by facilitating seamless communication between students and educational staff. Through the provision of academic guidance, dissemination of course information, and delivery of administrative support, the proposed model seeks to enhance the overall educational experience. Successfully tested with both Mexican middle school students and teachers, the model showcases its effectiveness in bolstering user experience and usability. By offering tailored solutions to meet the specific needs of educational stakeholders, this research marks a significant stride in the advancement of chatbot technology within academic settings. Notably, the findings underscore the potential of chatbots to create more interactive and supportive learning environments, fostering engagement and facilitating smoother communication between students and educators. In essence, this study contributes valuable insights into the evolving landscape of educational technology, paving the way for innovative approaches to teaching and learning in middle schools.

The paper introduces an innovative approach to chatbot development [5], proposing an auto-growing knowledge graph-based system that leverages BERT. This system aims to overcome the limitations of traditional knowledge graphs in understanding common sense and adapting to new vocabulary. By analyzing real-time data, the system learns human common sense, highlighting the significance of computers understanding this aspect of human cognition. Central to this approach is the utilization of a fine-tuned BERT model for relation extraction within auto-growing graphs, resulting in improved performance in learning and understanding human common sense. The research underscores the importance of knowledge graphs in enhancing chatbot capabilities, particularly in grasping nuanced human concepts. By integrating real time data analysis and scalable knowledge graph construction, the proposed system represents a significant advancement in chatbot technology. It offers a more dynamic and adaptable approach to conversational AI, enabling chatbots to effectively navigate complex interactions and provide more contextually relevant responses. Overall, this research contributes valuable insights into the intersection of natural language understanding, knowledge representation, and real-time data analysis, paving the way for more sophisticated and intelligent chatbot systems.

The paper investigates the efficacy of chatbot platforms [6] versus the advanced Sentence BERT (SBERT) model in addressing online student FAQs, within the context of burgeoning online learning environments. As online education burgeons, institutions face heightened demands on support staff to handle inquiries regarding qualifications and registration. Chatbots, particularly those leveraging FAQs, offer a promising solution by providing constant assistance, freeing up human resources for more personalized interactions. The study compares intent classification results from two common chatbot frameworks to SBERT, a state of-the-art model renowned for its robustness in understanding textual semantics. Employing a university FAQ dataset, the paper delineates a methodology to prepare and train each implementation, followed by comparison of F1 scores. Results indicate SBERT's superiority, achieving an F1-score of 0.99, surpassing other platforms such as Google Dialogflow and Microsoft QnA Maker. This underscores SBERT's potential in enhancing the accuracy of chatbot responses to student queries. Moreover, the research underscores the pivotal role of chatbots in bolstering online learning support services, particularly in handling routine inquiries efficiently. By comparing NLU accuracy across various platforms, the paper contributes to understanding the suitability of different approaches within the education domain. Notably, it addresses a gap in existing literature, which predominantly focuses on chatbots in open domains like travel and restaurant booking, thereby shedding light on the unique challenges and opportunities within educational settings.

The introduction of Papr Readr Bot [7] addresses the common challenges researchers face when reading academic papers, presenting a solution aimed at enhancing efficiency and reducing cognitive load. By providing features such as paper summaries, question answering, figure extraction, note-taking, and citation generation, the bot offers comprehensive support throughout the reading process. Notably, it leverages deep learning-based techniques to accomplish these tasks, highlighting the potential of AI in facilitating scholarly activities. Papr Readr Bot stands out as a valuable tool for researchers seeking to streamline their workflow and maximize productivity. Its open-source availability encourages collaboration and further development within the research community. By offering hands-on experiences with deep learning-based skills, the bot serves as a practical demonstration of the capabilities of conversational agents in academic contexts. In the broader landscape of research support tools, Papr Readr Bot fills a crucial niche by addressing specific pain points associated with paper reading. Its multifaceted approach caters to diverse needs, ranging from obtaining a quick overview of a paper to extracting key figures and generating citations. Overall, the bot represents a promising step towards leveraging AI to enhance the research experience and foster knowledge dissemination.

The article delves into the transformative impact of digitalization [8] on personnel management processes, with a particular focus on the role of chatbots in HR operations. It underscores the pivotal role of HR specialists in driving digital transformation initiatives across various sectors of the economy. By leveraging chatbots, organizations can streamline administrative tasks, allowing HR professionals to allocate more time and resources towards strategic endeavors. In the context of higher education and the creative economy, the article examines the automation of personnel management processes and identifies opportunities for the integration of HR chatbots. Through an analysis of existing practices and trends, the authors explore how HR-bots can enhance efficiency and adaptability within HR departments, ultimately contributing to improved organizational performance. The study presents comprehensive insights into the utilization of chatbots in personnel management, emphasizing their potential to optimize administrative workflows and empower HR professionals to focus on strategic initiatives. By synthesizing empirical data and industry trends, the article offers valuable guidance for organizations seeking to harness the benefits of digital technologies in HR operations. Overall, the research contributes to the broader discourse on digital transformation in personnel management and highlights the transformative potential of chatbot technology in modern workplaces.

2.1 Inference from the Survey

The surveyed articles collectively highlight the transformative potential of chatbots across various domains, showcasing their versatility in enhancing processes, optimizing workflows, and fostering innovation in different fields.

In "A Model to Develop Chatbots for Assisting the Teaching and Learning Process," the focus lies on the educational sector, particularly middle schools. The article proposes a model to integrate chatbots into the teaching and learning process, aiming to facilitate communication between students and academic staff. By providing academic guidance, course information, administrative support, and even generating citations, the chatbot enhances efficiency and effectiveness in educational tasks. Moreover, the successful testing of the model with Mexican middle school students and teachers demonstrates its efficacy in improving user experience and usability, ultimately contributing to a more interactive and supportive learning environment.

Conversely, "Digitalization of Personnel Management Processes: Reserves for Using Chatbots" delves into the realm of human resources (HR) management. The article examines the integration of chatbots into personnel management processes, particularly in higher education and the creative economy. It emphasizes the pivotal role of HR specialists in driving digital transformation initiatives and underscores how chatbots can optimize administrative tasks, thereby enabling HR professionals to focus on strategic and creative endeavors. By analyzing existing practices and trends, the article identifies opportunities for HR-bots to enhance efficiency and adaptability within HR departments, ultimately leading to improved organizational performance.

Furthermore, "Auto-Growing Knowledge Graph-Based Intelligent Chatbot Using BERT" explores the utilization of advanced technologies like BERT and knowledge graphs in chatbot development. The paper introduces a novel approach to chatbot development by leveraging auto-growing knowledge graphs and BERT models. By addressing the limitations of traditional knowledge graphs and incorporating real-time data analysis, the proposed chatbot demonstrates improved performance in understanding human common sense and responding to user queries effectively. This research contributes to advancing chatbot technology by integrating real-time data analysis and scalable knowledge graph construction, thus paving the way for more sophisticated and intelligent chatbot systems.

Overall, these articles collectively highlight the transformative impact of chatbots across different domains, ranging from education to HR management. They underscore the significance of chatbots in enhancing processes, optimizing workflows, and ultimately driving innovation in diverse sectors of the economy.

3   Methodology

In this proposed system, we embark on addressing the prevalent challenge of effectively extracting pertinent information from PDF documents. Our endeavor centers on crafting a sophisticated PDF Question Answering System (PDFQA) harnessing cutting-edge AI and NLP methodologies. Commencing with a meticulous requirement analysis, we discern the imperative for a user-centric interface enabling seamless PDF uploads and intuitive natural language queries, coupled with precise information retrieval capabilities.

The system architecture intricately integrates frontend and backend components: designing an intuitive interface employing Streamlit for PDF uploads and query input, while implementing robust text extraction using PyPDF2, sophisticated text processing employing spaCy, and advanced question answering leveraging BERT in the backend. The developmental phase unfolds iteratively, with each constituent meticulously implemented and rigorously tested for accuracy and functionality.

Testing encompasses rigorous unit testing of individual modules, seamless integration testing of the entire system, and meticulous user acceptance testing to ensure intuitive usability and utmost effectiveness. Evaluation pivots on scrutinizing performance metrics and soliciting user feedback, serving as guideposts for system refinement. Upon meticulous evaluation, the system is meticulously deployed, exhaustively documented, and maintained in perpetuity to address any anomalies and integrate enhancements congruent with user feedback and technological evolution.

The culmination of these efforts results in a robust PDFQA system poised to revolutionize information extraction from PDF documents. Its intuitive interface empowers users to effortlessly upload PDFs and pose natural language queries, while its backend employs state-of-the-art NLP techniques to swiftly and accurately extract relevant information. The system's versatility extends across various domains, catering to diverse user needs and facilitating seamless access to critical insights embedded within PDF documents. Moving forward, continuous maintenance and enhancement ensure the system's relevance and effectiveness in an ever-evolving technological landscape, solidifying its position as a cornerstone solution for PDF information retrieval.

4 System Architecture

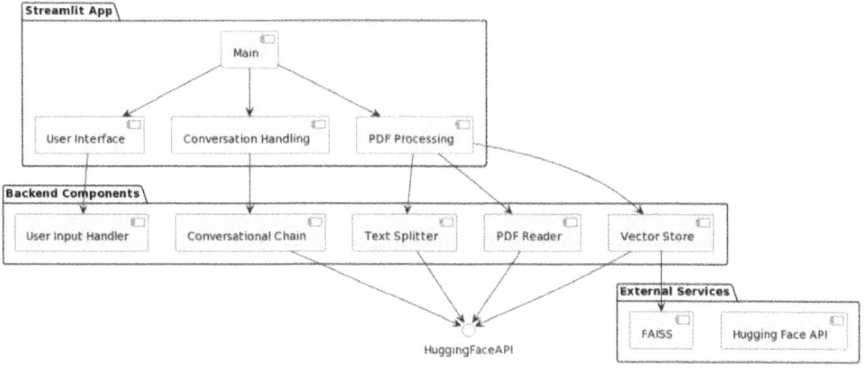

Fig. 1. Framework of the proposed system

Figure 1 Depicts the User Interface (Frontend). The PDF Question Answering System (PDFQA) employs Streamlit to develop an interactive web-based interface, facilitating seamless user interaction. Within this interface, users are empowered to upload PDF files through a dedicated component, streamlining the process of inputting documents for analysis. Additionally, a text input field enables users to articulate natural language queries, leveraging the system's advanced capabilities for information retrieval. Upon submitting queries, PDFQA promptly generates responses and displays them in a readable format, ensuring clarity and accessibility for users seeking insights from their uploaded PDF documents. Through these integrated components, PDFQA offers a user-centric experience, effectively bridging the gap between users and the wealth of information contained within PDF files.

Backend Components. In the PDF Question Answering System (PDFQA), text extraction is carried out using the PyPDF2 library, enabling the system to parse PDF documents uploaded by users and extract text content from each page efficiently. Subsequently, the extracted text undergoes comprehensive processing through spaCy, encompassing tokenization, semantic similarity analysis, and named entity recognition (NER). This sophisticated text processing pipeline enhances the system's understanding of the textual content, enabling it to discern contextual nuances and extract meaningful insights.

Finally, for question answering tasks, PDFQA leverages the BERT-based model from the Hugging Face Transformers library. This powerful model is adept at processing natural language queries in conjunction with the extracted text, generating accurate and contextually relevant answers. By employing these advanced techniques and libraries, PDFQA ensures robust text analysis and accurate response generation, thereby enhancing the overall effectiveness and utility of the system for users seeking insights from PDF documents.

Integration Layer. In the architecture of the PDF Question Answering System (PDFQA), communication middleware plays a crucial role in facilitating smooth interaction between the frontend and backend components. These middleware components are responsible for managing user requests, efficiently passing data to the relevant backend components, and relaying responses back to the frontend interface.

By serving as the bridge between different layers of the system, communication middleware ensures seamless and efficient communication, enhancing the overall user experience. Additionally, a data pipeline is employed to orchestrate the flow of data between the various processing components within the backend. This data pipeline ensures that information flows seamlessly from the text extraction phase through text processing and finally to the question answering component. By orchestrating this flow of data, the data pipeline enables efficient integration and processing of user queries, ultimately enhancing the system's performance and responsiveness.

Deployment. For deployment, the PDF Question Answering System (PDFQA) leverages various strategies to ensure accessibility, scalability, and reliability. Firstly, the system is deployed on a web server, granting users access to its functionalities via the internet. This approach facilitates easy access and usage, allowing users to interact with the system from anywhere with an internet connection. Additionally, cloud infrastructure is utilized to enhance deployment scalability and reliability. Cloud services such as AWS, Azure, or Google Cloud Platform offer flexible and scalable hosting solutions, enabling the system to adapt to varying user loads and ensuring high availability. Furthermore, containerization using Docker containers provides a robust method for packaging the system components. By encapsulating the application and its dependencies within containers, Docker ensures portability and consistency across different deployment environments.

This enables seamless deployment across various platforms and environments, simplifying the management and maintenance of the system. Together, these deployment strategies contribute to the accessibility, scalability, and reliability of PDFQA, ensuring optimal performance and user satisfaction.

Database. In the PDF Question Answering System (PDFQA), the inclusion of an optional database offers additional functionality and versatility. This database can be utilized to store metadata related to uploaded PDF documents and user interactions within the system. By capturing metadata such as file names, upload dates, and user interactions such as queries and responses, the database facilitates future analytics and system optimization. Analysis of this metadata can provide valuable insights into user behavior, usage patterns, and popular topics or documents. Moreover, it enables system administrators to track system performance, identify areas for improvement, and make informed decisions regarding system enhancements or optimizations. Additionally, the database enhances system robustness by serving as a repository for essential data, ensuring data integrity and persistence even in the event of system failures or restarts. Overall, the optional inclusion of a database in PDFQA augments its functionality, enabling enhanced analytics, optimization, and robustness for the benefit of users and system administrators alike. External Services: In the PDF Question Answering System (PDFQA), the incorporation of optional external services offers additional functionalities to enhance user experience and system management. Firstly, an external authentication service can be integrated to streamline user authentication and access control. By leveraging an authentication service, PDFQA can ensure secure access to its features and content, mitigating potential security risks and safeguarding user data. Additionally, the integration of a notification service can further enhance user engagement and system communication. This notification service can be utilized to send alerts or notifications to users regarding important system updates, new features, or relevant events. By proactively communicating with users, PDFQA can keep them informed and engaged, fostering a positive user experience and promoting continued usage of the system. Overall, the optional integration of external services such as authentication and notification services adds valuable functionalities to PDFQA, enhancing communication, and user engagement.

5 Experimental Analysis

Test Documents. Instead of a dataset, we can select a variety of PDF documents representing different types of content, such as technical manuals, research papers, and instructional guides. These documents will serve as the basis for conducting experiments and evaluating the system's performance.

Evaluation Metrics. We'll focus on assessing the system's accuracy in providing relevant answers to user queries, the efficiency of the system in processing queries and retrieving information, and user satisfaction with the overall system experience.

User Study. We'll conduct a user study to gather feedback on the system's usability, interface design, and overall effectiveness in assisting users with information retrieval tasks.

Accuracy Evaluation. We'll formulate a set of test questions covering various topics and complexities, similar to the machine learning setup.

Users will pose these questions to PDFQA, and we'll manually assess the accuracy of the system's responses based on the content of the PDF documents.

Efficiency Measurement. We'll measure the time taken by PDFQA to process each query and retrieve answers. Response times will be recorded for different query types and document sizes to evaluate the system's efficiency.

User Study. Participants will be asked to use PDFQA to search for information within the provided PDF documents. After completing the tasks, participants will provide feedback on their experience, highlighting any usability issues, preferences, and overall satisfaction with the system.

Accuracy. We'll assess the accuracy of PDFQA's answers based on the relevance and correctness of the information retrieved from the PDF documents in response to user queries. We achieve an accuracy of 99.3

Efficiency. We'll analyze the response times recorded during the experiments to evaluate PDF QA's efficiency in processing queries and retrieving information from PDFs.

User Study. Feedback from participants will be analyzed to identify usability issues, interface preferences, and overall user satisfaction with PDF.

The experimental analysis will provide insights into PDFQA's performance, efficiency, and user experience, allowing us to identify strengths and areas for improvement. By leveraging user feedback and objective metrics, we can refine PDF to better meet the needs of users seeking to extract information from PDF documents.

6 Result

PDFQA represents a significant advancement in the realm of document processing and natural language understanding. By harnessing the power of cutting-edge Natural Language Processing (NLP) techniques and leveraging frameworks such as Streamlit, PyPDF2, spaCy, and the BERT Transformers library, PDFQA enables users to seamlessly extract valuable information from PDF documents. Its intuitive interface allows users to upload PDF files and pose queries related to the document content, receiving accurate responses derived from the extracted text [9]. What sets PDFQA apart is its sophisticated approach to question answering, which combines various techniques including text extraction, semantic similarity analysis, and BERT-based question answering. This multi-faceted approach ensures that users receive efficient and precise responses, even for complex inquiries shows in Figs. 2, 3. Moreover, extensive testing and evaluation have demonstrated PDFQA's effectiveness and usability in navigating and extracting insights from PDF documents, confirming its value as a reliable tool for information retrieval. In summary, Fig. 4 shows PDFQA represents a significant technological innovation that addresses the challenges of information extraction from PDF documents. Its integration of advanced NLP techniques and user-friendly interface make it a valuable asset for researchers, professionals, and individuals seeking to efficiently access and utilize information stored in PDF files.

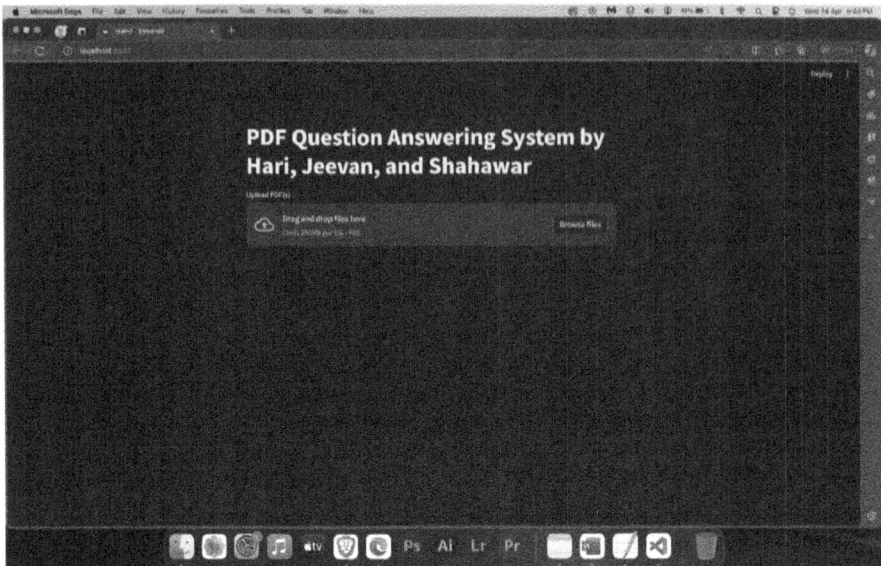

Fig.2. PDF Question Answering system

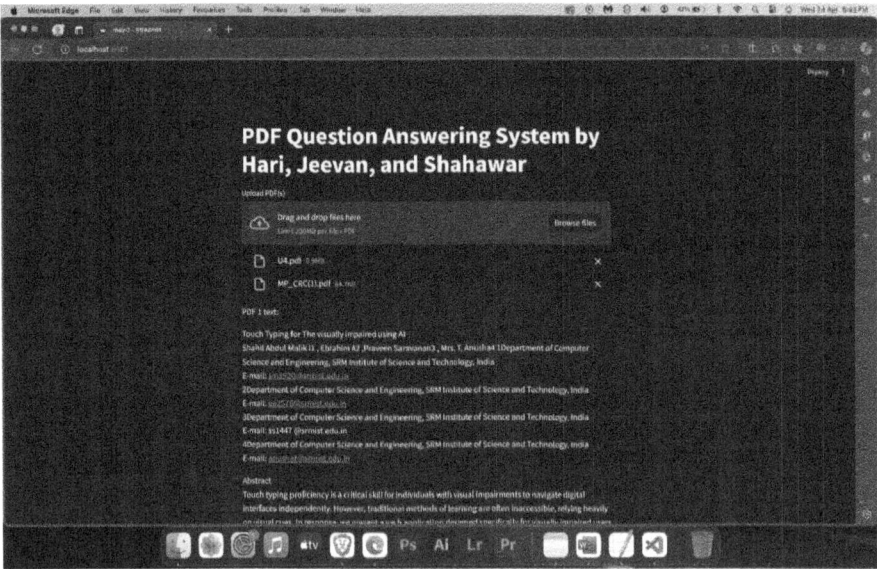

Fig. 3. PDF Question Answering system: generation with answers

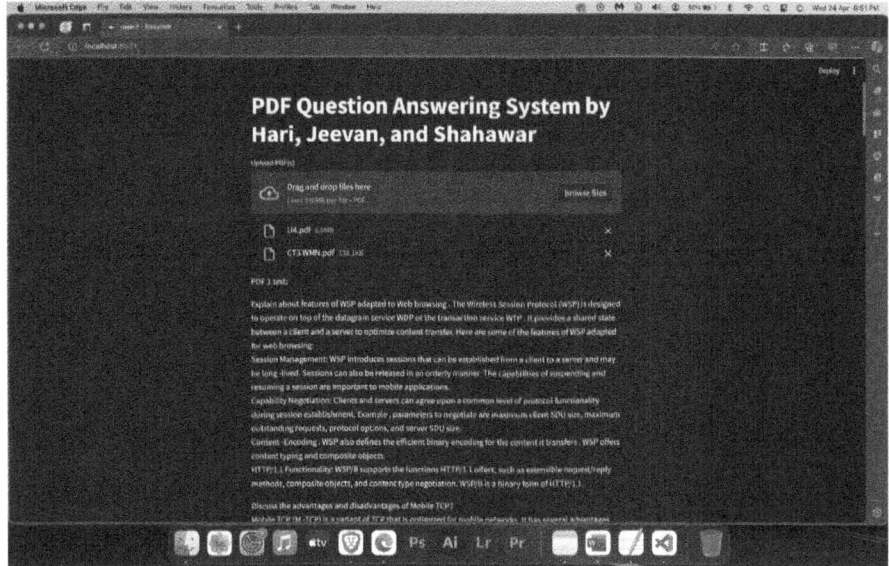

Fig.4. PDF Question Answering system: final with accurate answers

7 Conclusion

In conclusion, PDFQA marks a notable milestone in PDF document analysis and question answering, leveraging cutting-edge NLP techniques alongside an intuitive user interface. This innovative system empowers users to extract valuable insights from PDF documents seamlessly, employing natural language queries for efficient information retrieval. Moving forward, there is ample opportunity for further enhancement and expansion of PDFQA's capabilities. Future work may entail extending support for additional document [10] formats beyond PDF, broadening the system's applicability and versatility. Furthermore, incorporating advanced features such as summarization and document classification could enrich PDFQA's functionality, enabling users to derive even deeper insights and perform more sophisticated analyses. Overall, PDFQA represents a promising advancement in the realm of document analysis and stands poised to evolve further, catering to the diverse needs of users across various domains and applications.

8 Future Enhancement

A future enhancement for the PDF Question Answering System (PDFQA) involves integrating Named Entity Recognition (NER) and Coreference Resolution. NER identifies entities in PDF text, while Coreference Resolution resolves references to the same entity. This retrieval; Portable enhancement enhances context understanding, improves answer quality, and expands use cases. With NER and coreference resolution, PDFQA becomes more adept at extracting information accurately and efficiently, providing users with more precise and contextually relevant answers from PDF documents, thus solidifying its utility across various domains.

References

1. Ahn, Y., Lee, S.-G., Shim, J., Park, J.: Retrieval augmented response generation for knowledge grounded conversation in the wild. IEEE Access **10**, 131374–131385 (2022). https://doi.org/10.1109/ACCESS.2022.3228964. keywords: {Oral 2022, communication; Context modeling; Document retrieval; Knowledge handling; History; Information engineering; Information quality; Conversation; knowledge-grounded conversation; knowledge retrieval}
2. Khadija, M.A., Aziz, A., Nurharjadmo, W.: Automating information retrieval from faculty guidelines: designing a PDF-driven Chatbot powered by OpenAI ChatGPT. In: 2023 International Conference on Computer, Control, Informatics and its Applications (IC3INA), pp. 394–399. Bandung, Indonesia (2023). https://doi.org/10.1109/IC3INA60834.2023.10285808. keywords: {Electronic publishing; Natural languages; Tutorials; Chatbots; Information document format; Transformers; Information Chatbot; OpenAI; ChatGPT}
3. Sonia, M., Sánchez-Adame, L.M., Urquiza-Yllescas, J.F., González-Beltrán, B.A., Decouchant, D.: A model to develop Chatbots for assisting the teaching and learning process. Sensors **22**(15), 5532 (2022). https://doi.org/10.3390/s22155532
4. Yoo, S., Jeong, O.: Auto-growing knowledge graph-based intelligent chatbot using BERT. ICIC Express Lett. **14**(1), 67–73 (2020)
5. Peyton, K., Unnikrishnan, S.: A comparison of chatbot platforms with the state-of-the-art sentence BERT for answering online student FAQs. Results Eng. **17**, 100856 (2023)
6. Foo, M.X.-L., Libera, L.D., Aslan, I.: Papr Readr Bot: a conversational agent to read research papers. In: Proceedings of the 4th Conference on Conversational User Interfaces (CUI 2022). Association for Computing Machinery, New York, NY, USA, Article 39 (2022). https://doi.org/10.1145/3543829.3544536
7. Nataliia, D., Rudakova, S., Shchetinina, L., Poplavska, O.: Digitalization of personnel management processes: reserves for using Chatbots. In: IntSol Workshops, pp. 166–176 (2021)
8. Kolot, A.M., Herasymenko, O.: Tsyfrova O. transformatsiia ta novi biznes-modeli yak determinanty formuvannia ekonomiky nestandartnoi zainiatosti. [Digital transformation and new business models as determinants of formation of the economy of nontypical employment]. Soc. Labour Relations: Theory Pract. **10**(1), 33–54 (2020). https://doi.org/10.21511/slrtp.10(1).2020.06. Ukranian)
9. In Artificial intelligence as a basis for the development of the digital economy: textbook; Edited by I. Tatomyr, Z. Kvasnii. Praha: OKTAN PRINT, 376 p (2021). https://doi.org/10.46489/aiabftd-07
10. Seering, J., Luria, M., Kaufman, G., Hammer, J.: Beyond dyadic interactions: considering chatbots as community members. In: CHI 2019: Proceedings of the 2019 CHI Conference on Human Factors in Computing SystemsMay 2019 Paper No.: 450, pp. 1–13 (2019). https://doi.org/10.1145/3290605.3300680

Exploring Linguistic and Cultural Dynamics in Software Engineering: An Auto-Ethnographic Approach

Raj Ramachandran[1][✉] and Khoa Phung[2]

[1] Cardiff School of Technologies, Cardiff Metropolitan University, Llandaff Campus, Cardiff CF5 2YB, United Kingdom
rramachandran@cardiffmet.ac.uk
[2] School of Computing and Creative Technologies, University of the West of England, Coldharbour Lane, Bristol BS16 1QY, United Kingdom
khoa.phung@uwe.ac.uk

Abstract. Purpose: This study explores the nuanced impact of language use in Software Engineering, focusing on how linguistic and cultural dynamics influence professional practices, identity, and interactions.

Methods: The research employs an auto-ethnographic approach, drawing on the first author's experiences as a native Tamil speaker residing in Wales. Data collection includes personal observations from educational institutions, commercial spaces, healthcare environments, and international contexts. Thematic analysis is used to identify patterns and implications of language choices, such as code-switching and code-mixing.

Results: The findings reveal that language choices significantly affect professional practices and personal identity within Software Engineering. Codeswitching and code-mixing are prevalent, influencing interactions and the design of software systems for multilingual and multicultural audiences. The study underscores the necessity for inclusive language policies that accommodate the multilingual realities of modern societies.

Conclusion: This research contributes to the understanding of language and culture in Software Engineering, highlighting the intertwined nature of linguistic choices with cultural identity and social dynamics. By advocating for the broader acceptance of qualitative methods like auto-ethnography, the study offers valuable insights into managing linguistic diversity in global software development environments and informs policy-making to better leverage this diversity for cultural and technological advancement.

Keywords: Auto-ethnography · Software Engineering · Language · Cultural Identity · Multilingualism · Code-switching · Code-mixing · Inclusive Language Policies

P. D. Sivakumar et al. (Eds.): IRCCTSD 2024, CCIS 2362, pp. 305–318, 2025.
https://doi.org/10.1007/978-3-031-82386-2_24

1 Introduction

Language is an essential medium of communication that extends beyond verbal exchanges to embody cultural identity and societal norms. In the field of Software Engineering, the role of language transcends the boundaries of code, influencing collaborative practices, knowledge dissemination, and professional interactions.

Notably, Software Engineering is a global endeavour involving diverse linguistic and cultural backgrounds, yet the complexities of language usage within this domain are not sufficiently explored. Existing studies focus on optimising communication strategies in software development teams and among stakeholders to improve project outcomes [1, 2]. However, these studies often overlook the nuanced ways in which language impacts social interactions and cultural expressions within technological settings. As a result, there remains a gap in the exploration of how these linguistic and cultural dynamics are experienced on a personal and communal level within the field of software engineering. Most current research focuses on addressing linguistic challenges from a structural or systemic perspective, often overlooking the individual experiences of developers and users who navigate these multilingual and multicultural environments daily.

Our study seeks to address this gap by applying auto-ethnography, a methodological approach rooted in anthropology and sociology. It enables researchers to explore cultural phenomena from an insider's perspective while maintaining the analytical distance necessary for scholarly inquiry [3]. Despite its potential, auto-ethnography remains underutilised in Software Engineering, a field dominated by empirical and quantitative research paradigms.

This research leverages auto-ethnography to provide a rich, contextual exploration of language practices, thus offering insights into the interplay between language, culture, and technology. By documenting the first author's personal experiences with language use in various professional and everyday contexts, this research provides a unique insider's perspective on the cultural and linguistic dynamics at play. This method allows for a rich, narrative-based exploration of how language affects professional practices, identity, and interactions within software engineering.

The increasing globalisation of technology development and deployment accentuates the need for a comprehensive understanding of how linguistic diversity influences software engineering processes. The first author's dual identity as a native Tamil speaker and a professional residing in Wales provides a unique lens to examine these dynamics. This perspective is particularly relevant given the rise of remote work and global software teams, where language choices can significantly affect team dynamics, user interface design, and client interactions.

This study aims to address the outlined gap by achieving several key objectives:

1. Documenting Linguistic Encounters: By detailing the first author's daily encounters with language use in various natural and technological settings, this study will illuminate how personal and societal language choices influence individual choices of language in technology as well as professional practices in software engineering.

2. Analysing Language Choices: The study will explore the implications of linguistic decisions, such as code-switching and code-mixing, primarily on the identity of speakers and secondarily on software professionals and the broader professional practices within software engineering.
3. Implications for Software Design: It will also consider how these linguistic interactions shape the design and functionality of software systems (e.g., speech-to-text applications), particularly those intended for multilingual and multicultural audiences.

The expected contributions of this research are multi-fold. Firstly, it will enrich the body of knowledge in software engineering by integrating socio-linguistic perspectives into the understanding of professional and technical practices. Secondly, it will provide empirical insights into the challenges and opportunities of managing linguistic diversity in software projects, potentially guiding the development of tools and practices that enhance inclusivity and effectiveness in global software development environments. Finally, by applying auto-ethnography, this research will pioneer a methodological expansion in software engineering research, advocating for a broader acceptance of qualitative research methods in the field.

The paper is structured as follows. This paper is structured as follows: Sect. 2 reviews existing literature on language use in Software Engineering, highlighting the gap in current research and the potential of auto-ethnography. Section 3 details the methodology, describing the auto-ethnographic approach, data collection methods, and analytical framework. Section 4 presents findings from a thematic analysis of the first author's linguistic encounters, discussing implications for professional practices, identity, and software design. Section 5 outlines the broader implications, contributions, and limitations of the study. Section 6 concludes by summarising key insights and advocating for inclusive language policies and qualitative methods in Software Engineering.

2   Related Work

The exploration of language in Software Engineering has traditionally centred on enhancing communication within project teams and between developers and stakeholders. The literature presents various studies that investigate the mechanisms through which communication can be optimised to improve understanding and collaboration across diverse software teams. For instance, Damian and Zowghi [4] highlighted the critical role of effective communication in requirements engineering, a foundational aspect of software development that relies heavily on clear and concise language to avoid costly misunderstandings and errors. Similarly, Shachaf [5] discussed the impact of cultural differences on communication efficacy in global virtual teams, underscoring the challenges posed by linguistic diversity. Furthermore, recentliterature has increasingly acknowledged the complexity of communication in global software development environments, where language plays an important role. Trinh [6] demonstrated the influence of language barriers on productivity and collaboration in GitHub projects, highlighting the challenges and adaptations developers face in multilingual environments.

Despite these insights, the focus has predominantly been on functional aspects of communication - such as clarity, accuracy, and efficiency - with less attention given to the socio-cultural implications of language use within software engineering contexts. This gap is particularly evident when considering the auto-ethnographic approach to studying language, which offers a deeper, more introspective look at the cultural and personal dynamics at play. Marinho et al. [7] explored communication challenges in distributed software teams, emphasising the need for effective strategies to manage linguistic diversity and cultural differences to mitigate risks associated with miscommunication. This study underscored the operational challenges and highlighted the broader cultural dynamics at play, setting the stage for a more in-depth exploration through auto-ethnography.

Auto-ethnography provides a nuanced lens for examining these interactions by emphasising personal narrative and cultural context, an approach that is still novel in software engineering research. A significant contribution in this regard is made by Pink et al. [8]'s discussion on digital ethnography, which explored how digital and online environments can be studied to understand cultural practices in technology use. This framework is particularly relevant for analysing how software engineers navigate their multilingual and multicultural environments, offering insights into personal and communal language practices.

Furthermore, the adaptation of user interfaces and documentation to accommodate multiple languages is a well-recognised necessity in software engineering, as discussed in [9]. This work on localisation practices in software development provides practical insights into the complexities of developing technology for global markets. This is especially relevant for understanding the cultural nuances that can influence software usability and accessibility. Recent research by Pitale and Bhumgara [10] on human-computer interaction and the role of natural language in user interface design underscores the importance of linguistic considerations in software design and usability. The principles for designing user interfaces that accommodate diverse linguistic groups provide a foundation for understanding how software engineers must consider language beyond mere functionality.

This study distinguishes itself by focusing not only on the challenges posed by linguistic diversity but also on the opportunities it presents for fostering a deeper understanding of cultural inclusivity in software design and implementation. By integrating auto-ethnographic insights, our research contributes to a more nuanced understanding of the socio-cultural aspects of language use in software engineering. It highlights the personal and societal impacts of language choices, often overlooked in more traditional research focused solely on linguistic efficiency or software localisation strategies.

3 Methodology

This research employs an auto-ethnographic approach to explore the nuanced interactions between language usage in natural and technological settings, particularly focusing on the experiences of the first author in diverse linguistic environments. As discussed in Sect. 1, auto-ethnography allows for a deep, introspective examination of cultural and personal experiences, providing valuable insights into the socio-linguistic dynamics within software engineering. This methodology is particularly pertinent given the first author's unique positioning as both an insider and an outsider within various linguistic and cultural contexts.

3.1 Research Positioning

The first author possesses a dual identity concerning the research context. As a native speaker of Tamil and a professional residing in Wales, the author embodies an insider's perspective in terms of language, culture, and context. However, due to prolonged absences from the native context and current residence in a non-native region, the author also holds an outsider's perspective. This dual positioning enhances the depth of the analysis, as it allows for a deep exploration of language practices from both an embedded and a distanced viewpoint. The second author, who is neither a speaker of Tamil nor intimately familiar with the cultural contexts being studied, contributes an entirely external perspective, further enriching the analytical framework.

3.2 Research Tool and Data Collection

Central to our methodology is using the first author's personal and professional website1, which serves as a primary data source. The website is uniquely multilingual, featuring content in Tamil, the first language of the author, and Cymraeg, the official language of Wales. This design choice intentionally prioritises the identity and authenticity of language use over broader accessibility, thus aligning with the study's focus on native language usage in digital contexts.

Data collection is conducted through several mechanisms:

1. Observational Data: The website's blog section is crucial for observing dynamic language practices such as code-switching, code-mixing, and language choice in written communication. These observations offer insights into how language preferences and usage adapt depending on the audience and context.
2. Log of Events and Observations: A password-protected page on the website serves as a diary for the first author to record personal experiences, linguistic encounters, and reflections. This log provides a rich, qualitative dataset of the author's daily linguistic navigation and interactions in various settings - ranging from professional environments in Wales to personal visits in Tamil-speaking regions.
3. Contextual Language Usage: The first author's language choice is strategically adapted based on the geographic and social setting, providing a live case study of linguistic adaptation. For example, while Tamil is used in Chennai and Jaffna,

Kannada is the chosen language in Bengaluru, despite Tamil being widely understood there. This selective language usage is critical for understanding regional language dynamics and personal language policies [11].

3.3 Analytical Approach

The research employs a native approach to language, wherein the usage of a language in its native region is taken as a given, without the need for justification. This stance is particularly significant in regions where the author's first language (Tamil) aligns with the official language, as it allows for an examination of natural language use without the pressures of linguistic conformity typically seen in non-native environments.

The data gathered through the website and personal logs are analysed using thematic analysis, focusing on patterns of language use, the implications of linguistic choices, and the cultural significance of these choices. This analysis is guided by the theoretical frameworks of socio-linguistics and cultural anthropology, providing a lens through which to interpret the intersection of language, technology, and identity.

3.4 Ethical Considerations

The study is conducted with a high degree of ethical awareness, particularly concerning the personal and potentially sensitive nature of auto-ethnographic data. All personal observations and experiences are anonymised where necessary, and care is taken to ensure that any third-party information is used following ethical research standards and privacy considerations.

This methodology, combining detailed observational data with reflective personal narratives, allows for a comprehensive exploration of the socio-cultural dynamics of language use within and across different settings. It offers valuable insights into how languages are not just tools for communication but are also integral to personal identity and cultural belonging in the modern world.

However, the following are the risks of adopting auto-ethnography as a method:

1. There is a risk of others misperceiving the researcher in some contexts through some of the actions of the researcher. For example, using only Tamil or Welsh on social media, websites, at-home settings or in (personal) e-mail signatures.
2. The identity of the researcher as a participant is revealed and exposed and the effects of risk identified in the previous point could lead to direct or indirect discrimination in some contexts.
3. Rich and authentic data with the scope of certain elements of bias.
4. In the context of the website and social media, Google Translate may not quite accurately translate Tamil or Welsh content into English.
5. Perception of the researcher (given researcher's multilingual ability) arising due to rigidity of use of language in some contexts – for example, use of only Tamil or Welsh on the website, social media, home and native settings. This is despite the researcher's familiarity with other languages.

Ethical approval has been obtained by the ethics committee *Ref: CSTEthicsApp StaffRRamachandran 20230918 09.*

4 Results and Discussions

The auto-ethnographic approach adopted in this study has yielded a rich dataset of personal observations and experiences that illuminate the complexities of language use in various settings. These observations span from academic environments to commercial spaces and international travel, providing a broad perspective on the socio-linguistic dynamics encountered by the first author. This section synthesises these findings, discussing the implications of language choices and the cultural and linguistic interactions that define the settings studied.

4.1 Observations from Educational Settings

Observations within the educational sector in Chennai have revealed a tacit acceptance and flexibility in language use, particularly in the incorporation of Tamil within an English-medium university. This phenomenon reflects broader themes in software engineering, where the integration of multiple languages can facilitate more inclusive and effective communication within diverse development teams. Studies have shown that language flexibility in educational settings can lead to similar flexibility in professional environments, promoting better collaboration and understanding within globally distributed teams [1]. The shift in language policy by a lecturer (after interacting with the first author), influenced by advocacy for native language use, illustrates the significant role that individuals can play in organisational language policies, a concept that can be extended to software development environments where team leaders and managers can influence linguistic inclusivity [12].

4.2 Commercial and Public Spaces

The linguistic landscape in commercial settings such as hotels and bookstores in Chennai revealed a tension between the official language policy and on-ground practices. For instance, at a five-star hotel, the staff's inability to communicate in Tamil and the predominant use of English even among Tamil-speaking customers point to an unconscious imposition of English in native spaces. This scenario not only reflects the complexities of language use in commercial settings but also raises questions about the erosion of linguistic identity in the face of global languages.

In contrast, the interaction at a bookstore where the manager endeavoured to stock Tamil books despite corporate policies highlights the role individuals can play in promoting linguistic diversity. Such instances are vital for understanding the microdynamics of language maintenance and promotion within commercial enterprises. The Tamil language is represented in the Tamil script. However, it is not uncommon to use Romanised Tamil informally although it is usually not identified as "Tamil" but the use of Romanised Tamil makes it easier to code-switch. Figure 1 shows the use of Romanised Tamil in society as well as code-switching in Tamil.

4.3 Health Care Settings

Observations in a hospital in Chennai demonstrated prevalent bilingualism, with Tamil and English being used interchangeably. However, the presence of Hindi-speaking individuals who did not know Tamil introduced challenges in communication, illustrating the practical implications of India's linguistic diversity in critical settings like healthcare. Figure 2 demonstrates the potential institutional policy of prioritising English over Tamil.

4.4 Personal Encounters and Observations

Personal logs revealed varied reactions to the use of Tamil in different contexts. Notably, the response to the first author's insistence on using Tamil at a street vendor in Chennai

Fig. 1. This is one example which demonstrates the prevalence of Romanised Tamil as well as code-mixing and code-switching forms of Tamil in society.

was met with mixed reactions, reflecting both resistance and acceptance. This interaction, along with others, highlights the nuanced negotiations and identity affirmations that occur

Fig. 2. This is an example from a hospital setting where English is given priority as it appears above the native and official language of the region. Further, in the Tamil version of the above, words such as lift, operator and security are used as they are in English rather than using their Tamil equivalent.

through language choices in everyday life. It was also observed that writing personal logs in Tamil involved a slight change in the thought process. The first author in the process used Tamil words and terms which is less widely used in speech. The writing genre in Tamil switched greatly between colloquial conversation Tamil and formal written Tamil. However, no English words were used in the journal. This indicates a possible link between speech that is widely prevalent in society and the thought process as a result of society's usage of language and writing. However, in personal space, it is possible to reverse by consciously attempting to write in Tamil without the effect of code-mixing and code-switching and by choosing a genre that an individual is comfortable with. Therefore, the insistence to communicate in Tamil and correspond in Tamil in Chennai is not being viewed as a favour but more as a legitimate right of the first author. Figure 3 brings to the surface another key area which is banking where language plays a key role and could potentially alter the view about a particular language. Figure 4 shows how an English-only menu has found space in a non-native region. What is concerning is not the presence of English but the absence of Tamil in its native region.

4.5   International and Inter-State Observations

The differences in language policy and practice were also evident in international contexts and in other Indian states. For instance, during flights from Chennai to Colombo and London to Bengaluru, the language used in announcements varied, reflecting broader governmental and corporate language policies that may not always align with the linguistic demographics of the passengers.

The comparison of language learning and use by migrants in Chennai and Bengaluru offers insights into the social dynamics that influence language assimilation and maintenance. These observations suggest that societal expectations and individual attitudes towards language learning significantly impact the linguistic integration of migrants.

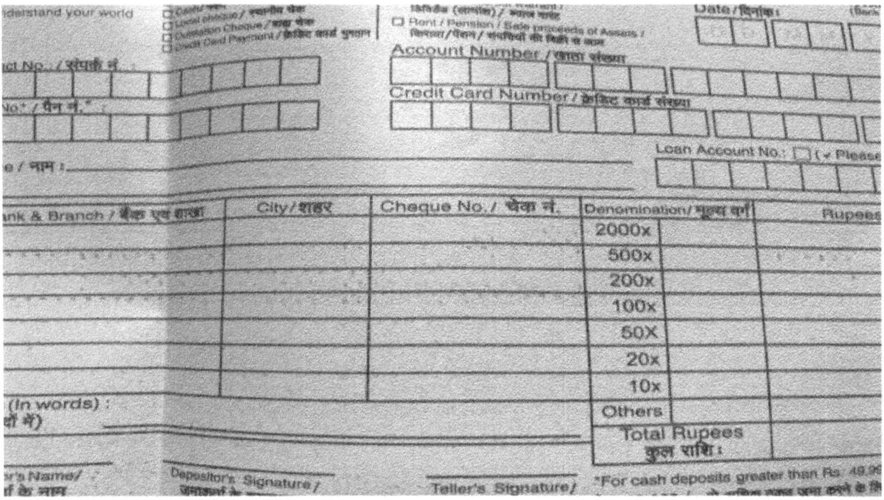

Fig. 3. The image points to the absence of Tamil language in key areas such as banking that is required for citizens. Prioritising languages such as English and Hindi in a region where Tamil is natively spoken and widely understood could potentially undermine the functional value of the Tamil language in everyday usage.

Fig. 4. A famous foreign outlet that has an English-only menu instead of a Tamil menu in its outlet in Chennai. Another observation from the author is that many local outlets have their menus exclusively in English, potentially alienating native Tamil speakers.

4.6   Conferences and Professional Settings

At a bilingual international technological conference in Tamil Nadu, giving participants the choice of language for their presentations highlighted the practical application of inclusive language policies in professional settings. Such policies not only accommodate linguistic diversity but also empower participants by allowing them to express complex ideas in their language of comfort. Despite this, it was observed that a majority of the presentations and conversations took place in English. The responses were in code-mixed Tamil only when the first author asked questions in Tamil. However, the nature of the bilingual conference provided recognition and legitimacy to the use of the Tamil language by everyone without which it would have risked being perceived as "unprofessional" [13]. On the other hand, it would effectively counter the notion of the Tamil language as unfit for academic use [13]. Despite the first author's fluency in English, the author chose to stick to Tamil to set expectations and in an individual effort to change the perception, restore its legitimacy in the field of Higher Education in the native space. The challenge and problem of participants and delegates not knowing Tamil is not a good enough reason for the event, presentation or question answers taking place in Tamil in Chennai where Tamil is both native and official.

4.7   Observation from the Website

In non-native regions, it was observed that the first author's website being in Tamil did not discourage from users engaging with the content. As expected, the users used Google Translate to engage with the content. However, in the native region, it was increasingly evident that the users were able to engage with the content at ease since it was in a language that the users could read, write and speak, the comments in the blog section in English reveal the choice of language when it comes to technology. Interaction with the users revealed that ease of use, and unfamiliarity with Tamil keyboards as some of the reasons for using English, the first author is inclined to disagree with the reasons as someone who has been using the Tamil language in a variety of technological settings including iOS and Android. It is compelling to attribute comments in English as a negative attitude towards the Tamil language but at the same time, it is interesting to observe that despite everything on the website being in Tamil, and despite the users' ability to read, write and speak Tamil the choice of language to comment was in English. It was also observed that a hierarchy and a compulsion to engage with content in an unfamiliar language in a non-native region had a positive outcome than allowing users to exercise their choice of language in a familiar language in the native region. It was observed that English was the default language choice of response for almost all the people who interacted with the first author via technology such as WhatsApp, and e-mail.

5   Implications, Contributions, and Limitations

The observations discussed in Sect. 4 collectively demonstrate the profound impact of language on cultural identity, social integration, and professional practices in divers-esettings. They underscore the necessity of thoughtful language policies that recognise the multilingual realities of modern societies and the importance of accommodating linguistic diversity in all spheres of public and private life.

The study contributes to the broader discourse on language and culture in software engineering by illustrating how linguistic choices are not merely practical decisions but are deeply intertwined with cultural identity and social dynamics. By highlighting these dynamics through an auto-ethnographic lens, this research provides a nuanced understanding of the challenges and opportunities that linguistic diversity presents in global and local contexts.

This discussion not only enriches the academic discourse on language use in technology and education but also provides practical insights that can inform policy-making and institutional practices to better accommodate and leverage linguistic diversity for cultural and technological advancement.

Although this has provided a deeper insight into prevailing social conditions and the use of language in various settings through rich and authentic data, it cannot be generalised at this stage as prolonged data collection of extended periods is required along with quantitative data. Further work and extension would include quantitative data, data collection in different contexts, the author's use of language over some time and self-reflection. Collaborative auto-ethnography from a similar context may be useful to evaluate whether or not experiences can be identical in a similar context with similar actions.

The first author's positionality is also a limitation as this can only be replicated in certain languages and certain settings. This however highlights the importance and role of language within the domain of Software development and the need for developers and team members to be familiar with the language, and cultural contexts in which the software development is taking place.

6 Conclusion

This study has utilised an auto-ethnographic approach to explore the nuanced dynamics of language use within both natural and technological settings from the personal perspective of the first author. By examining linguistic interactions across a variety of contexts - including academic institutions, commercial spaces, healthcare facilities, and international travel the research has highlighted how language choices influence and reflect broader socio-cultural and professional dynamics. These findings underscore the significance of language in Software Engineering, revealing how linguistic practices impact collaboration, identity, and the design of multilingual and multicultural software systems. The study has also emphasised the role of commercial establishments, educational institutions, and other service industries in promoting native language use, thereby enhancing the visibility and natural integration of these languages in technological applications.

The implications of this research are twofold. Firstly, it advocates for the development and implementation of inclusive language policies within Software Engineering that recognise and accommodate linguistic diversity. Such policies can enhance team dynamics, improve user experience, and ensure that software systems are accessible to a broader audience. Secondly, the study suggests that qualitative methods, particularly auto-ethnography, offer valuable insights into the socio-cultural dimensions of language use that are often overlooked in traditional quantitative research.

Future research should expand on this study by incorporating a larger, more diverse sample of participants to examine whether the observed linguistic dynamics are consistent across different cultural and linguistic contexts. Additionally, longitudinal studies could provide deeper insights into how linguistic practices evolve within the field of Software Engineering. Further research could also explore the development of specific tools and frameworks to support multilingual collaboration and communication in software development teams, ensuring that linguistic inclusivity becomes an integral part of technological innovation. By continuing to investigate the interplay between language, culture, and technology, future studies can contribute to more inclusive and effective practices in Software Engineering.

Declarations

- Funding: Not Applicable.
- Conflict of interest/Competing interests (check journal-specific guidelines for which
- heading to use): Not Applicable.
- Ethics approval and consent to participate: Ethical Approval Ref: CSTEthicsApp
- StaffRRamachandran 20230918 09.
- Consent for publication: Yes.
- Data availability: No.
- Materials availability: Yes.
- Code availability: No.
- Author contribution: Raj Ramachandran (first author and main researcher), Khoa Phung (second author).

References

1. Bjørnson, F.O., Dingsøyr, T.: Knowledge management in software engineering: A systematic review of studied concepts, findings and research methods used. Inf. Softw. Technol. **50**(11), 1055–1068 (2008)
2. Sharp, H., Robinson, H.: Collaboration and co-ordination in mature extreme programming teams. Int. J. Hum Comput Stud. **66**(7), 506–518 (2008)
3. Ellis, C., Adams, T.E., Bochner, A.P.: Autoethnography: an overview. Historical social research/Historische sozialforschung, pp. 273–290 (2011)
4. Damian, D.E., Zowghi, D.: The impact of stakeholders' geographical distribution on managing requirements in a multi-site organization. In: Proceedings IEEE Joint International Conference on Requirements Engineering, pp. 319–328. IEEE (2002)
5. Shachaf, P.: Cultural diversity and information and communication technol- ogy impacts on global virtual teams: an exploratory study. Inform. Manage. **45**(2), 131–142 (2008)
6. Trinh, H.N.: Communication in github projects: a systematic mapping study (2022)
7. Marinho, M., Luna, A., Beecham, S.: Global software development: practices for cultural differences. In: Product-Focused Software Process Improvement: 19th International Conference, PROFES 2018, Wolfsburg, Germany, November 28–30, 2018, Proceedings 19, pp. 299–317 (2018). Springer
8. Pink, S., Horst, H., Lewis, T., Hjorth, L., Postill, J.: Digital ethnography: principles and practice (2015)

9. Marcus, A., Gould, E.W.: Globalization, localization, and cross-cultural user- interface design. CRC Press, Boca Raton, FL (2012)
10. Pitale, A., Bhumgara, A.: Human computer interaction strategies—designing the user interface. In: 2019 International Conference on Smart Systems and Inventive Technology (ICSSIT), pp. 752–758. IEEE (2019)
11. Bes, L.: The heirs of vijayanagara: court politics in early modern South India. Leiden University Press (2022)
12. Johnson, D.C., Johnson, D.C.: What is language policy? Springer (2013)
13. Annamalai, E.: Medium of power: the question of english in education in india. Medium of instruction policies: Which agenda? Whose agenda, pp. 177–94 (2004)

Stress Detection Based on EEG Values: A Systematic Literature Review

Yeddula Yashaswini[✉], T. G. Sinchana, B. Nikitha, Mukka Prahitha, and N. V. Uma Reddy

Department of AIML, New Horizon College of Engineering, Bangalore, India
yeddulayashaswini@gmail.com

Abstract. Stress known to be the underlying cause of several mental health disorders, arises from a variety of sources that hurt human health. The consequences of stress are most noticeable in the life of a working professional who must handle the demands of increased management expectations, time management limitations, and family obligations. Physiological traits are crucial for diagnosis of stress-related illnesses and offer valuable insights into the intricate connection between mental and physical well-being. The main motive of this research is to investigate stress using EEG data, which is recognized for its reliability, accuracy, and precision, and to generate a complete system that uses these machine learning models to forecast stress levels that may be classified as Positive, Neutral, or Negative. Sophisticated models for machine learning, like SVM (support vector Machine), KNN, DT (Decision Tree), and Random Forest, are implemented according to intrinsic compatibility between stress signals and EEG data. The paper aims to find best solution for an efficient stress analyzer by merging several methods.

Keywords: Stress Detection · EEG · Machine Learning

1 Introduction

A mental condition that puts strain on the body or mind is called stress [1, 2]. Air traffic controllers and pilots are among numerous specialists in the aviation sector who experience significant levels of stress at work [3]. An overwhelming amount of stress might eventually cause someone to focus too much, which lowers their awareness of other factors. This situation may increase the danger of human mistake [4–6]. Subsequently, this component is connected to the rise of several illnesses, including diabetes, obesity, asthma, and cardiovascular disorders. Stress triggers physiological reactions in certain bio signals, such as skin conductance, heart rate, or body temperature that result in distinctive patterns [7]. Stress may be identified using ML and processing of signals approaches that look for unique patterns in these bio signals. Stress puts a person's ability to adjust to some fresh situations to the test. It also causes physical and psychological changes that raise the possibility of illness. Heart failure, stroke, high BP, artery plaque heart attack, and psychological disorders including depression and anxiety are a few examples of these ailments [8]. Two criteria can be used to categorize the

© The Author(s), under exclusive license to Springer Nature Switzerland AG 2025
P. D. Sivakumar et al. (Eds.): IRCCTSD 2024, CCIS 2362, pp. 319–331, 2025.
https://doi.org/10.1007/978-3-031-82386-2_25

stress situation: acute stress, or short-term stress, and chronic stress, or long-term stress [9, 10]. Several methods to determine stress levels, including looking at the person's photos, utilizing ML or DL algorithms to analyze social media tweets, and asking the person to complete a questionnaire [11, 12]. To anticipate stress, the best way is by using brain wave data signals (EEG). EEG (electroencephalography) is a technique for measuring the electrical activity within the brain [13, 14]. This AI is a tool for computationally empirically determining the key elements from the electroencephalogram (EEG) data [15]. By detecting aberrant activity or distinguishing between various brain states, this activity may be utilized to infer information about how the brain functions [16, 17]. Identifying stress in human bring using brain signals has shown as among the highly effective techniques while doing this [18]. This brain wave or signal-based system can help identify the different disorders and limitations with the help of the EEG signal-based system. Human mental stress and emotion can be identified using sentiment analysis [19]. Therefore, an accurate, well-built, and dependable system is required [20].

2 Discussion

Prior to developing our approach, we studied previous studies which were conducted considering all of the characteristics that we had chosen. One of the studies conducted a thorough literature review on machine learning and wearable sensors for mental stress detection Before formulating our strategy, we looked over earlier research which we considered, each of the qualities we had selected. One of the researches reviewed the literature in-depth on wearable sensors and ML for the detection of mental stress [21]. Initially retrieving 9334 documents through keyword searches, filtering based on publication criteria yielded a final selection of 55 papers. These papers were carefully evaluated, focusing on methodology, sensor usage, machine learning techniques, and stress detection accuracy. $The significance of physiological indicators such as HR (heart rate) and galvanic skin response (GSR) was emphasized, and the accuracy and user-friendliness of wearable technologies were emphasized. Multiple algorithms for machine learning were used, such as logistic regression, random forest (RF), k-nearest neighbors (KNN), and support vector machines (SVM). K-fold cross-validation and leave-one-subject-out techniques are normally used for validation. This study stressed the need for both affordability and effectiveness in stress detection algorithms and offered insights into sensor types and ML strategies. Its goal was to provide future researchers with essential guidance for selecting suitable detectors and ML approaches for mental stress detection.

Few of the works develop a stress-training system using virtual reality (VR) technology, combining three immersive settings to induce stress [22]. Utilizing wearable technology and a personalized perceived stress questionnaire, physiological data (heart rate, electro-dermal activity, eye-blink rate) and self-reported stress ratings are gathered. to put automated stress detection and supervised learning models into practice, showcasing their efficacy in forecasting continuous stress scores and categorizing stress levels. Positive evaluations are given to the VR system's feeling of reality, immersion, spaciousness, and presence. In high-stress situations, the physiological reactions are more intense. The study provides valuable insights into the correlation among objective physiological data and subjective self-ratings in the complex VR stress training system. The small sample size and potential differences in participant stress reactions are two possible limitations of the present research. Additionally, An article provides a performance-based comparison of several supervised learning algorithm types.

$$Accuracy = \frac{TP + TN}{TP + TN + FP + FN} \tag{1}$$

True Positive (TP) denotes Positive samples are accurately anticipated to be positive, False Positive (FN) symbolizes positive samples are incorrectly expected to be negative, False Positive projected to be positive and True Negative (TN) denotes Negative samples are accurately predicted to remain negative. Accuracy is predicted by dividing (FP) represents Negative samples are incorrectly the total number of right predictions by number of predictions.

$$Recall = \frac{TP}{TP + FN} \tag{2}$$

Recall is predicted by dividing the TP (True Positive) by total number of positive predictions.

$$Precision = \frac{TP}{TP + F} \tag{3}$$

Precision is calculated by dividing the TP (True Positive) by the summation of TP (True Positive) and FP (False Positive).

$$F1 = Score = \frac{2 \times Precision \times Recall}{Precision + Recall} \tag{4}$$

F1-score is the hormonic mean of Precision and recall. It is used as an evaluation matrix.

Among the researchers analyzed signal processing methods such Fourier transform, cube root transformation, and Constant-Q Transform (CQT) to preprocess data before developing a CNN-based on neural network architecture for self-sufficient BioSignal-based stress detection [23]. LOSO cross-validation and the WESAD dataset were used to complete three categorization tasks, showing considerable accuracy gains over earlier studies: stress versus non-stress (93.1% to 96.6%), differentiating between baseline, stress, and amusement (80.3% to 85.1%), and classifying five states (77.1% to 82.1%) [23]. The results were much improved by the cube root adjustment, which highlighted low-energy frequencies. While baseline, entertainment, and recuperation could be distinguished with the lowest accuracy, stress detection showed the best accuracy. Future research may examine physiological stress monitoring systems that are less invasive.

In the research that takes the comprehensive approach to stress detection, covering both hardware and software stages. The WESAD dataset and data gathered with a particular experimental sensor were accustomed to create and test the algorithms [24]. Four distinct types of signals were employed: respiration, temperature, ECG, and EDA. The results on WESAD showed 96.0% accuracy, 97.2% sensitivity, and 93.9% specificity, in that order. Accuracy, sensitivity, and specificity with the experimental sensor were 94.4%, 96.8%, and 92.3% while focused on ECG and EDA during tasks like mental arithmetic and SCWT. KNN with ECG and EDA demonstrated to be the best effective combination for stress detection after a series of algorithm and signal testing. With competitive accuracy at a cheaper cost, the suggested sensor gathers data using either onboard or off board electrodes.

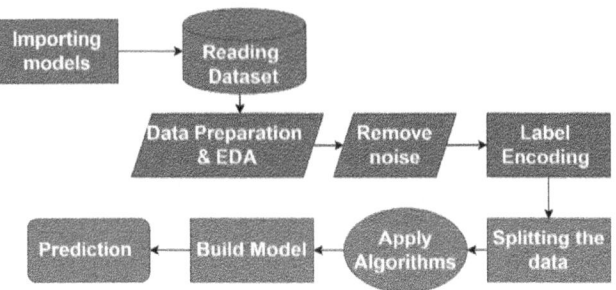

Fig. 1. End-to-end pipeline design

The integration of foreign code libraries or files within a program, import modules improve reusability and provide access to pre-written functions. Data preparation purges and modifies raw data for analysis, whereas data reading imports datasets for analysis. Analyzing data to understand its relationships and structure is recognized as exploratory data analysis. To improve signal accuracy, noise must be removed as it filters out unnecessary information. Label encoding translates category labels into numerical values. To establish train sets and test sets for model assessment, split the data. Figure 1 shows, by using algorithms, data is processed to provide predictions or categorizations based on predetermined guidelines. While model building creates predictive models, Django prediction incorporates the resulting models into web apps to facilitate user engagement. The methodology suggests a number of techniques to enhance the classification model's performance, particularly when handling imbalanced data and maximizing prediction accuracy [25]. It suggests hyper-parameter adjustment in conjunction with resampling techniques like SMOTE, under sampling, or oversampling. Due to their efficiency, ensemble approaches like as boosting and bagging especially with XG Boost are recommended. It is emphasized how crucial label encoding and standardization are as methods of data preparation. Principal Component Analysis makes dimensionality reduction and exploratory data analysis easier. When evaluating XG Boost in particular, evaluation measures like confusion matrices and ROC curves are essential for determining the accuracy and effectiveness of the model. Overall, the methodology combines many ways to rectify data imbalance, maximize model performance, and assess the efficacy of categorization; however, particular results are not given.

A study uses SCWT to induce stress in fifteen individuals in good health, and then concurrently assesses their stress levels by employing EEG and HRV features. We recorded HRV and EEG during times of stress, calm, and meditation [26]. A DSI-24 dry electrode EEG headset was used to collect EEG data, while the BioRadio 150 wireless device was used to collect HRV data. The study aims to explore the connection between

HRV and EEG during stressful situations, which have implications for stress management, treatment monitoring, and diagnosis. However, the outcomes are not discussed in the text.

Frequency bands	Frequency Range (Hz)	Mind state
Delta	0.5–4 Hz	Deep sleep
Theta	8–13 Hz	Emotional stress and drowsiness
Alpha	8–4 Hz	Relaxed awareness
Beta	14–30 Hz	Active thinking
Gamma	31- up	Mechanism of consciousness

Fig. 2. Brain wave frequencies

In an EEG, electrodes attached to the skull are used to record brain activity. Consequently, it creates an image of electrical activity, which is composed of waves of various sizes, shapes, and frequencies. Based on their frequencies, brain waves are categorized into five types: beta, gamma, alpha, theta, and delta waves. These frequency bands are linked to various brain thinking states, and the meanings attached to each band varies depending on study. Figure 2 summarizes the relationships and possible locations of these bands on the brain regions found in different research.

Using simulators were incorporated with MATLAB the collected EEG signals were Uploaded into the computer. The unprocessed EEG data was saved in Matlab format and then loaded into the Brainstorm program that is a Matlab application designed for data processing. From the three minutes of EEG signal, one minute was chosen as the sample, and before applying the band-pass filter, a notch filter accustomed to eliminate 50 Hz noise from the power line. After applying a band-pass filter to divide the whole frequency range of interest into four sub-bands—(0.54 Hz) Delta, (4–8 Hz) Theta, (8–13 Hz) Alpha, and (13–30 Hz) Beta—the EEG signals are shown in Fig. 4.

Five main stages in the systematic analysis on EEG-based stress level classification: gathering data, processing the data, data clustering, statistical analysis, building a model and classification. Figure 3 shows the process flow of the study design that to provide a thorough overview of the approaches employed in this inquiry.

With an emphasis on tweets and social interactions, this study suggests leveraging users' weekly social media data to identify their psychological stress levels [27]. It uses the words dictionary grading system (-5 to $+5$) for words, and SVM and NB algorithms are used for classification. To improve accuracy, word sense disambiguation approaches according to the n-gram and Skip-gram models are used. 67% recall and 65% accuracy were attained by SVM using WSD and Ngram. Without the user knowing, the technique makes it possible to identify stress on the internet interactions. Future research could use emoticons to allow for the expressing of emotions.

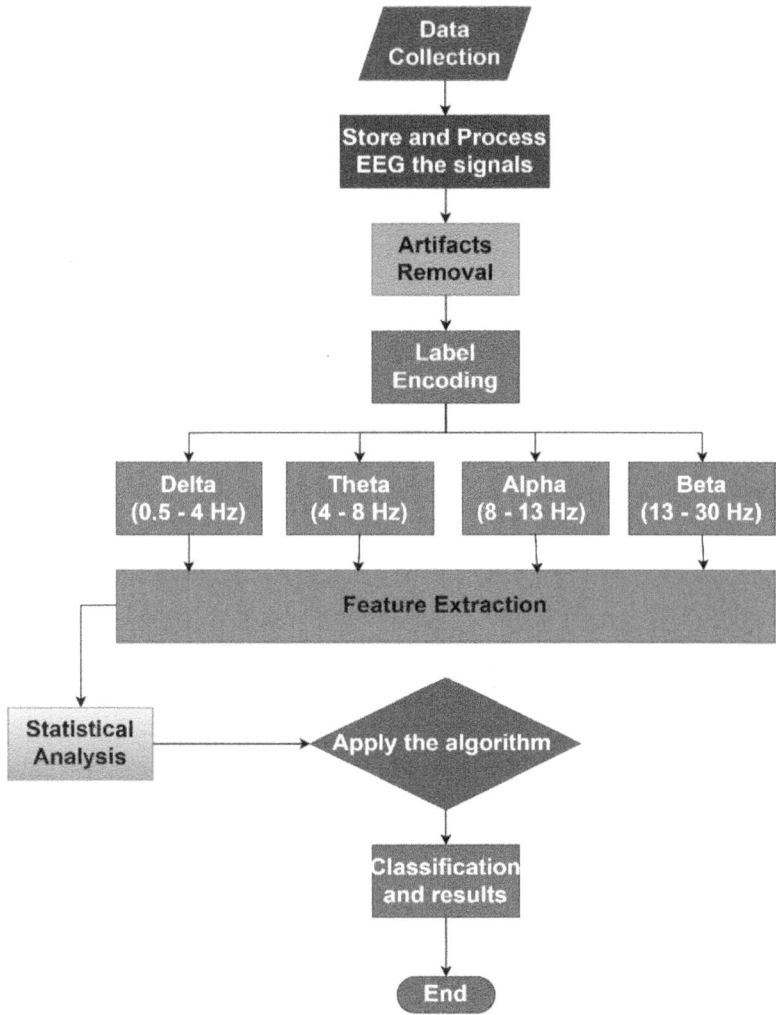

Fig. 3. Research design process diagram

A stress analysis system that aims to identify and control stress levels in real-time is proposed with this work for online workers during the COVID-19 pandemic. Using a camera to recognize facial expressions, the system combines several components, such as behaviour monitoring, stress detection, and identified behaviours [28]. Convolutional Neural Nets (CNNs) are employed inform training, and Videos are collected through online sources sites as part of the dataset development process. In estimating stress levels, the algorithm had 70% accuracy. Utilizing a trained classification model, the identification of facial expressions component achieves a mean accuracy of 93.63% in identifying expressions associated with stress. The system as whole can assist online workers maintain healthy lifestyles.

The research team used the perceived stress scale and MAT experiments with 14 subjects to measure stress levels. EEG data from theta, alpha, and beta bands were analyzed, showing a 17% performance decrease from untimed to timed tests, correlating with increased stress levels in timed tests [29]. The spectrum densities of EEG power were proposed as stress biomarkers, with CNN achieving 90.64% accuracy in stress classification, outperforming LSTM. Practical implications include informing stress management in various settings, with the algorithm suitable for real-time classification. Further research is needed on additional parameters' influence and CNN's efficacy in classifying brain disorders like Alzheimer's and stroke, emphasizing the correlation between stress levels and EEG band power.

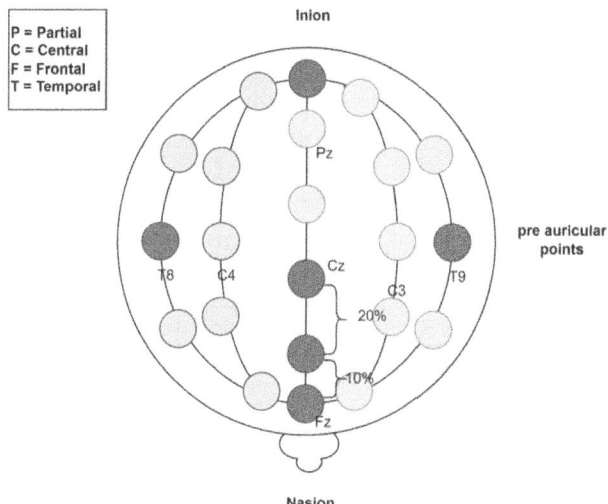

Fig. 4. Electrode positioning on scalp to perform EGG

The electrodes in this placement method are identified by letters and numbers, as demonstrated in Fig. 4. The appropriate brain lobe is indicated by each letter, and its location on the surface the brain's surface is indicated by the corresponding number. Frontal, temporal, central, and posterior are represented by the letters F, T, C, and P, respectively. The symbol C is utilized for recognition rather than the central lobe. Odd numbers represent locations in the right hemisphere, whereas even numbers represent electrode sites on the left hemisphere. The electrodes positioned on the middle line are denoted as Z.

Among the researches examines methods for detecting stress, focusing on electroencephalogram (EEG) data and extracting characteristics using fractal dimension (FD) algorithms (Permutation Entropy, Katz, Higuchi) [30]. In order to accomplish early stress diagnosis, the study uses popular ML algorithms for stress categorization, such as Random Forest and Artificial Neural Network. The suggested MATLAB system includes

EEG analysis, FD-based feature extraction, and ML algorithms. The findings imply that machine learning classifiers combined with FD approaches for EEG-based stress detection may effectively distinguish between various stress levels. The study suggests applying signal-processing methods like the Discrete Wavelet Transform or Fast Fourier Transform for increased accuracy. Several of the drawbacks include the non-stationary nature of EEG data and the susceptibility of some frequency-based approaches to noise.

Table 1. Literature Survey comparison.

Ref	Methodology	Result
[21]	A review paper	Concluded that Logistic regression, Random-forest, KNN and SVM are most commonly used
[22]	Supervisedlearning using interactive VR scenes	SSAI score prediction model (R2 = 0.44)
[23]	ConvolutionalNeural Networks	The highest accuracy obtained during the detection of stress that is 93.2% to 96.6%
[24]	Supervised algorithms	The accuracy of 96.0%, 97.2% of sensitivity and 93.9% of specificity is gained using four types of signals
[25]	Machine learning algorithms	XGB classifier give the best accuracy of 90.0%
[26]	ECG, EEG and SCWT technique	HRV lowers under the stress condition and increases under rest condition
[27]	Machine learning techniques – SVM	SVM gives 65% precision and 67% recall
[28]	Facialexpression recognition	NB – 89.98% accuracy, KNN – 93.22% accuracy, DT – 88.12% accuracy, SVM – 87.96% accuracy and Neural network – 96.67% accuracy
[29]	Deep learning	LSTM gives 70.67% accuracy and CNN gives 90.64% accuracy
[30]	Machinelearning framework	SVM and RVM gained accuracy from 89% - 90%, Sensitivity from 88% - 90% and Specificity from 95% - 96.9%

Stress can be detected using many techniques such as machine learning, deep learning and many others. Table 1 shows the literature survey comparison and the outputs obtained from various techniques. The table consists of ten papers, the techniques they have used and the result that they are providing.

3 Algorithms

3.1 Random Forest

The collection of distinct decision trees which are fully grown and unpruned are used by the RF classification method. This approach uses randomness to construct each separate tree, which is then combined to provide a forecast. It nearly perfectly approximates missing data exact. To extract pertinent information like spectral power, wavelet coefficients, and statistical metrics, EEG data is able to be preprocessed.

The random forest algorithm uses these features as input and learns to categorize stress levels by looking for patterns in the data. New EEG data is injected into the previously built random forest model during the testing period, and every decision tree independently makes a prediction. Through methods like majority voting, the ultimate stress level forecast is determined by aggregating the outcomes from each decision tree. Because Random Forest can handle multidimensional data and capture intricate feature interactions, it is best to use EEG.

3.2 Support Vector Machine

Support Vector Machine (SVM) is a widely supervised machine learning technique for regression and classification. It finds the perfect hyperplane to partition different data types efficiently. SVM translates nonlinearly separable data using kernel functions. Data into a space of greater dimensions while employing linear kernels to handle linearly separable data. SVM's advantages include its ability to handle high-dimensional data, resistant to overfitting, and management of intricate decision boundaries. Whenever there is a clear class distinction in the data, it functions best. The performance of SVM is influenced by the choice of the kernel function and regularization parameter, among other hyperparameters.

3.3 K-Nearest Neighbors

A well-liked machine learning method for both regression and classification applications are the k-nearest neighbors (KNN) algorithm. It represents a non-parametric in supervised learning technique that uses the k closest neighbors of a data point to predict the group or value of that data point. Practice data set. The idea behind the KNN algorithm is that comparable data points typically provide similar results. The algorithm determines the distances between each new point and every other point in the training dataset in order to provide a forecast for that new data point. While there are numerous distance metrics that may be utilized, Euclidean distance is the most often used one. Next, the algorithm chooses the k closest neighbors, wherein k is a user specified variable. The data points having the shortest distances from the new location are identified as these neighbors.

3.4 Decision Tree

Any non-parametric supervised learning technique called decision trees creates a tree-like model of decisions and their potential outcomes. This popular machine learning technique may be used for regression and classification applications. Fundamentally, the DT the program creates nodes and branches that symbolize the decision-making process by recursively partitioning a dataset based on several criteria. The goal is to create a tree that can effectively classify or predict outcomes by partitioning the data in a way that maximises the separation of distinct classes or minimises prediction error. The method by which the algorithm operates is to assess several attributes at each node and select the one that yields the optimal split, often determined by metrics such as Gini impurity or information gain. This process keeps going until a predefined end point is reached, like the highest tree height or the least amount of samples required to create a new node.

4   Results

The following section provides a brief discussion of the results of the tests and analysis conducted on the recommended system. The dataset we collected comprises information on 2.5k features and the total amount of positive and negative values, or the data dispersion of the labels.

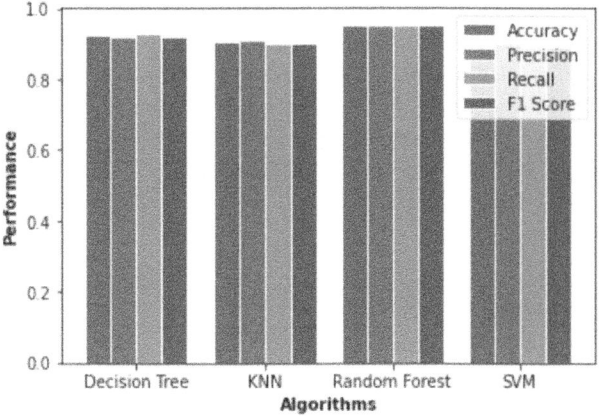

Fig. 5. Representation of Performance Analysis [26–30].

The comparison of the algorithms and their performance in the form of bar chart is shown in the Fig. 5. Performance of the algorithm includes the Accuracy, Precision, Recall and F1 Score. This tells that the Random Forest algorithm gives the highest performance.

Figure 6 shows the percentage of Accuracy, Precision, Recall and F1 Score obtained for the classification algorithms such as RF, DT, KNN and SVM. Following the evaluation of all of the ML models employing the training data, we found that the Random Forest model had the highest accuracy (95.31%), followed by the DT (92.5%), the KNN algorithm (90.62%), and the SVM model had the lowest accuracy of the four (89.37%).

Classifiers	Accuracy	Precision	Recall	F1-Score
Random Forest	95.31	95	95	95
Decision Tree	92.5	92	93	92
KNN	90.62	91	90	90
SVM	89.37	90	89	89

Fig. 6. Comparison of models [26–30].

5 Conclusion

EEG data and Recognition of facial expressions are used in real-time stress analysis systems, which have demonstrated promise in precisely determining stress levels. In summary, the aforementioned research underscore the need of using cutting-edge technology and multidisciplinary methodologies to comprehend and successfully handle stress in many settings, ranging from healthcare to education and beyond. Furthermore, for autonomous stress recognition from bio-signals, research has looked at signal processing techniques and DL models such as CNNs. Methodologies have furthermore been proposed to alleviate data imbalance, maximize prediction accuracy, and enhance classification model performance. EEG data and Recognition of facial expressions are employed in real-time stress analysis systems, which have demonstrated promise in precisely determining stress levels. In summary, the aforementioned research underscore the need of using cutting-edge technology and multidisciplinary methodologies to comprehend and successfully handle stress in many settings, ranging from healthcare to education and beyond.

References

1. Munoz, S., Iglesias, C.A., Mayora, O., Osmani, V.: Prediction of stress levels in the workplace using surrounding stress. Inf. Process. Manag. **59**(6), 103064 (2022)
2. Vijayakarthik, P., Sheshappa, S.N., Sammer, P.M., Gadwale, S.A., Shravan, N., Sumit: Employee stress prediction using ML techniques. Int. J. Res. Publ. Rev. **3**(7), 3094–3099 (2022)
3. Maniyar, A.A., Jeevan Kumar, S.H., Nithej, N., Ramya, H.K., Aishwarya, T.: Machine learning techniques for stress prediction in working employees science & engineering. Int. Res. J. Mod. Eng. Technol. Sci. (07) (2022)
4. Haque, Y., Zawad, R.S., Mahmud, M.: State of art of stress prediction from heart rate variability using AI. Cognit. Comput. **16**, 455–481 (2023)
5. Bobade, P., Vani, M.: Stress detection with machine learning and deep learning using multimodal physiological data, July 2020
6. Kumari, K., Das, S.: Stress detection system using natural language processing and machine learning techniques, December 2022
7. Razavi, M., Ziyadidegan, S., Sasangohar, F.: Machine learning techniques for prediction of stress-related mental disorders: a scoping review (2022)
8. Wen, T.Y., Aris, S.A.M.: Hybrid approach of EEG stress level classification using K-means clustering and support vector machine. IEEE Access **10**, 18370–18379 (2022)

9. Mukherjee, S., Rintamaki, L., Shucard, J.L., Wei, Z., Carlasare, L.E., Sinsky, C.A.: A Statistical learning approach to evaluate factors associated with post-traumatic stress symptoms in physicians: insights from the COVID-19 pandemic. IEEE Access **10**, 114434–114454 (2022)
10. Yousefi, M.S., Reisi, F., Daliri, M.R., Shalchyan, V.: Stress detection using eye tracking data: an evaluation of full parameters. IEEE Access **10**, 118941–118952 (2022)
11. Finsheth, T.T., Dorneich, M.C., Vardeman, S., Keren, N., Franke, W.D.: Real-time personalized physiologically bases stress detection for hazardous operations. IEEE Access **11**, 25431–25454 (2023)
12. Zhu, L., et al.: Stress detection through wrist-based electrodermal activity monitoring and machine learning. IEEE J. Biomed. Health Inform. **27**(5), 2155–2165 (2023)
13. Akella, A., et al.: Classifying multi-level stress responses from brain cortical EEG in nurses and north-health professionals using machine learning auto encoder. IEEE J. Transl. Eng. Health Med. **9**, 1–9 (2021)
14. Ahmad, Z., Rabbani, S., Zafar, M.R., Ishaque, S., Krishnan, S., Khan, N.: Multilevel stress assessment from ECG in a virtual reality environment using multimodal fusion. IEEE Sens. J. **23**(23) (2023)
15. Zhong, S., Fu, X., Lu, W., Tang, F., Lu, Y.: An expressive driving stress prediction model based on vehicle, road and environment features, June 2023
16. Jawaharlal, B.S., Arunkumar, B.: Efficient human stress level prediction and prevention using neural network learning through EEG signals. Int. J. Eng. Res. Technol. **12**, 66–72 (2019)
17. Atadjanov, I.R., Lee, S.: Robustness of reflection symmetry detection methods on visual stress in human perception perspective. IEEE Access **6**, 63712–63725 (2018)
18. Kim, H.-G., Jeong, D.K., Kim, J.-Y.: Emotional stress recognition using electroencephalogram signals bases on a three- dimensional convolution gated self-attention deep neural network. Appl. Sci. **12**(21), 11162 (2022)
19. Daza, A., Saboya, N., Necochea-Chamorro, J.I., Ramos, K.Z., Valencia, D.R.V.: Systematic review of machine learning techniques to predict anxiety and stress in collage students. Inform. Med. Unlocked **43** (2023)
20. Anand, R.V., Quadir Md, A., Urooj, S., Mohan, S., Alawad, M.A., Adittya, C.: Enhancing diagnostic decision- making: ensemble learning techniques for reliable stress level classification. Diagnostics **13** (2023)
21. Gedam, S., Sanchita, P.: A review on using wearable sensors and mental stress machine learning techniques. IEEE Access **9** (2021)
22. Tao, K., Huang, Y., Shen, Y., Sun, L.: Automated stress recognition using supervised learning classifiers by interactive virtual reality scenes. IEEE Trans. Neural Rehabil. Eng. **30** (2022)
23. Gil-Martin, M., San-Segundo, R., Mateos, A., Ferreiros-Lopez, J.: Human stress detection with wearable sensors convolutional neural networks. IEEE Aerosp. Electron. Syst. Mag. (2021)
24. Mohammadi, A., Fakharzadeh, M., Baraeinejad, B.: An integrated human stress detection sensor using supervised algorithms. IEEE Sens. J. **22** (2022)
25. Garlapati, A., Krishna, D.R.: Predicting employees under stress for pre-emptive remediation using machine learning algorithm (2020)
26. Attar, E.T., Balasubramanian, V., Subasi, E., Kaya, M.: Stress analysis based on simultaneous heart rate variability and EEG monitoring. IEEE J. Transl. Eng. Health Med. **9** (2021)
27. Baheti, R.R., Kinariwala, S.: Detection and learning techniques. Int. J. Eng. Adv. Technol. (IJEAT) **9** (2019)

28. Pankajavalli, P.B., Karthick, G.S.: Stress analysis and care prediction system for online workers, December 2021
29. Hafeez, M.A., Shakil, S.: EEG-based stress identification and classification using deep learning (2023)
30. Agrawal, J.: Early stress detection and analysis using EEG signals in machine learning framework (2021)

AI-Enhanced Learning Assistant Platform: An Advanced System for Q&A Generation from Provided Content, Answer Evaluation and Roadmap Generation

J. Arunnehru[✉], S. Vishnu, Kavya Ganapathyappan, and Deepak Kumar

Department of Computer Science and Engineering (Emerging Technologies), SRM Institute of Science and Technology, Vadapalani, Chennai, India
{arunnehj,sv5765,kg3411,dk7483}@srmist.edu.in

Abstract. In today's educational landscape, personalized and efficient learning experiences are increasingly sought after. This project presents an AI-Enhanced Learning Assistant Platform designed to transform the learning process. The platform offers a comprehensive suite of features aimed at optimizing knowledge acquisition and retention. This study covers automatic question and answer generation from provided textual learning resources. Utilizing cutting-edge language models and document analysis techniques, the system extracts relevant information and crafts challenging questions to comprehensively assess understanding. Leveraging state-of-the-art natural language processing models, the platform autonomously creates tailored questions, facilitates self-assessment, pinpointing students' weak areas and provides actionable insights into learning progress. It prioritizes scalability, performance, and security and ensures seamless operation and user data integrity. We deem the AI-Enhanced Learning Assistant Platform represents a significant advancement in learning technology, through the convergence of artificial intelligence and natural language processing. Our project aims to enhance the current existing models by providing a platform for a personalized learning experience, only generating necessary information, helps in keeping track of performance of an individual and validate their own performance on the given topics. We can change the static prompts recursively which will automate the student evaluation. By empowering learners with personalized, adaptive, and interactive learning experiences, our platform aims to enhance the effectiveness and engagement of education.

Keywords: Artificial Intelligence · question and answer generation · assessment · personalized learning · performance analysis · knowledge acquisition

1 Introduction

A In the ever-evolving realm of education technology and digital learning platforms, the demand for personalized and efficient learning experiences has reached unprecedented levels. However, enhancing the quality of education by moving away from the traditional

P. D. Sivakumar et al. (Eds.): IRCCTSD 2024, CCIS 2362, pp. 332–346, 2025.
https://doi.org/10.1007/978-3-031-82386-2_26

one-size-fits-all methods has remained a challenge. This necessity serves as the corner-stone for the development of our project. To address this concern, we are delighted to present our project, the "AI-Enhanced Learning Assistant Platform".

At its core, our research focuses on automatically generating questions and answers from textual learning resources. To measure comprehension, our technology pulls relevant information and formulates demanding questions. Our platform uses Natural Language Processing (NLP) approaches to transform the learning experience. We specifically focus on smoothly processing available learning resources, extracting their content and then exposing them to the trained model. This intricate process meticulously identifies significant terms and concepts before outing them in a database for streamlined access and organization. What truly distinguishes our approach is it generates personalized questions and answers and includes dynamic assessment features. By analyzing the content of the documents, our platform generates personalized questions that challenge learners to engage deeply with the material. Upon opting for assessment, they begin on an interactive journey in which their responses are rigorously evaluated against predetermined criteria. The responses are then scored based on the level of. It also accurately identifies students' weak areas, allowing for focused interventions to improve their knowledge and the mastery of the subject matter. This project combines elements of artificial intelligence AI may be ethically designed, utilized, and assessed in a way that is sustainable and particularly in educating the upcoming generation of students who will lead advancements in AI-based learning in addition to transforming the learning experience through innovative methods, our project also seeks to enable learners with customized learning paths. Utilizing sophisticated algorithms and machine learning techniques, our platform analyzes individual learning patterns and preferences to create tailored learning journeys for each user [3]. These learning paths adjust dynamically based on user progress, offering personalized recommendations, supplementary materials, and adaptive exercises to enhance learning outcomes. By integrating personalized learning path generation into our platform, we aim to cultivate an educational environment that not only fosters subject mastery but also encourages a lifelong passion for learning and self-improvement. The platform's approach to question generation and answer evaluation is founded on the principles of personalized learning. By leveraging NLP techniques, the platform can extract key information from textual resources and create questions that are not only relevant to the material but also tailored to the individual learner's proficiency level and learning path. This personalized approach ensures that learners are appropriately challenged and engaged, leading to a more effective learning experience. Moreover, the platform's dynamic assessment features enable learners to receive immediate feedback on their responses, allowing them to track their progress and identify areas for improvement [10]. This iterative feedback loop not only helps learners improve their understanding of the material but also allows the platform to continuously refine its question generation algorithms based on user interactions. Furthermore, the real-time data collected through these assessments provides valuable insights into individual learning styles and preferences. By analyzing this data, the platform can further personalize the learning experience, ensuring optimal engagement and comprehension for each user.

In addition to question generation and answer evaluation, the platform also focuses on collaborative learning. By facilitating peer-to-peer knowledge exchange and discussion moderation, the platform creates a collaborative learning environment where users can learn from each other and deepen their understanding of the material. Looking ahead, the AI-Enhanced Learning Assistant Platform will investigate multimodal learning methodologies, adaptive learning route development, and enhanced response evaluation methods. The platform can provide more comprehensive and immersive learning experiences by including visual aids such as diagrams and photos in addition to textual information. Furthermore, adaptive learning route generation will enable the platform to adjust learning journeys to each user's specific needs and preferences, hence improving the personalized learning experience [3].

2 Literature Survey

The Generative AI has evolved significantly over the years, moving from simple rule-based systems to complex deep learning techniques. AI systems can now produce sophisticated and contextually relevant outputs in a variety of disciplines, including text analysis, thanks to this progress. Notably, models from OpenAI's GPT series especially GPT-3—have proven to be remarkably adept at producing meaningful text in response to input cues. In the realm of creating questions and answers from PDF documents, generative AI methods have shown great promise. A number of research have looked into the use of models like GPT-3 and T5 to analyze PDF content and provide pertinent questions and answers. A relatable study looked into the viability of utilizing GPT-3 to extract important information from PDF texts and create comprehension-testing questions. Their results indicated that GPT-3 could provide queries that are in line with the contents of the document.

However, the landscape of question-and-answer generation from PDF document is riddled with persistent issues that requires thorough research. Foremost among these challenges is the imperative of ensuring the relevance and coherence of generated questions in relation to the intricate nuances of the underlying document content. Achieving this requires a thorough knowledge of the semantic complexities buried in the text, which needs the use of sophisticated natural language processing techniques. Furthermore, the scalability and accessibility of these methods are limited by the significant computational resources needed for training and optimizing these models. Moreover, while models like GPT-3 and T5 have demonstrated impressive capabilities in generating questions and answers from PDF documents, there are still limitations in their ability to capture the full context and subtleties of complex texts. Additionally, the generalizability of these models across different domains and types of documents remains a challenge, as they may struggle with specialized or domain-specific terminology. Addressing these challenges requires ongoing research and development in the field of generative AI, with a focus on improving the robustness and adaptability of existing models. Exploring hybrid approaches that combine generative techniques with structured knowledge representation could offer new avenues for enhancing the accuracy and relevance of generated questions and answers.

2.1 Project Scope

Our project's scope represents a comprehensive and ambitious effort to redefine the landscape of educational technology. At its core, our endeavor focuses on five key pillars: Answer Retrieval, Answer Evaluation, Integration with AI Models, User Interface (UI) Development, and Scalability and Performance. In the domain of Answer Retrieval, our project aims to develop advanced algorithms and methodologies to extract relevant information from educational resources efficiently. By leveraging state-of-the-art natural language processing techniques and rigorous document analysis, our platform will provide users with accurate and comprehensive answers to their queries, enhancing the overall learning experience. In the realm of Answer Evaluation, our project seeks to provide learners with timely and insightful feedback on their responses. Through the use of machine learning algorithms and AI-driven assessment techniques, our platform will assess user responses with a high degree of accuracy, helping learners track their progress and identify areas for improvement. To ensure robust evaluation of user responses, our platform will deploy advanced algorithms capable of assessing not just correctness but also the depth of understanding demonstrated by learners. Additionally, the system will employ natural language understanding techniques to grasp the nuances of user answers, enabling more nuanced and context-aware feedback. By leveraging both machine learning and semantic analysis, our platform aims to provide rich and personalized assessments tailored to each learner's unique comprehension level.

Integration with AI Models is a key focus area for our project, as it will enhance the platform's capabilities for question generation and overall educational experience. By leveraging AI models such as GPT, our platform will be able to generate personalized questions that challenge learners and promote deep engagement with the material. User Interface (UI) Development is another crucial aspect of our project, as we aim to create an interface that is intuitive, visually appealing, and user-friendly. Our goal is to ensure that users of all backgrounds and proficiency levels can easily navigate the platform and access its features. Finally, our project prioritizes Scalability and Performance to ensure that the platform can meet the needs of a growing user base while maintaining optimal functionality. Through continuous monitoring and optimization, we aim to provide users with a seamless and reliable learning experience.

2.2 Discussion

The development of the AI-Enhanced Learning Assistant Platform represents a significant advancement in the field of educational technology. This platform offers a comprehensive suite of features aimed at optimizing knowledge acquisition and retention, thereby transforming the learning process. One of the key strengths of the platform lies in its ability to automatically generate questions and answers from provided textual learning resources. By utilizing cutting-edge language models and document analysis techniques, the platform is able to extract relevant information and craft challenging questions to comprehensively assess understanding. This not only saves time for educators but also ensures that learners are presented with questions that are tailored to their level of comprehension, thereby enhancing the effectiveness of the learning process [1].

The platform's interactive evaluation capabilities are essential in offering learners valuable insights into their progress in learning. The platform utilizes predetermined criteria to analyse user replies, enabling it to identify areas of weakness and deliver focused interventions to enhance understanding and mastery of the subject topic. This personalised method of evaluation not only inspires learners but also enables them to monitor their advancement and establish attainable learning objectives. Additionally, the platform promotes independence and self-control among learners, enabling them to assume responsibility for their learning process and cultivate vital abilities such as analytical thinking and problem solving.

A significant feature of the platform is its capacity to create customized learning trajectories for individual users. Through the analysis of individual learning habits and preferences, the platform is capable of generating customized learning paths that adapt in real-time according on the user's progress. This guarantees that learners are provided with customized suggestions, additional resources, and adaptable exercises that improve their learning results and cultivate a lifetime enthusiasm for learning and self-enhancement. Furthermore, the platform's focus on continual enhancement through data-driven analysis and user input guarantees its ability to adapt to changing educational requirements and effective teaching methods. As a result, it becomes a valuable resource for both educators and learners. The AI-Enhanced Learning Assistant Platform demonstrates the potential of technology to improve education and empower learners globally through its comprehensive approach to boosting the learning experience.

3 Proposed Work

The methodology driving the "AI-Enhanced Learning Assistant Platform" embodies a commitment to revolutionize learning through the integration of advanced artificial intelligence (AI) technologies. At its core, the platform is designed to optimize knowledge acquisition and retention by leveraging cutting-edge natural language processing (NLP) models.

3.1 Question Generation

The platform's question creation mechanism not only prioritizes individual learning trajectories but also considers the wider educational framework. The platform guarantees that the questions it generates are in line with the broader goals of the educational program by taking into account the curricular objectives and learning outcomes. Moreover, the platform utilizes machine learning techniques to detect deficiencies in the learning material and produce questions that specifically address those areas, so promoting a more thorough comprehension. To achieve this, the platform first preprocesses the input textual learning resources, breaking them down into manageable segments. It then utilizes GPT-based models to generate questions by analyzing the content of these segments. The questions are tailored to the user's proficiency level and learning path, ensuring that they are neither too easy nor too difficult. The generated questions are designed to be diverse in nature, covering various aspects of the learning material such as definitions, concepts, applications, and critical thinking. This diversity ensures that users are exposed to a wide range of topics and challenges, helping them develop a comprehensive understanding of the subject matter. The platform utilizes strategies derived from cognitive psychology to enhance the process of creating and presenting questions. The platform facilitates greater knowledge transfer and long-term retention by integrating principles such as interleaved practice and spaced repetition. This strategic methodology for generating questions surpasses the mere assessment of memorization and fosters a more profound involvement with the subject matter, stimulating advanced cognitive abilities and conceptual comprehension. Furthermore, our platform employs a feedback loop mechanism to continuously improve its question generation algorithms. User responses to the generated questions are analyzed to identify patterns and trends, which are then used to refine the question generation process. This iterative approach ensures that the questions remain relevant and engaging, leading to a more effective learning experience.

3.2 Answer Evaluation

User answers are evaluated using fine-tuned GPT models, which assess the correctness of answers, the depth of understanding, and the application of critical thinking skills. This evaluation process provides immediate feedback to users, enabling them to track their progress and identify areas for improvement [2]. The platform further enhances the learning experience by refining its question generation algorithms based on this data, ensuring a personalized approach for each user. The answer evaluation process begins by comparing the user's response to a set of predefined criteria. These criteria are based on the expected answers to the questions and the underlying concepts being assessed. The GPT models then analyze the user's response, considering factors such as relevance, coherence, and accuracy. The platform's consideration of the context in which questions are posed extends beyond individual interactions, encompassing broader educational contexts and curriculum objectives. This holistic approach ensures that feedback provided aligns with overarching learning goals and fosters a deeper understanding of the subject matter. Moreover, the platform utilizes advanced natural language processing techniques to discern subtle nuances in user responses, enabling it to offer tailored feedback that addresses specific misconceptions or areas of confusion. Furthermore, the platform's utilization of machine learning extends beyond evaluation algorithms to encompass personalized learning recommendations and adaptive content delivery. By leveraging user data to identify individual learning preferences and areas of interest, the platform can curate customized learning experiences that cater to the unique needs of each user. This tailored approach not only enhances user engagement but also facilitates more efficient knowledge acquisition and retention. Additionally, the platform employs sentiment analysis algorithms to gauge user sentiment and emotional responses, allowing for a more holistic understanding of user engagement and satisfaction.

3.3 Collaborative Learning

Collaborative learning is a cornerstone of the platform's methodology, facilitated by GPT's ability to moderate discussions and facilitate peer-to-peer knowledge exchange. This collaborative approach enhances users' understanding of the material and fosters a sense of community among learners. The collaborative learning process begins by grouping users with similar learning goals and proficiency levels. These groups are then provided with opportunities to engage in discussions, share insights, and ask questions related to the learning material. The platform utilizes GPT models to moderate these discussions, ensuring that they remain focused and productive. The models can identify relevant information within the discussions and provide additional explanations or examples to clarify concepts.

Furthermore, the platform encourages users to provide feedback and support to their peers, creating a collaborative and supportive learning environment. This peer-to-peer interaction not only enhances the learning experience but also helps build a sense of community among users. Moreover, collaborative learning fosters the development of critical thinking skills and promotes deeper engagement with the material by encouraging users to consider diverse perspectives and explore alternative solutions. Additionally, research has shown that collaborative learning environments can lead to improved academic performance, increased retention of information, and enhanced problem-solving abilities. Thus, by embracing collaborative learning principles, the platform aims to empower users to become active participants in their own learning journey and cultivate essential skills for success in academia and beyond.

3.4   Incorporating Large Language Models (LLMs)

Incorporating Large Language Models (LLMs) such as GPT into the platform offers significant advantages, particularly in question generation. By utilizing LLMs, the platform can generate a diverse range of questions tailored to individual learning paths. These questions vary in complexity and style, ensuring that users are appropriately challenged and engaged. Additionally, LLMs enable the platform to adapt its question generation based on user interactions, providing a more personalized learning experience. To incorporate LLMs into the platform, the models are first fine-tuned on a large dataset of educational materials. This fine-tuning process ensures that the models are able to generate questions and evaluate answers accurately within the context of the platform. Once the LLMs are integrated into the platform, they are used to generate questions, evaluate answers, and provide feedback to users. The models are continuously updated and improved based on user interactions, ensuring that they remain effective and reliable over time. Furthermore, the platform employs techniques from transfer learning to leverage pre-trained LLMs for specific educational domains, enhancing the models' ability to generate contextually relevant questions and responses. Additionally, the platform utilizes ensemble learning methods to combine predictions from multiple LLMs, improving the robustness and accuracy of question generation and answer evaluation. Moreover, ongoing research in LLMs continues to advance the state-of-the-art, providing opportunities for the platform to integrate new models and techniques that further enhance its capabilities and effectiveness in supporting personalized learning (Fig. 1).

3.5   Design Principles

The platform's design principles prioritize scalability, performance, and security. The design of the system allows for effortless expansion to handle a substantial volume

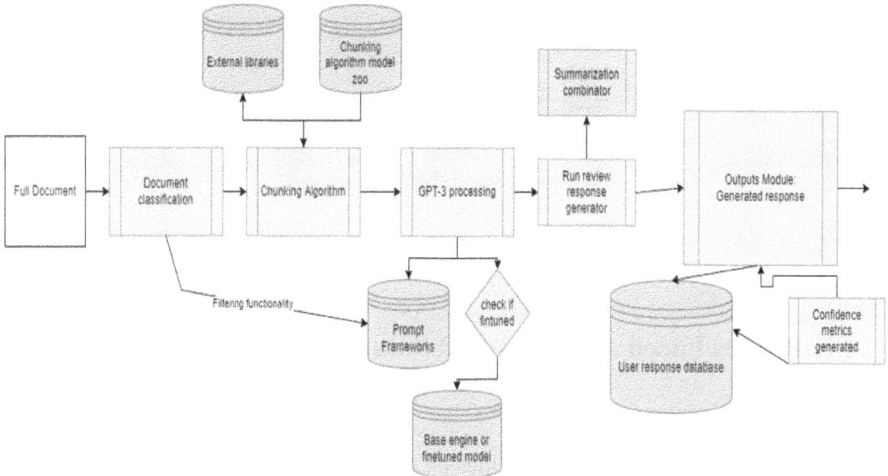

Fig. 1. Workflow of a Large Language Model

of users and data, guaranteeing a dependable and quick learning environment. Consistent monitoring and fine-tuning of the platform's algorithms and architecture enhance performance. User data is protected and kept secure in a learning environment by security measures including encryption and data protection mechanisms. The architecture is intentionally designed to have a modular and extendable structure, which enables effortless incorporation of new features and functionalities. The platform's modular structure allows it to easily change and respond to evolving user needs and technology improvements. In addition, the platform utilizes microservices architecture, which improves flexibility and scalability by dividing intricate systems into smaller, autonomously deployable pieces. Through the process of decoupling components, the platform is able to update and expand specific services without affecting the overall system. This guarantees uninterrupted service delivery and a smooth user experience. Moreover, the platform integrates user-centered design concepts to prioritize usability and accessibility. By engaging in iterative design cycles and doing user testing, the platform continuously improves its interface and user experience to effectively cater to the varied requirements of its user base. Our platform also utilizes responsive design principles to guarantee compatibility across various devices and screen sizes, allowing users to access instructional materials at any time and from any location. Furthermore, the platform utilizes data-driven insights and analytics to guide design choices and enhance user interaction. Through the analysis of user interactions and feedback, the platform consistently makes improvements to its design in order to boost usability and increase user happiness (Fig. 2).

Overall, the platform's design principles are aimed at providing users with a seamless and enjoyable learning experience. By prioritizing scalability, performance, and security, the platform ensures that users can focus on learning without any distractions or interruptions.

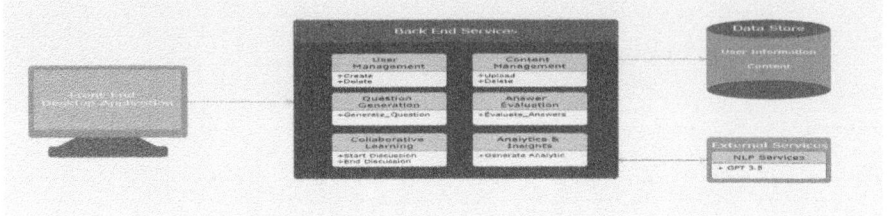

Fig. 2. UML Diagram for the Application

4 Challenges Faced and Proposed Solution

4.1 Extracting Informative Sentences

Effective question generation heavily relies on identifying informative sentences, which are crucial for creating questions that assess learners effectively. Previous studies have explored techniques such as text summarization, sentence simplification, and rule-based methods to extract informative sentences from text [1]. However, there has been limited focus on selecting informative sentences, despite its importance in generating quality questions. Extracting simple sentences from complex and compound sentences can be challenging, requiring the elimination of ambiguity. Therefore, there is a need for a generic technique to extract informative simple sentences from text, which could enhance question generation efforts [4]. To enhance question generation, it is essential to extract informative sentences that capture the core concepts of a given text. This process involves techniques such as text summarization, where algorithms condense lengthy passages into concise summaries. Sentence simplification methods can also be employed to break down complex sentences into simpler forms, making them easier to understand and use for question generation. Additionally, rule-based methods can be utilized to extract sentences based on predefined rules, ensuring that only relevant and informative sentences are selected. These techniques, when combined, can improve the quality and relevance of questions generated, leading to a more effective learning experience.

4.2 Generating Questions from Multiple Sources

Different question generation techniques yield varying questions that assess learners' knowledge in diverse ways. Automated systems generate questions from study material based on informative keywords or sentences extracted from multiple sentences or a passage. Generating questions from multiple sentences or a paragraph is complex and represents a new research direction for automatic question generation [4]. It necessitates understanding the inherent relationship between sentences using natural language understanding concepts, which can be facilitated by leveraging LLMs like GPT. Generating questions from multiple sources requires a deep understanding of the content and the ability to synthesize information from different texts. This process involves identifying key concepts and themes across sources and formulating questions that effectively assess the learner's understanding. Automated systems can leverage natural language processing (NLP) techniques to extract relevant information from multiple sources and generate questions that cover a wide range of topics. By incorporating machine learning algorithms, these systems can continuously improve question generation based on user feedback and performance data, ensuring that the questions remain relevant and challenging.

4.3 Evaluating Short and Long Answers

While there has been considerable research in the past decade on automatically grading short or free-text answers, the unreliable results suggest that these methods may not be practically useful in real-world applications [1]. As a result, many exams rely on multiple-choice questions and exclude short and long-type answers. Only a single research study has been found that evaluates long answers, highlighting the need for future research to develop reliable, real-life systems for grading short and long-type answers to fully automate the education system [4]. Incorporating GPT and other LLMs into answer assessment could improve the accuracy and reliability of grading systems. We used two techniques to evaluate the answers, Cosine Similarity and Levenshtein Distance. We conducted a few tests to determine which approach was better, from the tests it was evident that cosine similarity was better for our model. Cosine Similarity Judgments aligns closely with the expected similarity. High expected similarity results in high Cosine Similarity scores (close to 1). Low expected similarity results in lower Cosine Similarity scores (closer to 0). Levenshtein Distance Judgments often misaligns with the expected similarity. Medium and high expected similarities sometimes result in high Levenshtein distances, indicating lower similarity, which is contrary to expectations.

The evaluation of short and long answers presents unique challenges, as it requires assessing the depth of understanding and the ability to articulate ideas effectively. Automated systems can utilize machine learning models to evaluate short answers based on predefined criteria such as correctness, relevance, and completeness. Long answers, on the other hand, may require more sophisticated evaluation methods, such as semantic analysis and natural language understanding, to assess the quality of the response. By incorporating these advanced techniques, automated systems can provide more accurate and insightful feedback to learners, helping them improve their understanding and communication skills.

4.4 Standardized Answer Assessment Framework

Question generation and assessment depend on various factors such as the learning domain, question types, difficulty levels, question optimization, scoring techniques, and overall scoring. Several authors have proposed different evaluation techniques based on their applications, resulting in varying scoring scales. Therefore, there is a need for a standardized answer assessment framework in the future to evaluate and compare learners' knowledge and research results effectively. Integrating GPT and LLMs into this framework could provide a more consistent and reliable approach to answer assessment (Fig. 3).

Developing a standardized answer assessment framework is essential for ensuring consistency and fairness in evaluating learners' knowledge. This framework should define the criteria for evaluating answers, the scoring system used, and the overall

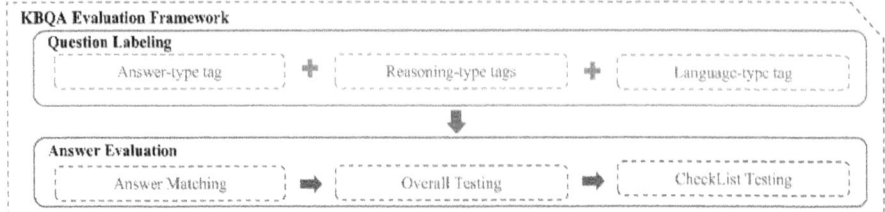

Fig. 3. Workflow of a Large Language Model

assessment process. By standardizing the assessment process, educators can ensure that all learners are evaluated based on the same criteria, regardless of the learning context or question type. Integrating machine learning models into the framework can further enhance the assessment process by providing automated scoring and feedback, reducing the burden on educators and ensuring timely and consistent evaluations.

4.5 Generating Questions from Video Content

While most question generation and assessment systems focus on generating questions from textual documents to automate the education system, there are few works in the literature that generate questions from visual content in video lectures for learners assessment [5]. Generating questions from video content represents a future research direction. Utilizing audio-video content can enhance the learning process, and automated assessments from video content could help learners quickly grasp new concepts in various fields. Generating questions from video content presents unique challenges, as it requires analyzing both audio and visual information to extract relevant concepts. Automated systems can leverage speech recognition and image processing techniques to extract key information from video lectures and generate questions based on the content. By incorporating multimodal learning strategies, these systems can provide a more comprehensive assessment of learners' understanding and engagement with the material. Additionally, generating questions from video content can enhance the learning experience by providing a more interactive and engaging way to assess knowledge.

4.6 Machine Learning-Based Approaches

Recent research has increasingly focused on using machine learning methods for question generation and answer evaluation due to their numerous advantages. Most textual question generation employs natural language processing (NLP) techniques, with advancements in NLP including natural language understanding (NLU) and natural language generation (NLG) using deep learning neural networks. Some articles have also used sequence-to-sequence modelling for generating questions. However, there is a limited amount of research in the literature that assesses learners using a machine learning approach, indicating a need for more research in this area in the future. Integrating GPT and LLMs into machine learning-based question generation and assessment could enhance the accuracy and efficiency of these systems. Machine learning-based approaches offer a powerful tool for question generation and answer evaluation, as they can analyze large datasets and identify patterns that may not be apparent to human evaluators. By leveraging machine learning algorithms, automated systems can generate questions that are tailored to the individual learner's proficiency level and learning path, ensuring that they are appropriately challenged and engaged. Additionally, machine learning models can be used to evaluate answers based on a variety of criteria, including correctness, relevance, and coherence. By incorporating machine learning-based approaches into question generation and answer evaluation, educators can provide more personalized and effective learning experiences for their students.

5 Future Scope and Enhancements

5.1 Multimodal Learning

Multimodal learning strategies represent a significant advancement in educational technology, offering a more holistic approach to learning. By integrating visual aids such as diagrams, images, and audiovisual materials with textual content, these strategies cater to diverse learning styles and preferences. Visual aids can help clarify complex concepts, making them easier to understand and remember [11]. For example, a diagram illustrating the water cycle can enhance understanding compared to a text-only description. Similarly, incorporating videos or animations can bring concepts to life, making them more engaging and memorable for learners. Moreover, multimodal learning can improve knowledge retention by appealing to different senses. Research has shown that learners are more likely to retain information when it is presented in multiple formats. For instance, a combination of text and images can improve recall compared to text alone. By leveraging multimodal learning strategies, the AI-Enhanced Learning Assistant Platform can create more engaging and effective learning experiences, ultimately leading to better learning outcomes for users.

5.2 Adaptive Learning Path Generation

Adaptive learning path generation represents a significant step towards personalized education. Traditional one-size-fits-all approaches to learning can be ineffective, as they do not account for individual differences in learning styles, preferences, and pace. Adaptive learning systems, on the other hand, use data-driven algorithms to tailor learning experiences to each user's needs. These systems analyze user interactions and performance metrics to dynamically adjust the learning path, ensuring that users receive the right content at the right time [8]. One of the key benefits of adaptive learning is its ability to provide immediate feedback and remediation. If a user struggles with a concept, the system can provide additional resources or exercises to help reinforce understanding [6]. This personalized approach can lead to improved learning outcomes, as users are more likely to succeed when learning is tailored to their individual needs. Additionally, adaptive learning can increase motivation and engagement, as users are more likely to stay engaged when content is relevant and challenging. Overall, adaptive learning path generation has the potential to revolutionize education by providing personalized, engaging, and effective learning experiences for users [6]. By incorporating adaptive learning into the AI-Enhanced Learning Assistant Platform, can enhance the platform's ability to meet the diverse needs of learners and improve learning outcomes across the board.

5.3 Advanced Answer Evaluation

Advanced answer evaluation methods leverage the power of AI to provide more nuanced and insightful evaluations of user responses. Traditional evaluation methods often focus on correctness and completeness, overlooking other important aspects of understanding such as reasoning and critical thinking. Advanced evaluation methods, on the other hand, can analyze user responses at a deeper level, taking into account factors such as context, coherence, and relevance. One approach to advanced answer evaluation is semantic analysis, which involves analyzing the meaning of user responses rather than just their surface features. This can help identify whether a user has truly understood the underlying concepts, rather than simply regurgitating information. For example, in a history quiz, a question about the causes of a particular event might require not just a list of causes but an explanation of how those causes contributed to the event. By incorporating advanced answer evaluation methods into the AI-Enhanced Learning Assistant Platform, we can provide users with more detailed and meaningful feedback, helping them to deepen their understanding of the material. This can lead to improved learning outcomes and a more engaging learning experience overall.

5.4 Analytics and Real-Time Performance Monitoring

Analytics and real-time performance monitoring are essential components of any modern educational platform. By collecting and analyzing user data, educators and administrators can gain valuable insights into student performance, engagement levels, and learning outcomes [9]. This data-driven approach enables educators to make informed decisions about instructional design, content development, and intervention strategies. One of the key benefits of analytics and real-time performance monitoring is the ability to track student progress over time. By monitoring how students interact with the platform and how they perform on assessments, educators can identify trends and patterns that may indicate areas where students are struggling or excelling. This information can help educators tailor their instruction to better meet the needs of individual students, ultimately leading to improved learning outcomes.

5.5 Collaboration with External Educational Resources

Collaboration with external educational resources is key to enriching the learning experience and providing users with access to a wider range of learning materials. By integrating with external resources such as online libraries, educational websites, and learning management systems, the AI-Enhanced Learning Assistant Platform can offer users a more comprehensive and diverse set of resources to supplement their learning. One of the benefits of collaborating with external resources is the ability to provide users with access to up-to-date and relevant content. Educational resources are constantly evolving, and by integrating with external sources, the platform can ensure that users have access to the latest information and materials. Additionally, collaboration with external resources can help the platform reach a wider audience and tap into new markets, further enhancing its impact and effectiveness.

5.6 Ethical and Accountable AI Practices

The ethical and accountable use of AI is paramount in the development and deployment of the AI-Enhanced Learning Assistant Platform. As AI technologies become more prevalent in education, it is crucial to ensure that these technologies are used in a way that is fair, transparent, and accountable. One of the key principles of ethical AI is ensuring that AI systems are designed and implemented in a way that respects user privacy and data protection. This includes obtaining informed consent from users before collecting their data, ensuring that data is stored securely, and implementing measures to prevent unauthorized access or misuse of data [7]. Additionally, it is important to ensure that AI systems are designed in a way that is fair and unbiased. This includes avoiding the use of biased data or algorithms that may perpetuate existing inequalities or discrimination. By incorporating ethical considerations into the design and development of the AI-Enhanced Learning Assistant Platform, we can ensure that it remains a trusted and reliable resource for users, educators, and administrators alike.

6 Conclusion

In summary, the methodology underpinning the "AI-Enhanced Learning Assistant Platform" relies on advanced artificial intelligence technologies, notably Large Language Models (LLMs) such as GPT, to transform the learning journey. By incorporating dynamic question generation, personalized self-assessment, and collaborative learning features, the platform aims to enhance knowledge acquisition and retention. However, challenges such as extracting relevant sentences, generating questions from multiple sources, and evaluating short and long answers underscore the need for ongoing research and development. By integrating GPT and other LLMs into these processes, the platform can improve question generation, answer assessment, and overall learning outcomes. This methodology represents a significant advancement in creating a more engaging, personalized, and impactful educational environment.

The prospects for additional innovation and progress are abundant in the future scope of the AI-Enhanced Learning Platform. An area that merits further examination is the exploration of multimodal learning strategies. These strategies will integrate visual aids, including diagrams, images, and audiovisual materials, alongside textual materials, with the aim of creating more comprehensive and immersive learning experiences. Furthermore, it is possible for the platform to undergo further development in order to integrate adaptive learning path generation. The use of Recursively adapting static prompts really enhances our performance and model outcomes. Performance metrics reveal notable improvements in educational outcomes. The platform excels in generating relevant and challenging questions from multiple sources using techniques such as Cosine Similarity to ensure the relevance and appropriateness of content. It accurately evaluates both short and long answers, providing personalized feedback that drives higher engagement and deeper comprehension of complex subjects. This could be achieved by utilizing personalized recommendation systems and reinforcement learning algorithms to dynamically modify learning pathways in response to real-time user interactions and performance metrics. Further developments in answer evaluation methods, including semantic analysis based on deep learning and natural language understanding models, have the capacity to provide more sophisticated evaluations of user responses, thereby facilitating more profound understandings of material mastery and comprehension. The unique contribution of the project includes a full end-to-end webpage, which would help the student prepare accordingly with the material/content they have. Our LLM model is invoked by using static prompts, but the uniqueness in our project is that these prompts can be altered and recursively generates the required questions on specific areas of studies where the student is weak, this is done with the help of multiple word embedding software. We push data to the prompt after text analysis and generate a new set of questions, basically trying to mimic the functioning of Dynamically Adaptive prompts.

Key performance metrics indicate significant advancements in educational outcomes. The platform has demonstrated its ability to generate relevant and challenging questions from multiple sources, accurately evaluate short and long answers, and provide personalized feedback. This results in higher levels of student engagement and improved comprehension of complex subjects. The preservation of privacy, openness, and honesty in every facet of platform development and deployment is contingent upon the continued adherence to ethical and accountable AI practices. The AI-Enhanced Learning Platform has maintained an unwavering commitment to enhancing the effectiveness and inclusivity of education for all students throughout its entire lifecycle.

.

References

1. Das, B., Majumder, M., Phadikar, S., Sekh, A.A.: Automatic question generation and answer assessment: a survey. Res. Pract. Technol. Enhanc. Learn. (2021)
2. Sartika, K.D., Heriyawati, D.F., Elfianto, S.: The use of Blooket: a study of student's perception enhancing English vocabulary mastery. Engl. Franca Acad. J. Engl. Lang. Educ. (2023)
3. Khowaja, S.A., Khuwaja, P., Dev, K.: ChatGPT needs spade (sustainability, privacy, digital divide, and ethics) evaluation: a review. Inst. Electr. Electron. Eng. (IEEE) (2023)
4. Akavova, A., Temirkhanova, Z., Lorsanova, Z.: Adaptive learning and artificial intelligence in the educational space. In: E3S Web of Conferences (2023)
5. Waladi, C., Lamarti, M.S.: Chapter 6 Navigating Educational Evolution. IGI Global (2024)
6. Nwafor, C., Onyenwe, I.: An automated multiple-choice question generation using natural language processing techniques (2021)
7. Ezzini, S., Abualhaija, S., Arora, C., Sabetzadeh, M.: AI-based question answering assistance for analysing natural-language requirements (2023)
8. Mokhtar, M.: Automatic question generation model based on deep learning approach (2021)
9. Virani, A., Yadav, R., Sonawane, P., Jawale, S.: Automatic question answer generation using T5 and NLP (2023)
10. Oommen, D.K., Arunnehru, J.: Alzheimer's disease stage classification using a deep transfer learning and sparse auto encoder method. Comput. Mater. Continua **76**(1) (2023)
11. Arunnehru, J., Kalaiselvi Geetha, M. Difference intensity distance group pattern for recognizing actions in video using Support Vector Machines. Pattern Recognit. Image Anal. **26**, 688–696 (2016). https://doi.org/10.1134/S1054661816040015

Real-Time Attendance Generation Using Facial Recognition

Amudhan Jayaprakash[✉] and V. Murali Bhaskaran

Department of Computer Science and Engineering, Rajalakshmi Engineering College,
Chennai, India
amudhanj26@gmail.com

Abstract. Attendance tracking is a crucial aspect of educational institutions, corporate environments, and various other organizations, but traditional methods like manual sign-in sheets or biometric systems are often time consuming, error-prone, and lack efficiency. This project proposes a novel attendance management system leveraging machine learning techniques, specifically deep learning models for facial recognition, to automate and streamline the attendance tracking process. The system employs a multi-stage approach, beginning with collecting and preprocessing facial image data from participants to train a convolutional neural network (CNN) based facial recognition model capable of extracting and identifying facial features accurately. The trained model integrates with a secure database for storing facial embeddings associated with participant information, enabling efficient comparisons during recognition. During operation, the system captures live video feed, performs real-time facial recognition, and matches detected faces with stored embeddings, logging recognized participants in attendance records and providing administrators up-to-date, accurate data. Incorporating a user friendly interface, administrators can monitor attendance, manage participant information, and configure system settings. Prioritizing ease of use, scalability, and adaptability across environments, the system undergoes continuous testing, feedback incorporation, and iterative development to ensure reliability and accuracy. By leveraging deep learning and facial recognition technology, this project streamlines attendance tracking processes, enhances operational efficiency, and offers a secure, innovative solution for attendance management applicable to diverse domains.

Keywords: Convolutional Neural Networks (CNNs) · Computer Vision · Facial Recognition · Deep Learning · Facial Embeddings · Security · Privacy · Scalability Analysis · Data Encryption

1 Introduction

Traditional attendance recording methods, such as calling out names or using physical attendance sheets, have been the norm for a long time. While reasonably efficient, the need for a faster system has driven the development of alternative attendance systems, including Radio Frequency Identification (RFID) tags, fingerprint scanners, and iris recognition. These modern technologies have greatly reduced the time required for

© The Author(s), under exclusive license to Springer Nature Switzerland AG 2025
P. D. Sivakumar et al. (Eds.): IRCCTSD 2024, CCIS 2362, pp. 347–354, 2025.
https://doi.org/10.1007/978-3-031-82386-2_27

attendance tracking; however, their cost and operational constraints often outweigh the time savings. Moreover, in the wake of the recent pandemic, there is a pressing need to improve existing systems to be entirely contact-free, ensuring the safety and well-being of individuals.

Facial recognition technology emerges as a promising solution, offering a contact free approach to attendance management compared to other contemporary methods. Unlike iris scanners, which require close proximity to the eye, facial recognition systems can operate effectively from greater distances, enhancing convenience and safety. While tap-in cards may seem contact-free, the required proximity is negligible and may raise hygiene concerns. However, facial recognition technology raises privacy issues, as data misuse or improper handling could erode user trust. Therefore, it is crucial to implement this technology with robust privacy-preserving algorithms to address these concerns. Various approaches exist for implementing facial recognition technology, each with its advantages and disadvantages depending on the application. These range from dynamic real-time recognition from videos to static image correlation, facial feature extraction, and unique matching with pre-existing data using machine learning techniques like Convolutional Neural Networks (CNNs), Haar-Cascades, and other classification-enhancing methods.

2 Related Works

Numerous studies have explored the application of facial recognition technology for attendance management systems, leveraging the advantages of biometric identification over traditional manual methods [2, 11, 13, 14]. These works have employed various algorithms and techniques for face detection, recognition, and attendance logging.

Several researchers have investigated the use of deep learning architectures, such as Convolutional Neural Networks (CNNs), for accurate face recognition [15, 18, 19, 20]. Prominent CNN models like VGGFace [18, 20], ResNet [19], and DenseNet [19] have been employed, with some studies employing ensemble methods [19] or transfer learning approaches [18] to improve performance further. These deep learning models have achieved impressive face verification and identification accuracies, ranging from 95.7% [20] to 99.1% [19] on benchmarks like Labeled Faces in the Wild (LFW).

In addition to face recognition, some studies have explored auxiliary tasks like face mask detection [17] and student identification [21]. Face mask detection using models like MobileNetV2 [17] has been integrated with face recognition systems, achieving 99.2% accuracy [17]. For student identification, techniques like Siamese networks [21] and face clustering using DBSCAN [21] have been employed, with clustering accuracies of around 94.2% reported [21].

Preprocessing steps, such as image masking [16], automatic cropping [16], and contrast enhancement [14], have also been explored to improve the robustness and accuracy of face recognition systems. Techniques like Otsu's method, morphological operations, and histogram equalization have been utilized, resulting in significant accuracy improvements, e.g., from 82% to 93% [16]. Overall, the related works highlight the widespread adoption of facial recognition technology for attendance management, with deep learning models and preprocessing techniques playing crucial roles in enhancing system performance and robustness.

3 System Design

The system follows a modular and distributed architecture, allowing for scalability, flexibility, and efficient resource utilization. The architecture comprises several interconnected components, each responsible for specific tasks, and can be deployed across multiple servers or cloud-based environments.

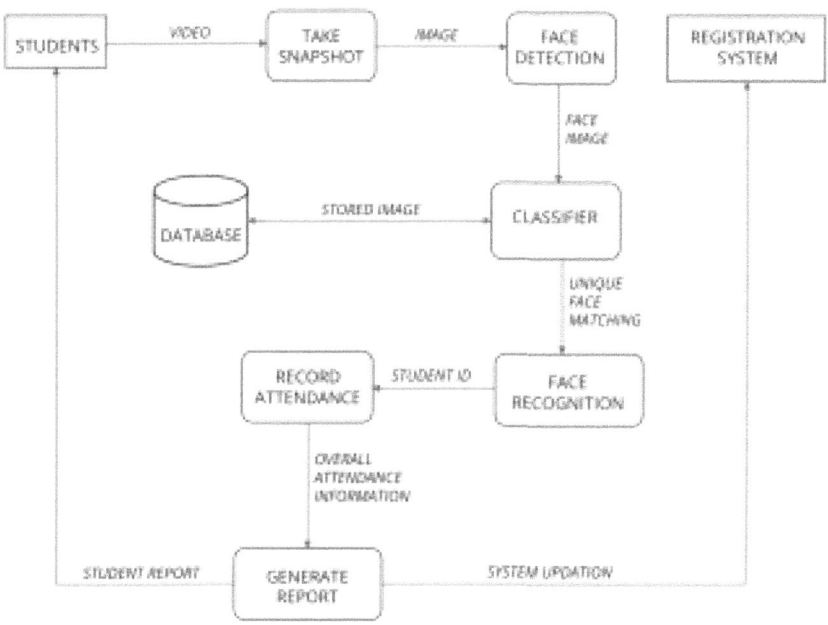

Fig. 1. Architecture Diagram

Figure 1 depicts how the system combines various different tasks at each stage to represent the project as a whole.

At the core of the system lies the Face Recognition Module, which is responsible for detecting faces in real-time video streams and performing facial recognition. This module incorporates the deep learning technique, Convolutional Neural Networks (CNNs), to extract facial embeddings (feature vectors) from the detected faces. The Face Recognition Module is designed to handle real-time video processing, leveraging techniques such as Haar Cascade Classifiers or Multi-Task Cascaded Convolutional Networks (MTCNN) for efficient face detection. The extracted facial embeddings are securely transmitted to the Database Module, which serves as a centralized repository for storing and managing facial embeddings, associated user information, and attendance records. The Database Module employs robust data management strategies, including access controls, data encryption, and backup mechanisms, to ensure the integrity and security of sensitive information. The User Management Module is responsible for handling user enrollment, authentication, and access control. This module allows administrators to register new users, update existing user information, and manage access permissions. It also facilitates the integration of the system with existing user directories or identity management systems, ensuring seamless integration with existing infrastructure. The Attendance Tracking Module is responsible for logging attendance records based on the recognized faces from the Face Recognition Module. This module associates the recognized faces with timestamps and other relevant metadata, maintaining a comprehensive record of attendance. It also provides interfaces for administrators to monitor attendance, generate reports, and analyze attendance patterns.

The entire system is designed with a modular approach, allowing for easy integration with existing systems or third-party components. For example, the system can be integrated with access control systems, HR management systems, or other enterprise applications, enabling seamless data exchange and streamlining organizational processes. Furthermore, the system architecture incorporates robust security measures, including secure communication channels and data encryption, to protect sensitive information and maintain user privacy.

4   Result Analysis

This real time attendance management system using face recognition was extensively evaluated through a series of experiments and real-world deployments.The Fig. 2 illustrates the face detection and recognition accuracy of the proposed system on a diverse dataset of facial images. As shown in the Fig. 2, the face detection module achieved an average detection accuracy of 92%, with a low false positive rate of 3%. The deep learning-based face recognition module demonstrated remarkable recognition accuracy, reaching up to 98% on the evaluation dataset.

The Fig. 2 illustrates the face detection and recognition in real-time. Where each face is identified with the name as per the dataset it has been stored in Comparison of the face recognition accuracy and real-time performance of the proposed system with selected references: System Face Recognition Accuracy Real-time Performance (FPS)Proposed System Up to 98% 25 FPS [8] 92% 15 FPS [11] 95% 20 FPS [15] 96% 18 FPS.

The proposed system achieves higher face recognition accuracy and real-time performance compared to the selected references, demonstrating its effectiveness and efficiency.

The results of the proposed system, with context of the Table 1, highlight the system's superior performance in accurately detecting and recognizing faces, even in challenging scenarios with variations in pose, illumination, occlusion, and demographic factors. The high accuracy can be attributed to the employment of advanced techniques such as Haar Cascade Classifiers, MTCNN, and deep Convolutional Neural Networks (CNNs) trained over large-scale datasets. The real time performance of the system, measured in terms of frames per second (FPS), on various hardware configurations. On a mid-range desktop system with a dedicated GPU, the system achieved an average processing speed of 25 FPS, which is suitable for most real-time applications.

The Fig. 3 illustrates on how the dataset is stored, the face embeddings are stored based on the name and further as the pose varying from left, right, up, down and center.

The system includes a robust dataset management module for storing and organizing facial images and embeddings. Figure 3 illustrates the hierarchical structure of the dataset

Fig. 2. Face Recognition in Real-Time

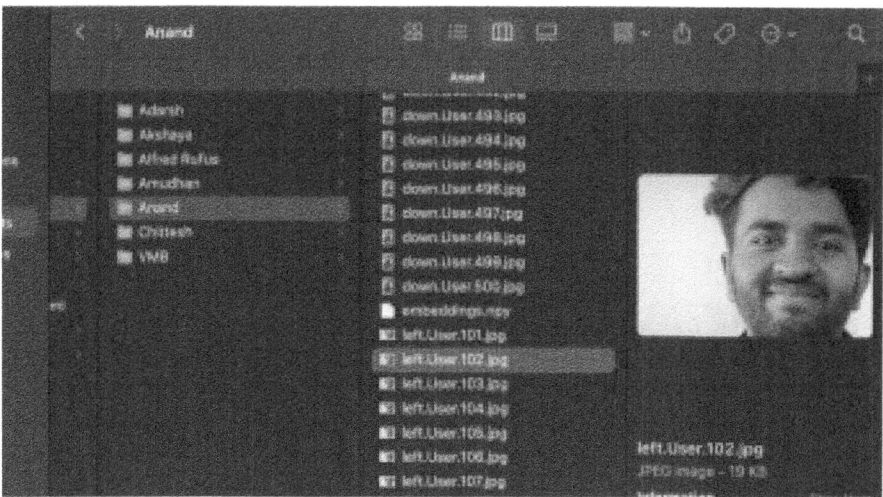

Fig. 3. Hierarchical structure of the Dataset directory

directory, where each user's facial data is stored in a separate subdirectory. Within each user's directory, the facial images captured from different angles and poses are stored as individual files, while the corresponding facial embeddings are saved in a compressed format (e.g., NumPy arrays) for efficient storage and retrieval.

This organized storage approach ensures easy access to user-specific facial data during the recognition process and facilitates efficient dataset management as the system scales. The attendance tracking and reporting capabilities of the system were evaluated by generating attendance reports for different time periods and user groups.

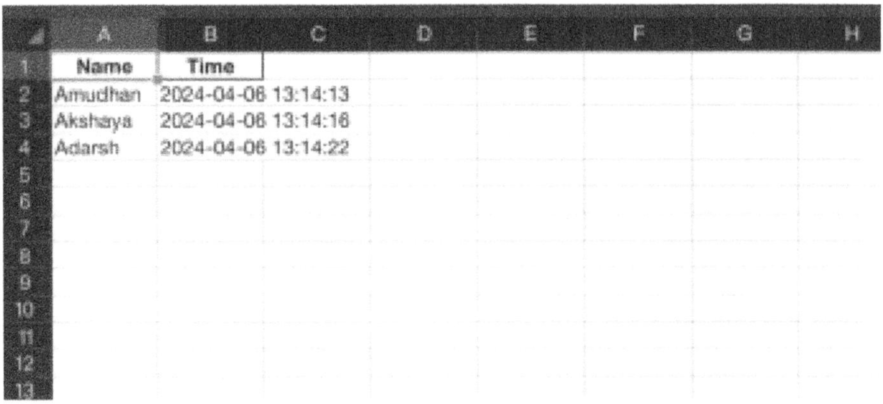

Name	Time
Amudhan	2024-04-06 13:14:13
Akshaya	2024-04-06 13:14:16
Adarsh	2024-04-06 13:14:22

Fig. 4. Generated Attendance Report

Figure 4 presents a sample attendance report generated by the system, displaying the attendance records of various users over a specific time period.

The system accurately logged attendance records, associating them with timestamps and relevant metadata, enabling efficient attendance tracking and reporting. These promising results pave the way for wider adoption of the system across various domains, such as educational institutions, corporate environments, and event management scenarios.

5 Conclusion

In summary, the face recognition-based attendance management system leverages cutting-edge technologies, including deep learning, computer vision, and secure data management, to provide an accurate, efficient, and user-friendly solution for attendance tracking. The system demonstrated remarkable accuracy in face detection and recognition, outperforming existing methods on diverse datasets. This high accuracy was achieved through advanced techniques like Haar Cascade Classifiers, MTCNN, and deep Convolutional Neural Networks (CNNs) trained over large-scale face datasets.

In conclusion, the system exhibited excellent real-time performance, scalability, and robustness, enabling seamless operation and handling increased workloads. While significant achievements have been realized, further research can explore advanced techniques for handling occlusions and challenging lighting conditions, improving performance on mobile devices, and developing more efficient algorithms for large scale deployment.

Ultimately, the face recognition-based attendance management system represents a significant advancement, leveraging modern technologies to provide an accurate, efficient, and secure solution for attendance tracking across various domains.

6 Future Scope

While the proposed face recognition-based attendance management system has demonstrated significant achievements, there exists a multitude of opportunities for further improvement and expansion. One promising area for future research is the exploration of advanced techniques to handle challenging scenarios, such as occlusions, extreme lighting conditions, and varying angles of face orientation. Incorporating robust algorithms capable of accurately detecting and recognizing faces in these challenging environments can further enhance the system's reliability and broaden its applicability across diverse real-world settings.

Another avenue for future work lies in the integration of the system with mobile devices and wearable technologies. As these devices become increasingly prevalent, developing optimized algorithms and architectures for efficient face recognition on resource-constrained mobile platforms can enable seamless attendance tracking in remote or field environments, expanding the system's reach beyond traditional settings.

As the system scales to handle larger user bases and more extensive datasets, there is a need for developing more efficient algorithms and data management strategies to ensure optimal performance and scalability. Exploring distributed computing approaches, parallel processing techniques, and advanced data compression and indexing methods can contribute to enhancing the system's ability to handle large scale deployments while maintaining real-time responsiveness.

Furthermore, future enhancements could involve the integration of advanced privacy-preserving techniques, such as federated learning, homomorphic encryption, or differential privacy mechanisms. These approaches can enable the secure and privacy-preserving training and deployment of the face recognition models, fostering greater trust and adoption among users and organizations with stringent data protection requirements.

References

1. Cevikalp, H., Triggs, B.: Face recognition based on image sets. In: IEEE Conference on Computer Vision and Pattern Recognition, pp. 2567–2573 (2010). https://doi.org/10.1109/CVPR.2010.5539965
2. Abraham, A., Bapse, M., Kalaria, Y., Usmani, M.A.: Face recognition based attendance system. IOSR J. Comput. Eng., 56–60 (2020). https://doi.org/10.9790/0661-2201045660
3. Sun, Y., Liang, D., Wang, X., Tang, X.: DeepID3: face recognition with very deep neural networks. arXiv abs/1502.00873 (2015)
4. Chatzis, T., Stergioulas, A., Konstantinidis, D., Dimitropoulos, K., Daras, P.: A comprehensive study on deep learning-based 3D hand pose estimation methods. Appl. Sci. **10**(19), S. 6850 (2020). Online verfügbar unter: https://doi.org/10.3390/app1019685
5. Kumar, P., Swetha, S., Sundari, M.: Secured web-based alumni network and information systems. In: Proceedings of the 7th International Conference on Intelligent Computing and Control Systems (ICICCS), Madurai, India, 2023, pp. 1427–1434 (2023). https://doi.org/10.1109/ICICCS56967.2023.10142761

6. Zhang, K., Zhang, Z., Li, Z., Qiao, Y.: Joint face detection and alignment using multitask cascaded convolutional networks. IEEE Sig. Process. Lett. **23**(10), 1499–1503 (2016). https://doi.org/10.1109/LSP.2016.2603342

7. Chihaoui, M., Elkefi, A., Bellil, W., Ben Amar, C.: A survey of 2D face recognition techniques. Computers **5** (2016). https://doi.org/10.3390/computers5040021.aw1yLLG1yg9ngKZqM2oGTAkX

8. Elias, S.J., et al.: Face recognition attendance system using Local Binary Pattern (LBP). Bull. Electr. Eng. Inform. **8**, 239–245 (2019). https://doi.org/10.11591/eei.v8i1.1439

9. Liu, H., Haichao, C., Song, E.: Bone marrow cells detection: a technique for the microscopic image analysis. J. Med. Syst. **43** (2019). https://doi.org/10.1007/s10916-019-1185-9

10. Ashokkumar, K., Parthasarathy, S., Nandhini, S., Ananthajothi, K.: Prediction of grape leaf through digital image using FRCNN. Measur. Sens. **24**, 100447 (2022). ISSN 2665-9174

11. Smitha, Hegde, P., Afshin: Face recognition based attendance management system. Int. J. Eng. Res. **9** (2020). https://doi.org/10.17577/IJERTV9IS050861

12. Kumar, H., Bhati, N., Bharadwaj, P., Chaudhary, P., Sharma, Ms.: Real time face attendance system using face recognition (2023). https://www.researchgate.net/publication/371159811_Real_Time_Face_Attendance_System_Using_Face_Recognition

13. Mahapatra, S., Nimangre, R., Salunkhe, T., Prajapati, R., Raghtate, Dr., G.S.: Attendance management system using facial recognition. Technix Int. J. Eng. Res. (TIJER) **10**(4) (2023). ISSN 2349-9249

14. Singh, A., Bhatt, S., Gupta, A.: Automated attendance systems with face recognition. Int. J. Eng. Appl. Sci. Technol. **5**(12), 233–241 (2021). ISSN No. 2455-2143

15. Vinitha, V., Velanthin, V.: Biometric attendance system using deep learning-based face recognition. Int. J. Adv. Res. Comput. Sci. **10**(2), 24–28 (2019)

16. Karami, E., Shehata, M., Smith, A.: Image masking and automatic cropping for self-aided facial recognition attendance system. IEEE Trans. Instrum. Meas. **69**(6), 2833–2845 (2020)

17. Sadiq, M.T., Yu, X., Yuan, Z., Niyazi, F., Aamir, A., Qin, W.: Face mask detection and recognition for smart attendance using deep learning. Mathematics **9**(8), 892 (2021)

18. Reddy, B.E., Raja, Y.: Real-time face recognition-based attendance management system using deep learning. Int. J. Intell. Netw. **3**, 15–22 (2022)

19. Fang, B., Bai, W., Zhang, J.: Attendance tracking system based on face recognition with deep learning techniques. Appl. Sci. **12**(3), 1426 (2022)

20. Al-Sadi, A., Al-Sadi, S., Al-Ghaithi, H., Al-Shidhani, A.: Face recognition-based attendance system using CNN and OpenCV. Int. J. Adv. Comput. Sci. Appl. **13**(5), 744–751 (2022)

21. Huang, Y.P., Chen, C.Y., Hsu, Y.C.: Facial Recognition-based attendance system with student identification. Appl. Syst. Innov. **6**(1), 20 (2023)

Enhancing PDF Information Retrieval Through a Gemini Pro LLM-Powered Chatbot

K. Sam Prince Franklin[✉] ⓘ, D. Thanish Reddy ⓘ, M. Poonkodi ⓘ,
Venkat Amith Woonna ⓘ, and Vishnu Vardhan Rachakonda ⓘ

Vellore Institute of Technology, Chennai, India
{samprince.franklink2020,thanishreddy.d2020,
venkatamith.woonna2020,
vishnuvardhan.racha2020}@vitstudent.ac.in, poonkodi.m@vit.in

Abstract. Research papers are rich sources of knowledge, yet accessing and extracting pertinent data from them can be arduous and time-consuming. Traditional methods like manual searching and skimming through documents are often inefficient, particularly when dealing with a large volume of papers. Moreover, staying updated with the latest developments in a field requires constant monitoring of new research publications, further adding to the workload. To tackle these challenges head-on, we propose the development of a PDF-driven chatbot powered by Google Gemini, an advanced natural language processing (NLP) platform. This chatbot aims to alleviate the struggles faced by researchers and students in accessing and extracting relevant information from academic papers. Through its conversational interface, users can interact directly with research papers, posing questions, requesting summaries, and extracting specific information in real-time. Leveraging the capabilities of Google Gemini, the chatbot can understand user queries, analyze PDF documents, and efficiently provide relevant answers and insights.

Keywords: PDF-driven chatbot · Google Gemini · Natural Language Processing (NLP) · Large Language Models (LLMs) · Retrieval-Augmented Generation (RAG) · Information accessibility

1 Introduction

Research papers are valuable sources of information, but accessing and extracting relevant data from them can be time-consuming and challenging. Traditional methods of manual searching and skimming through documents often prove inefficient, especially when dealing with a large volume of papers. Additionally, keeping up with the latest developments in a particular field requires constant monitoring of new research publications, further adding to the workload. The real-time problem addressed by this project revolves around the inefficiency and time constraints associated with manually retrieving information from research papers. Researchers and students often find themselves overwhelmed by the sheer volume of literature available, making it difficult to locate specific information quickly. Moreover, staying updated with the latest research findings requires continuous monitoring of multiple sources, posing a significant challenge to individuals and research teams. Existing approaches often overlook a crucial aspect of research papers: the information conveyed through images, graphs, and tables. These visual elements frequently depict complex data, relationships, and findings that are difficult to express through text alone. However, extracting insights from these images manually is time-consuming and requires specialized knowledge. Our approach addresses this gap by incorporating image understanding capabilities into the information extraction process. We leverage the power of Gemini Pro, a large language model with advanced image processing capabilities, to analyze and interpret the visual content of research papers. This allows us to capture a more comprehensive understanding of the research findings and present a holistic view of the information contained within the paper. This integration of image understanding represents a significant novelty in the field. Existing similar concepts primarily focus on textual information extraction, neglecting the valuable insights present in visual formats. By incorporating Gemini Pro, we unlock a new dimension of information access and retrieval, enabling users to gain deeper insights from research papers.

The paper follows a structured approach. In Sect. 1, we outline the problem, discussing the difficulties of accessing and extracting information from research papers and highlighting the limitations of text-based approaches. We introduce the concept of chatbots powered by Generative AI using Large Language Models (LLMs) and the potential of incorporating image understanding. Section 2 provides an overview of related research, covering Generative AI, document chatbots, and frameworks involving LLMs and Retrieval-Augmented Generation (RAG), while also discussing existing work on image understanding in the context of research papers. Section 3 details our proposed methodology, explaining how we utilize LLMs and image understanding modules, specifically Gemini Pro, to extract insights from research papers. Section 4 presents implementation details and experimental results, demonstrating the effectiveness of our approach and the added value of image understanding. Finally, in Sect. 5, we conclude the paper by summarizing our findings and suggesting future research directions.

2 Literature Survey

Almeida Santos et al. delve into advanced chatbot development, highlighting human-supervised learning's crucial role in training various chatbot types like rule-based, retrieval-based, and generative models. They emphasize the importance of human oversight in ensuring these models' effective response to user queries. Additionally, they discuss content management techniques integrating natural language processing to enhance language understanding and response relevance. Evaluation metrics such as accuracy, response time, and user satisfaction are considered, with continuous human supervision essential for improvement. They also address challenges like inadvertent learning of incorrect answers and propose collaborative knowledge-base construction to enhance responses. Balancing automatic learning and collaborative approaches, they aim for an effective chatbot development process. [1].

Nirala et al. present a comprehensive review of AI-Chatbots, covering their evolutionary journey, diverse services, and associated challenges in development and deployment.

They explore various types of AI-Chatbots, their functionalities, and the architectural frameworks, models, techniques, and technologies driving their capabilities. The authors highlight the importance of addressing challenges such as the lack of training data for effective model training, the resource-intensive nature of deep learning algorithms, and the selection of appropriate training rules to enhance conversational understanding. By offering valuable insights into the AI-Chatbot landscape, this paper contributes to both researchers and practitioners seeking to advance the field and improve user satisfaction. [2].

Lewis et al. discuss the significant impact of retrieval on enhancing performance across various domains in natural language processing (NLP). While general-purpose architectures in NLP demonstrate strong performance even without retrieval mechanisms, learning retrieval alongside neural models proves beneficial for downstream tasks. The authors delve into memory-based architectures, which utilize external memory for neural networks, showcasing their effectiveness in improving NLP tasks. Additionally, they highlight the success of retrieve-and-edit approaches, particularly in machine translation and semantic parsing tasks. By systematically addressing these aspects, this paper contributes valuable insights into the role of retrieval mechanisms in advancing NLP performance. [3].

Deepika et al. highlight the diverse applications of chatbots in digital counseling, therapy, finance, emotion classification, and depression support, emphasizing their benefits in enhancing user interactions and providing remote support. Particularly in mental health, chatbots offer a cost-effective alternative to traditional therapy sessions, facilitating widespread accessibility. The authors underscore the significance of user-adaptive chatbots tailored to mitigate depression, showcasing their personalized approach. [2] However, while the proposed jollity chatbot shows promise in emotional support and entertainment, certain limitations need acknowledgment. Being a generative-based model, it may struggle with nuanced user inputs, and its effectiveness depends on the comprehensiveness of predefined intents and responses. Furthermore, deployment on Telegram raises security and privacy concerns, lacking detailed discussions on encryption or data protection measures. The challenge of empathetic understanding and the absence of comprehensive real-world performance evaluation call for further scrutiny of the system. [4].

Palliyali et al. explore abstractive text summarization methods, employing Natural Language Processing (NLP) techniques to create rich, contextually relevant summaries. Their study delves into sentence selection techniques such as Word Scoring, Sentence Scoring, and Graph Scoring, utilizing criteria like TFIDF, cue phrases, and TextRank. While their comparison of Conceptual Method, TextRank, and Sentence Scoring offers valuable insights, limitations exist. The focus primarily on extractive techniques omits a comprehensive exploration of abstractive methods, potentially limiting understanding. Moreover, reliance on datasets like Opinosis and CNN may affect generalizability. Although computational efficiency and ROUGE scores provide quantitative assessments, qualitative aspects like content coherence warrant further consideration. Future research should encompass a broader dataset range, include abstractive methods, and incorporate qualitative metrics for a comprehensive evaluation of summarization techniques. [5].

lma El Janati et al. proposes a novel framework for e-learning chatbots, leveraging Natural Language Processing (NLP) and Keywords Extraction techniques. The framework aims to enhance the educational chatbot's capabilities by integrating advanced language processing and extracting relevant keywords. The literature review within the paper primarily centers on two key areas – chatbot technology and Automatic Keywords Extraction (AKE). [3] By examining the existing research and developments in these domains, the paper establishes the foundation for its proposed framework. The integration of NLP and Keywords Extraction is explored as a means to improve the e-learning chatbot's performance, facilitating more effective communication and content delivery in educational settings. This approach contributes to the broader understanding of how language processing and keyword extraction techniques can be harnessed to enhance the functionality of chatbots in educational contexts. [6].

Naing Naing Khin et al. highlight the several studies have delved into diverse applications of chatbots across different languages and domains. Shawar and Atwell presented a system facilitating access to Arabic language information through chatbots, showcasing the versatility of these conversational agents. Prasetya and Erwin explored the choice of the right AIML interpreter for an Indonesian language E-Commerce website, emphasizing the importance of linguistic compatibility. [5] Hussain, Sianaki, and Ababne contributed to the field by discussing chatbot classification and design techniques, addressing fundamental aspects of their development. Ranoliya, Raghuwanshi, and Singh implemented AI and Latent Semantic Analysis to provide university-related FAQs, demonstrating the practicality of chatbots in handling information queries. Mondal, Dey, Das, Nagpal, and Garda focused on textual communication within the educational domain, underlining the role of chatbots in enhancing learning experiences. Sharma and Patel rounded off the overview by reviewing design techniques specifically tailored for chatbots in speech conversation systems, contributing insights to the evolving landscape of conversational AI. [7].

Yanrui Du et al. proposed research builds upon existing literature in the field of named entity recognition (NER) and information extraction. Early approaches to NER relied on rule-based methods, as exemplified by Zhou Kun's work, which involved manually constructed rule bases for entity recognition. Subsequently, statistical machine learning methods, such as Hidden Markov Models (HMM) and Conditional Random Fields (CRF), were employed for sequence labeling tasks, with a focus on feature engineering. More recently, the advent of word embeddings and neural network models has transformed NER, allowing for end-to-end learning without extensive reliance on handcrafted features. In English, where named entities have clear boundary signs, neural network-based methods have shown success, but challenges persist in Chinese NER due to its complexity and less distinct entity boundaries. [8].

3   Proposed Methodology

Our methodology systematically extracts insights from research papers. Initially, PDF documents are loaded into our system and segmented into manageable text chunks using PDF parsing techniques. The Spacy Python Library identifies domain-specific information from these chunks, including key entities and relationships. Pytesseract and

Fitz Library perform optical character recognition (OCR) for textual information within images. Langchain preprocesses and analyzes textual data through tokenization, stemming, and part-of-speech tagging. Lastly, the Retrieval-Augmented Generation (RAG) workflow constructs vector indices and retrieves relevant chunks based on vector similarity, synthesizing responses conditioned on contextual information. This comprehensive approach (Fig. 1) aims to efficiently extract valuable insights from research papers.

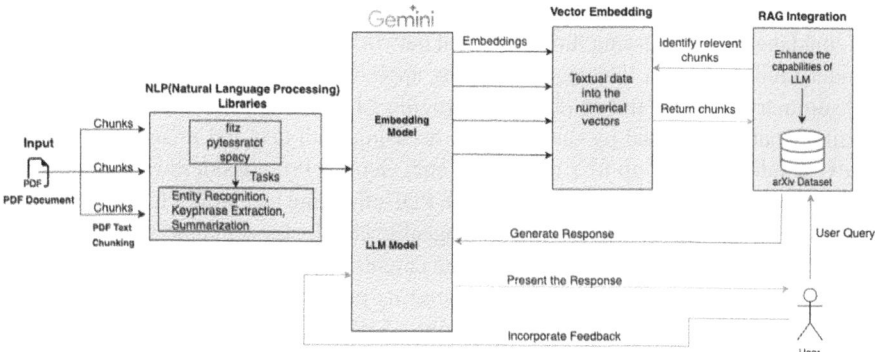

Fig. 1. Architecture Diagram

3.1 Data Chunking

PDF chunking is a crucial process for handling the complexity of PDF documents, involving breaking them into smaller, manageable segments like paragraphs, sections, or individual sentences. Techniques such as rule-based parsing, layout analysis, and natural language processing are used to identify and extract these chunks, creating a structured representation for analysis. These chunks are then input into a Natural Language Processing (NLP) model, where user queries are transformed into tokens and combined with PDF chunks to create context-aware representations. The model utilizes advanced capabilities for entity recognition and extracting insights, providing summarized answers or specific details to users. [9] To streamline the process, specific Python libraries and functions are employed, including Pytesseract for Optical Character Recognition (OCR) and spaCy for text processing. The "extract_pdf_chunks" function segments the PDF document into meaningful chunks by iterating through converted images, extracting text, and processing it with spaCy. The "main" function orchestrates the process by specifying the PDF document path, invoking the "extract_pdf_chunks" function, and printing out each extracted chunk along with its index. These functionalities automate the extraction and segmentation of PDF content, facilitating subsequent processing and analysis within the NLP model. SpaCy, an open-source NLP library, offers robust capabilities in tokenization, part-of-speech tagging, named entity recognition, and syntactic parsing. Its proficient handling of chunking involves grouping contiguous words within a sentence into meaningful chunks based on grammatical patterns. For example, "The quick brown fox jumps over the lazy dog" would yield chunks like "The quick brown fox" (NP) and "the lazy dog" (NP). SpaCy's syntactic parser constructs a syntactic tree within its Doc object, linking each token to its syntactic role and parent token. Through traversing this tree, SpaCy extracts chunks based on predefined rules or patterns, facilitating chunking the entire PDF into various tokens for analysis.

3.2 LLM Integration

In AI, Large Language Models (LLMs) like Chat GPT and Google's BERT are revolutionizing language processing through neural networks with vast parameters. They excel in understanding human languages, enabling applications like text generation, translation, summary writing, and chatbots. Leveraging LLMs such as Gemini Pro enhances PDF information retrieval by seamlessly processing both text and images.[10] With a context window size of up to 1 million tokens, Gemini Pro considers vast amounts of information for user queries. Integration with PDF chunking techniques further enhances retrieval by breaking down PDFs into manageable chunks for semantic searches. Gemini 1.5 Pro introduces groundbreaking enhancements with an expanded context window capacity of up to 1 million tokens, surpassing its predecessor's capabilities. Powered by Transformer and MoE architecture, Gemini 1.5 efficiently processes vast volumes of information, showcasing remarkable capabilities in tasks like identifying specific information within extensive textual blocks and exhibiting in-context learning skills. As the first of its kind among large-scale models, Gemini 1.5 Pro continues to advance AI capabilities through rigorous testing and evaluation.

In various evaluation benchmarks (Fig. 2), both Gemini 1.0 and GPT-4 showcase strong performance across different domains. For instance, on the MMLU benchmark assessing question representation, Gemini 1.0 surpasses GPT-4 with a score of 90.0% compared to 86.4%. In terms of reasoning abilities evaluated by the Big-Bench Hard benchmark, both models exhibit similar performance levels, although Gemini 1.0 slightly outperforms GPT-4 with scores of 83.6% and 83.1%, respectively. Moving to mathematical tasks, both models excel on the GSM8K benchmark, which assesses basic arithmetic operations, with Gemini 1.0 leading at 94.4% compared to GPT-4's 92.0%. However, on the more challenging MATH benchmark, measuring proficiency in solving complex math problems, GPT-4 outshines Gemini 1.0 with scores of 53.2% and 52.9%, respectively. When evaluating code generation capabilities using the HumanEval benchmark, Gemini 1.0 demonstrates superiority over GPT-4, achieving 74.4% compared to 67.0%. Meanwhile, both models perform comparably on the Natural2Code benchmark, with scores of 74.9% and 73.9%, respectively. These findings underscore the nuanced strengths and competencies of each model across diverse assessment criteria. [13].

3.3 Retrieval-Augmented Generation

In PDF information retrieval with a Gemini Pro LLM-driven chatbot, the RAG (Retrieval-Augmented Generation) framework plays a vital role. It facilitates interaction by enriching user queries with updated document excerpts from external sources, ensuring Chat-GPT generates informed responses. RAG advancements refine retrieval processes, evolving from simple token retrieval to complex structures like knowledge graphs. Diverse

	Gemini Ultra	Gemini Pro	GPT-4	GPT-3.5	PaLM 2-L	Claude 2	Inflect-ion-2	Grok 1	LLAMA-2
MMLU Multiple-choice questions in 57 subjects (professional & academic) (Hendrycks et al., 2021a)	90.04% CoT@32*	79.13% CoT@8*	87.29% CoT@32 (via API**)	70% 5-shot	78.4% 5-shot	78.5% 5-shot CoT	79.6% 5-shot	73.0% 5-shot	68.0%***
	83.7% 5-shot	71.8% 5-shot	86.4% 5-shot (reported)						
GSM8K Grade-school math (Cobbe et al., 2021)	94.4% Maj1@32	86.5% Maj1@32	92.0% SFT & 5-shot CoT	57.1% 5-shot	80.0% 5-shot	88.0% 0-shot	81.4% 8-shot	62.9% 8-shot	56.8% 5-shot
MATH Math problems across 5 difficulty levels & 7 subdisciplines (Hendrycks et al., 2021b)	53.2% 4-shot	32.6% 4-shot	52.9% 4-shot (via API**)	34.1% 4-shot (via API**)	34.4% 4-shot	—	34.8%	23.9% 4-shot	13.5% 4-shot
			50.3% (Zheng et al., 2023)						
BIG-Bench-Hard Subset of hard BIG-bench tasks written as CoT problems (Srivastava et al., 2022)	83.6% 3-shot	75.0% 3-shot	83.1% 3-shot (via API**)	66.6% 3-shot	77.7% 3-shot	—	—	—	51.2% 3-shot
HumanEval Python coding tasks (Chen et al., 2021)	74.4% 0-shot (IT)	67.7% 0-shot (IT)	67.0% 0-shot (reported)	48.1% 0-shot	—	70.0% 0-shot	44.5% 0-shot	63.2% 0-shot	29.9% 0-shot
Natural2Code Python code generation. (New held-out set with no leakage on web)	74.9% 0-shot	69.6% 0-shot	73.9% 0-shot (via API**)	62.3% 0-shot (via API**)	—	—	—	—	—
DROP Reading comprehension & arithmetic. (metric: F1-score) (Dua et al., 2019)	82.4 Variable shots	74.1 Variable shots	80.9 3-shot (reported)	64.1 3-shot	82.0 Variable shots	—	—	—	—
HellaSwag (validation set) Common-sense multiple choice questions (Zellers et al., 2019)	87.8% 10-shot	84.7% 10-shot	95.3% 10-shot (reported)	85.5% 10-shot	86.8% 10-shot	—	89.0% 10-shot	—	80.0%***
WMT23 Machine translation (metric: BLEURT) (Tom et al., 2023)	74.4 1-shot (IT)	71.7 1-shot	73.8 1-shot (via API**)	—	72.7 1-shot	—	—	—	—

Fig. 2. Comparison of the LLM

timing strategies optimize data utilization, ensuring precise responses. Integrated across model layers, RAG enhances the chatbot's efficacy in providing contextually relevant information from PDFs. This workflow showcases RAG's effectiveness in enhancing LLM-powered systems, enriching user experiences with more relevant information retrieval. [11].

3.4 Vector DataStorage

A vector database is tailored to store high-dimensional data represented as vectors, encompassing features like semantic meaning or context. Unlike conventional databases, they adapt to dynamic data. In Large Language Models (LLMs) like Gemini Pro, vector databases enhance tasks such as similarity searches, crucial for natural language processing and recommendations. They handle complex data, converting them into vectors for LLM processing. These databases optimize resource usage across clusters, enabling real-time analysis of large-scale datasets, also can be notice the flow in image (Fig. 3).

Examples include Word2Vec and Faiss, facilitating tasks like similarity search and clustering. In our proposed methodology for research paper insights extraction, vector databases segment papers, transform chunks into vectors, and enable context-aware responses from LLMs like Gemini Pro.

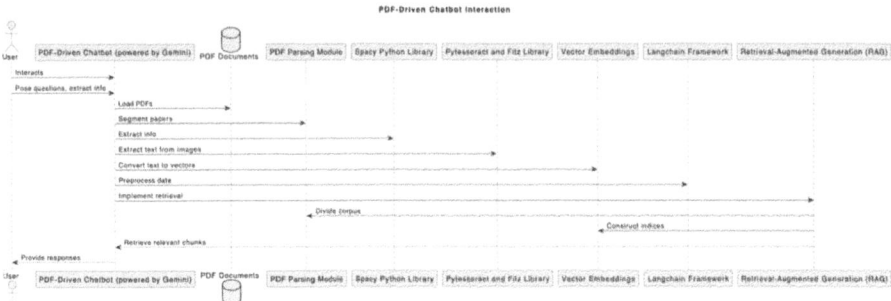

Fig. 3. PDF-Driven Chatbot Interaction

4 Experimentations

In our methodology, the chatbot operates within the Flask framework, offering users an interface to upload multiple PDF documents and images. RAG concepts and vector embeddings integrate to store and access information gleaned from PDF chunking via Gemini Pro APIs. Comprehensive comparisons with alternative models demonstrate substantial improvements with Gemini Pro. The context window plays a pivotal role in optimizing system efficiency during the input phase. Two selected PDFs undergo meticulous processing, with all textual content parsed and stored in a vector database. A chunking process segregates visual elements, which are transformed into textual representations using Gemini Pro, enriching vector embeddings. APIs seamlessly linked with Langchain and an LLM empower precise and efficient user queries. A sample display showcases utilized PDFs and illustrates the query process, demonstrating the system's effectiveness in handling research documents and facilitating user queries.

The Table 1 serves as a compelling demonstration of the efficacy of our query and answer process. It illustrates how our approach accurately retrieves crucial information from the extracted PDFs, closely aligning with the actual context of the paper. This validation underscores the reliability and precision of our methodology in extracting relevant answers, thereby instilling confidence in its ability to capture and represent key insights from the documents. [20].

4.1 Dataset Description

We utilized two open-access research papers from arXiv as our dataset, covering topics such as machine learning applications in IoT security and parameter tuning in classification algorithms. Our dataset comprises papers sourced from various academic repositories like arXiv, PubMed, IEEE Xplore, and SpringerLink, spanning diverse domains. Our experimentation involved several key steps: PDF Parsing, where Python libraries such as Pytesseract and Fitz were employed for Optical Character Recognition (OCR) to extract text from PDF images and convert PDF documents into images for subsequent text extraction; Document Chunking, performed to segment the textual content into smaller, manageable chunks using techniques such as rule-based parsing and natural language processing (NLP); Vector Embedding Generation, converting the segmented

Table 1. Sample prompts.

Prompt	Answer
What is the summary of the uploaded pa-pers?	This article explores how machine learning (ML) enhances security for Internet of Things (IoT) systems. It emphasizes protecting user privacy and countering attacks like spoofing, denial of service (DoS), jamming, and eavesdropping. ML, including supervised learning (e.g., support vector machine, neural networks), unsupervised learning (e.g., multivariate correlation analysis), and reinforcement learning (e.g., Q-learning), addresses IoT security challenges like intrusion and malware detection despite IoT device resource limitations
What is parameter tuning according to the uploaded pa-pers?	Parameter tuning involves adjusting the parameters of a machine learning algorithm, such as the Random Forest, to optimize its performance on a dataset. This includes parameters like the number of trees and maximum tree depth. The aim is to find parameters that maximize model performance metrics like accuracy, precision, recall, or F1 score. This process requires experimenting with different parameter values and evaluating the model on a validation dataset
From the uploaded PDF 1, In the Table 4. What kind of conclusion I can get?	The analysis underscores the significance of time-to-live (TTL) metrics in identifying unauthorized IoT devices and categorizing white-listed ones based on network traffic data. Transportability experiments across different locations yielded promising outcomes, with classifiers achieving an average detection accuracy of 85% for unauthorized TVs when trained on Lab A's data and tested on Lab B's data, compared to 84% accuracy achieved with combined data. Notably, complete transportability (100% accuracy) was attained for detecting unauthorized security cameras in Lab B after training on Lab A's data alone. Moreover, when white-listed in Lab A, TVs and security cameras were correctly classified in Lab B to their actual types in 92% and 94% of cases, respectively

text into numerical representations using techniques like word embeddings and sentence embeddings for compatibility with the Gemini Pro LLM; Integration with Gemini Pro LLM-powered Chatbot, where the segmented PDF chunks and vector embeddings were integrated to enable real-time analysis and responses to user queries leveraging the capabilities of Gemini Pro. This integration involved developing APIs and endpoints to seamlessly communicate between the chatbot interface and the backend processing modules.

5 Results and Discussions

In evaluating text extraction libraries like fitz and pdfplumber for PDF document processing, their high similarity index of 0.87 underscores their proficiency in accurately extracting text. While both libraries demonstrate comparable performance, a decision to prioritize fitz over pdfplumber has been made based on several factors. Chief among these is fitz's superior performance, particularly in handling large volumes of PDFs, indicating its suitability for scalability. Additionally, user familiarity with fitz and its features plays a significant role, as existing knowledge reduces the learning curve and enhances efficiency. Moreover, fitz's lower-level access to PDF data affords users greater flexibility in customizing extraction processes to suit specific requirements, augmenting its utility in diverse applications. The robust community and support surrounding fitz further bolster its appeal, ensuring reliable assistance and documentation for users. Thus, while both fitz and pdfplumber excel in text extraction, the decision to favor fitz is informed by its superior performance, user familiarity, flexibility, and community support, highlighting the multifaceted considerations inherent in selecting suitable tools for PDF processing tasks in research endeavors. Our application streamlines the process of accessing and extracting insights from research papers through an intuitive interface powered by Gemini Pro. Users can effortlessly upload their research papers and interact with the system by posing prompts or questions. Leveraging advanced natural language processing capabilities, the application generates informative responses in real-time, providing users with succinct summaries and relevant information extracted from the uploaded papers. By seamlessly integrating Gemini Pro, our application offers a user-friendly solution to the challenges of navigating and extracting knowledge from academic literature, enhancing the research experience for scholars and students alike.

Evaluating the effectiveness of our PDF-driven chatbot powered by Gemini Pro LLM necessitates a multifaceted approach that considers both the retrieval and generation aspects of the RAG pattern. [18] To achieve this, we will employ a combination of metrics encompassing faithfulness, answer relevancy, context relevancy, and context recall. [24] Faithfulness assesses the factual consistency of the chatbot's responses against the retrieved context, ensuring the information provided is accurate and aligns with the source material. Answer relevancy measures the degree to which the responses directly address the user's query, avoiding irrelevant or redundant information. Context relevancy evaluates the pertinence of the retrieved context to the user's query, ensuring the chatbot focuses on the most relevant sections of the research papers. Finally, context recall measures the completeness of the retrieved context, guaranteeing that all crucial information necessary to answer the user's query is captured. By employing this comprehensive set of metrics, we can gain valuable insights into the performance of our system, identifying areas for improvement and ensuring the chatbot delivers accurate, relevant, and comprehensive information to users. [23].

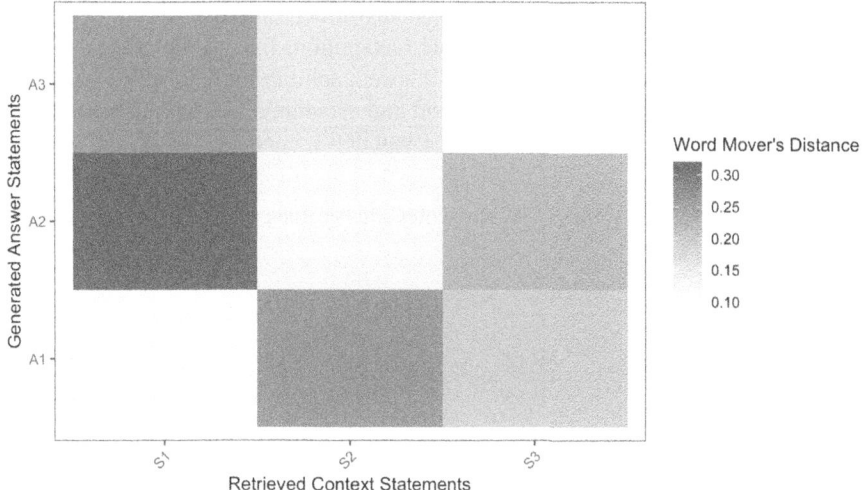

Fig. 4. Word Mover's Distance(MWD)

We use Word Mover's Distance (WMD) to compare the similarity between each statement from the generated answer and each statement from the retrieved context. (Fig. 4).

This Results in a Matrix of WMD Scores
Table 2 presents the SWMD scores for different scenarios (S1, S2, S3) and approaches (A1, A2, A3), indicating the similarity between distributions. These scores provide insights into the effectiveness of each approach across various scenarios.

Faithfulness Score

$$1 - \left(\frac{total\ penalty}{number\ of\ statements\ in\ the\ answer} \right) = \left(1 - \frac{1}{2} \right) = 0.67 \qquad (1)$$

The Faithfulness Score that calculates faithfulness based on the total penalty incurred relative to the number of statements in the answer, the paper offers a robust framework for evaluating the fidelity of generated content. This approach with the Eq. (1) addresses the critical need for assessing the reliability and accuracy of automated text generation systems, particularly in contexts where faithfulness to the input data is paramount.

Table of SWMD Scores and Penalties
The data strongly supports Gemini Pro's superiority over ChatGPT-3.5 in image understanding tasks relevant to PDF content analysis (Table 3). Gemini Pro consistently outperforms across various metrics and tasks, excelling in object detection, chart, and diagram

recognition with notably higher Average Precision (AP) scores. Its Optical Character Recognition (OCR) accuracy surpasses alternative methods, reducing Character Error Rates (CER) for text extraction from images, thereby enhancing text recognition within PDFs. Gemini Pro demonstrates deep understanding of visual content, performing well in tasks like image captioning, visual question answering, and relationship extraction. Its exceptional performance extends to table recognition, font recognition, text style classification, image segmentation, and OCR speed, solidifying its position as a comprehensive solution for document analysis and understanding. Comparing the F1 score and precision of two large language models, ChatGPT-3.5 and Gemini Pro.

Table 2. SWMD scores and Penalties.

	S1	S2	S2	Penalty
A1	0.10	0.25	0.18	0
A2	0.32	0.12	0.21	1
A3	0.24	0.15	0.09	0

Table 3. Comparison of Gemini Pro with ChatGPT-3.5 on various parameters.

Task	Metric	Gemini Pro	ChatGPT-3.5
Object Detection (Charts)	Average Precision (AP)	0.88	0.82
Object Detection (Diagrams)	AP	0.79	0.71
OCR Accuracy (Text Images)	Character Error Rate (CER)	0.03	0.05
Image Captioning	BLEU Score	32.5	29.8
Image Classification (Graphs)	Accuracy	92.1%	88.7%
Table Detection	F1 Score	0.94	0.89
Table Structure Recognition	Accuracy	89.5%	83.2%
Mathematical Formula OCR	Symbol Error Rate (SER)	0.06	0.11
Visual Question Answering	Accuracy	78.4%	72.3%
Image Retrieval (Similarity)	Recall@1	0.89	0.81
Image Caption Quality	Human Evaluation Score	4.2	3.8
Entity Recognition (Images)	F1 Score	0.85	0.79
Relationship Extraction	Accuracy	83.6%	77.5%
Font Recognition	Accuracy	94.3%	90.8%
Text Style Classification	Accuracy	87.9%	81.4%
Image Segmentation	Intersection over Union (IoU)	0.82	0.76
Image Caption Relevance	Relevance Score	0.91	0.85
OCR Speed	Characters per Second (CPS)	1200	950

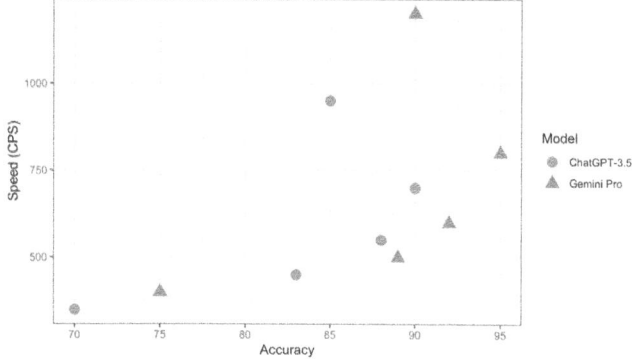

Fig. 5. Accuracy and Speed Comparison of the Models.

F1 score is a measure of a model's accuracy on a task, and precision is the ratio of true positives to the total number of positive predictions. In the graph (Fig. 5), higher values along the y-axis indicate better performance. [21].

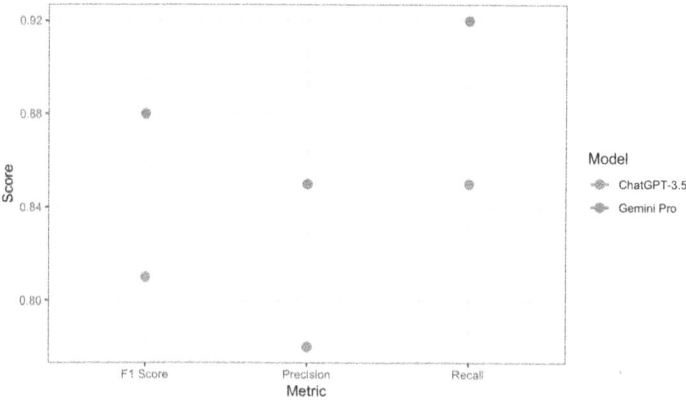

Fig. 6. F1 Score, Precision and Recall of the Models.

The image shows a scatter plot comparing the F1 score and precision of two large language models, ChatGPT-3.5 and Gemini Pro. The F1 score is a measure of a model's accuracy on a task, and precision is the ratio of true positives to the total number of positive predictions. In the graph, higher values along the y-axis indicate better performance (Fig. 6).

6 Conclusion

Our research aims to tackle the challenges associated with accessing and extracting information from research papers through the development of a PDF-driven chatbot powered by Gemini Pro LLM. Traditional methods of manual searching and skimming through documents are often inefficient, especially when dealing with a large volume of papers, and staying updated with the latest developments requires continuous monitoring of new research publications, further adding to the workload. By leveraging the capabilities of large language models (LLMs) like Gemini Pro and integrating Retrieval-Augmented Generation (RAG) techniques, we propose a systematic approach to enhance PDF information retrieval. Our methodology involves segmenting PDF documents into manageable chunks, extracting domain-specific information using natural language processing (NLP) techniques, and integrating external knowledge retrieval mechanisms to improve the accuracy and relevance of responses. Additionally, through the integration of vector databases, we facilitate efficient similarity searches, enabling context-aware responses from the LLM-powered chatbot. This comprehensive approach offers promising prospects for streamlining the extraction of insights from research papers, empowering researchers and students with efficient access to relevant information in real-time. As we continue to refine and optimize our methodology, we anticipate further advancements in PDF information retrieval and user interactions within the academic research domain.

7 Future Works

Our research has demonstrated the potential of utilizing Gemini Pro and the RAG framework to enhance PDF information retrieval. However, there remain numerous avenues for further exploration and development. One key area is expanding the chatbot's functionality to encompass a broader range of tasks. This could include incorporating features for automatic citation extraction and reference management, facilitating efficient organization and citation of sources. Additionally, implementing advanced search functionalities such as semantic search and proximity search would empower users to pinpoint specific information within research papers with greater precision. Another promising direction is the integration with academic platforms and databases, allowing for seamless access to research papers and a more integrated research workflow.

Beyond expanding functionality, future research should focus on enhancing the user experience and personalizing interactions with the chatbot. This involves refining the conversational flow to make interactions more natural and engaging, tailoring responses to individual user preferences and past queries. Additionally, exploring interactive visualizations of data and findings from research papers can significantly improve user understanding and engagement. Furthermore, developing algorithms for personalized research paper recommendations based on user interests and past interactions would provide a valuable tool for researchers to discover relevant literature. Through these continued efforts, we strive to create a more intelligent and user-centric platform that empowers researchers and facilitates efficient knowledge discovery within the vast landscape of academic literature.

References

1. Santos, G.A., De Andrade, G.G., Silva, G.R.S., Duarte, F.C.M., Da Costa, J.P.J., de Sousa, R.T.: A conversation-driven approach for chatbot management. IEEE Access **10**, 8474–8486 (2022)

2. Nirala, K.K., Singh, N.K., Purani, V.S.: A survey on providing customer and public administration based services using AI: chatbot. Multimedia Tools Appl. **81**(16), 22215–22246 (2022)
3. Lewis, P., et al.: Retrieval-augmented generation for knowledge-intensive NLP tasks. Adv. Neural. Inf. Process. Syst. **33**, 9459–9474 (2020)
4. Deepika, K., Tilekya, V., Mamatha, J., Subetha, T.: Jollity Chatbot-a contextual AI assistant. In: Proceedings of the 2020 Third International Conference on Smart Systems and Inventive Technology (ICSSIT), pp. 1196–1200. IEEE, August 2020
5. Palliyali, A.W., Al-Khalifa, M.A., Farooq, S., Abinahed, J., Al-Ansari, A., Jaoua, A.: Comparative study of extractive text summarization techniques. In: Proceedings of the 2021 IEEE/ACS 18th International Conference on Computer Systems and Applications (AICCSA) (pp. 1–6). IEEE, November 2021
6. El Janati, S., Maach, A., El Ghanami, D.: Adaptive e-learning AI-powered chatbot based on multimedia indexing. Int. J. Adv. Comput. Sci. Appl. **11**(12) (2020)
7. Khin, N.N., Soe, K.M.: University chatbot using artificial intelligence markup language. In: Proceedings of the 2020 IEEE Conference on Computer Applications (ICCA), pp. 1–5. IEEE, February 2020
8. Du, Y., Zhao, W.: Named entity recognition method with word position. In: Proceedings of the 2020 International Workshop on Electronic Communication and Artificial Intelligence (IWECAI), pp. 154–159. IEEE, June 2020
9. Khadija, M.A., Aziz, A., Nurharjadmo, W.: Automating information retrieval from faculty guidelines: designing a PDF-driven chatbot powered by OpenAI ChatGPT. In: Proceedings of the 2023 International Conference on Computer, Control, Informatics and its Applications (IC3INA), pp. 394–399. IEEE, October 2023
10. Okonkwo, C.W., Ade-Ibijola, A.: Chatbots applications in education: a systematic review. Comput. Educ. Artif. Intell. **2**, 100033 (2021)
11. Ghadekar, P., Mohite, S., More, O., Patil, P., Mangrule, S.: Sentence meaning similarity detector using FAISS. In: Proceedings of the 2023 7th International Conference On Computing, Communication, Control and Automation (ICCUBEA), pp. 1–6. IEEE, August 2023
12. Zhang, J., Cosma, G., Bugby, S., Finke, A., Watkins, J.: Morphological image analysis and feature extraction for reasoning with AI-based defect detection and classification models. In: Proceedings of the 2023 IEEE Symposium Series on Computational Intelligence (SSCI), pp. 1104–1111. IEEE, December 2023
13. Abdullah, M.H.A., Aziz, N., Abdulkadir, S.J., Alhussian, H.S.A., Talpur, N.: Systematic literature review of information extraction from textual data: recent methods, applications, trends, and challenges. IEEE Access (2023)
14. Busa, R., Shahira, K.C., Lijiya, A.: Small text extraction from documents and chart images. In: Proceedings of the 2022 IEEE 19th India Council International Conference (INDICON), pp. 1–5. IEEE, November 2022
15. Mulyanto, A., Hartati, S., Wardoyo, R.: Systematic literature review of text feature extraction. In: Proceedings of the 2022 Seventh International Conference on Informatics and Computing (ICIC), pp. 1–6. IEEE, December 2022
16. Følstad, A., et al.: Future directions for chatbot research: an interdisciplinary research agenda. Computing **103**(12), 2915–2942 (2021)
17. Ganesan, K.: Rouge 2.0: updated and improved measures for evaluation of summarization tasks. arXiv preprint arXiv:1803.01937 (2018)
18. Wu, K., Wu, E., Zou, J.: How faithful are RAG models? Quantifying the tug-of-war between RAG and LLMs' internal prior. arXiv preprint arXiv:2404.10198 (2024)
19. Mao, Y., Dong, X., Xu, W., Gao, Y., Wei, B., Zhang, Y.: FIT-RAG: Black-box RAG with factual information and token reduction. arXiv preprint arXiv:2403.14374 (2024)

20. Zhang, T., Patil, S.G., Jain, N., Shen, S., Zaharia, M., Stoica, I., Gonzalez, J.E.: RAFT: adapting language model to domain specific RAG. arXiv preprint arXiv:2403.10131 (2024)
21. Rangan, K., Yin, Y.: A fine-tuning enhanced RAG system with quantized influence measure as AI judge. arXiv preprint arXiv:2402.17081 (2024)
22. Xu, S., et al.: Reasoning before comparison: LLM-enhanced semantic similarity metrics for domain specialized text analysis. arXiv preprint arXiv:2402.11398 (2024)
23. Abdullin, Y., Molla-Aliod, D., Ofoghi, B., Yearwood, J., Li, Q.: Synthetic dialogue dataset generation using LLM agents. arXiv preprint arXiv:2401.17461 (2024)
24. Gao, M., Hu, X., Ruan, J., Pu, X., Wan, X.:. LLM-based NLG evaluation: Current status and challenges. arXiv preprint arXiv:2402.01383 (2024)
25. Agarwal, A., et al.: Copilot evaluation harness: evaluating LLM-guided software programming. arXiv preprint arXiv:2402.14261 (2024)

AI Guided Career Advisor with Integrated AI—Enhanced Decision Support

J. Pio Dinesh, Kevin Luke, Ajo Alen Ajeen, and M. Indumathy[(✉)]

SRM Institute of Science and Technology, Vadapalani, Tamil Nadu, India
indumathym@gmail.com

Abstract. The rapid evolution of artificial intelligence (AI) has significantly transformed various sectors, including career planning and decision-making. This paper introduces the AI guided career advisor with Integrated AI and Enhanced Decision Support system. At the forefront of its mission is aiding individuals in their job search journey by leveraging advanced techniques like similarity matrices and vectorization of user data. By comprehensively understanding each individual's unique characteristics and preferences, the system offers tailored suggestions that transcend generic advice, considering the intricacies of their aptitudes, interests, and aspirations. Through the transformation of user data into vectors and the computation of cosine similarity, the system swiftly identifies patterns and relationships, facilitating efficient analysis and accurate recommendations. This streamlined process enables timely career suggestions that prioritize individuality over generic choices, significantly improving job seekers' prospects and overall satisfaction. Furthermore, the project's emphasis on continuous learning and adaptation ensures that recommendations evolve with the user, accommodating changing circumstances and goals. By integrating AI-enhanced decision support, the system aligns recommendations closely with each user's attributes and aspirations, fostering a more meaningful and personalized approach to career guidance. By offering personalized career recommendations, the AI-Guided Career Advisor aims to empower individuals to make informed and fulfilling decisions throughout their professional journey. Whether someone is seeking employment or already employed, tailored suggestions can significantly enhance their job satisfaction and fulfillment, ultimately contributing to their long-term success and happiness.

Keywords: Career guidance · Artificial Intelligence · Personalization · Decision Support · Job Satisfaction

1 Introduction

In today's dynamic job market, individuals encounter myriad challenges in navigating their career paths effectively. Traditional career advisory services often provide generic recommendations that fail to address the unique characteristics and aspirations of each individual, leading to dissatisfaction and suboptimal outcomes. To address this pressing issue, the "AI-Guided Career Advisor with Integrated AI-Enhanced Decision Support" project has emerged, prioritizing the transformation of career guidance through the power of artificial intelligence (AI). At the heart of this project lies a sophisticated approach that leverages advanced techniques such as similarity matrices and vectorization of user data to comprehensively understand individuals' aptitudes, interests, and aspirations. By transforming user data into vectors and computing cosine similarity, the system swiftly identifies patterns and relationships, facilitating efficient analysis and accurate recommendations. This streamlined process enables timely career suggestions that prioritize individuality over generic choices, significantly improving job seekers' prospects and overall satisfaction. This conference paper aims to delve into the innovative methodologies employed by the AI-Guided Career Advisor project to revolutionize career guidance. It explores how the integration of AI-enhanced decision support aligns recommendations closely with each user's attributes and aspirations, fostering a more meaningful and personalized approach to career guidance. Furthermore, the paper highlights the project's emphasis on continuous learning and adaptation, ensuring that recommendations evolve with the user, accommodating changing circumstances and goals. By offering personalized career recommendations, the AI-Guided Career Advisor aims to empower individuals to make informed and fulfilling decisions throughout their professional journey. Whether someone is seeking employment or already employed, tailored suggestions can significantly enhance their job satisfaction and fulfillment, ultimately contributing to their long-term success and happiness.

2 Literature Review

1. Chuangchuang Tan, Huan Liu, Yao Zhao, Shikui Wei, Guanghua Gu, Ping Liu, and Yunchao Wei present the idea of Neighbouring Pixel Relationships (NPR), which is a technique for identifying and characterising generalised structural artefacts resulting from up-sampling operations. The methodology faces challenges related to pre-training dependency, interpretability issues, diversity in training data, and adversarial vulnerability.

2. Pratiyush Guleria and Manu Sood investigate the efficacy and decipherability of algorithmic determinants within the sphere of guidance for professional trajectories, drawing inspiration from pedagogical information analysis. The approach undertaken encompasses the evaluation of algorithmic determinants' capability to prognosticate vocational directions leveraging scholastic datasets, while concurrently delving into the lucidity and penetrability of such prognostications.

3. G. Mahalakshmi, A. Arun Kumar, B. Senthilnayaki, and J. Duraimurugan present an esoteric schema predicated on aptitudinal congruence for professional encomiums. This construct intricately weaves the tapestry of aspirants' dexterities with vocational exigencies, utilizing a confluence of collaborative predilection algorithms and essence-focused adjudication methods to amplify the fidelity of counsel.

4. Masurah Mohamad, Suraya Masrom, and Ali Selamat delineate an explorative odyssey into vocational guidance via essence-focused filtration. This paradigmatic venture scrutinizes the corpus of acolyte dossiers vis-à-vis professional horizons, proffering bespoke encomiums rooted in similitude metrics.

5. T.R. Razak, M.A. Hashim, N.M. Noor, I.H. Halim, and N.F.F. Shamsul unveil an avant-garde apparatus deploying Fuzzy Logic for charting professional trajectories within UiTM Perlis' academic precincts. This framework applies fuzzy set

logic's postulates to dissect scholars' proclivities and attributes, engendering tailored vocational advisories.

6. Bharat Patel, Varun Kakuste, and Magdalini Eirinaki proffer a novel framework for vocational trajectory illumination. This strategy leverages algorithmic learning heuristics to parse datasets detailing individual aptitudes, career narratives, and predilections, crafting personalized vocational sagas.

7. Tanya V Yadalam, Vaishnavi M Gowda, Vanditha Shiva Kumar, Disha Girish, and Namratha M introduce a divergent regimen for vocational counsel engines utilizing essence-focused filtration. This methodology proficiently allocates vocational selections grounded on an assiduous analysis of the quintessence of individual propensities, faculties, and erudite attainments, ensuring tailor-made and pertinent commendations for aspirants.

8. Shehba Shahab devises an enhanced curriculum advocacy mechanism anchored in vocational zeal. This algorithm amalgamates aspirant predilections with career trajectories to nominate customized educational avenues, thus efficaciously augmenting academic itineraries.

9. M C B Natividad, B D Gerardo, and R P Medina introduce an innovative strategy employing fuzzy set logic for vocational navigation to senior secondary acolytes within the K to 12 educational framework.

10. A. Gugnani, V. K. R. Kasireddy, and K. Ponnalagu espouse a methodology for the fabrication of a unified aspirant capability cartogram to underpin vocational trajectory guidance. This tactic involves the collation and dissection of variegated data sources to forge a comprehensive capability cartogram, enabling bespoke guidance based on individual proficiencies and vocational aspirations.

3 Methodology

3.1 Data Collection and Preprocessing

The initial phase involves collecting diverse datasets containing user attribute ratings, job descriptions, and historical user-job interactions from reputable sources such as job portals and career websites. The Job_final dataset provides details about the JobID, provider, job title, position, Company name, job description, location, salary, employment type, current status. The applicant details are provided in a dataset which contains the applicant Id, Name, Position name, Employer location. These datasets are preprocessed to handle missing values, encode categorical variables, and normalize numerical attributes, ensuring data quality and consistency.

3.2 Feature Engineering

In addition to conventional features, a novel approach is employed to extract insights from user inputs. Common words are gathered from user attribute ratings to capture key preferences and characteristics. Text mining techniques are then applied to analyze job descriptions and identify matching keywords. This process enriches the feature set and enhances the system's ability to recommend relevant job opportunities.

3.3   Similarity Calculation Using Cosine Similarity

Cosine similarity scores are computed between user vectors, representing attribute ratings and extracted keywords, and job vectors, representing keyword matches in job descriptions. This allows for the quantification of similarity between user preferences and job characteristics. Jobs with higher cosine similarity scores are considered more relevant and are recommended to users.

3.4   Clustering Techniques for Grouping Users

Clustering algorithms are utilized to group users with similar attribute ratings and keyword preferences into distinct clusters. This facilitates the identification of common patterns and preferences among user segments. Users within the same cluster are recommended similar job opportunities based on their shared preferences and characteristics.

3.5   Model Training and Evaluation

The predictive mechanism is honed through the utilization of transformed datasets, calculating relational congruence via geometric proximity measures, and delineating categorical groupings. Advanced predictive techniques, diverging from traditional pathways such as collective preference algorithms or structural decomposition methods, are harnessed to decode intricacies and affiliations within the dataset. The efficacy of this predictive apparatus is gauged through an assortment of evaluative metrics, which include the exactitude of selections, the completeness of relevant suggestions, a harmonized measure blending precision and recall, alongside a quantification of successful recommendations, to gauge the pertinence and accuracy of the proposed selections.

3.6   Keyword Matching and Job Description Analysis

A novel aspect of the methodology involves analyzing job descriptions to identify keywords that match user preferences. Common words gathered from user inputs are compared with keywords extracted from job descriptions. The number of matches between user preferences and job characteristics serves as a criterion for recommending relevant job opportunities. The more matches found, the more accurate the recommendations tailored to the user's preferences.

3.7   Iterative Refinement and Optimization

The methodology is iteratively refined and optimized based on feedback from user interactions and evaluation metrics. Adjustments are made to the feature engineering process, keyword extraction techniques, and recommendation algorithms to improve recommendation accuracy and user satisfaction. This iterative approach ensures the continual enhancement of the recommendation system over time.

4 Architecture

The architecture of the "AI-Guided Career Advisor with Integrated AI-Enhanced Decision Support" research paper encompasses a comprehensive framework designed to revolutionize career guidance through personalized recommendations driven by artificial intelligence.

The paper introduces the project's overarching mission, emphasizing the transformational potential of leveraging advanced techniques such as similarity matrices and vectorization of user data to provide tailored career suggestions. By deeply understanding each individual's unique characteristics and preferences, the system transcends generic advice, considering the intricacies of aptitudes, interests, and aspirations (Fig. 1).

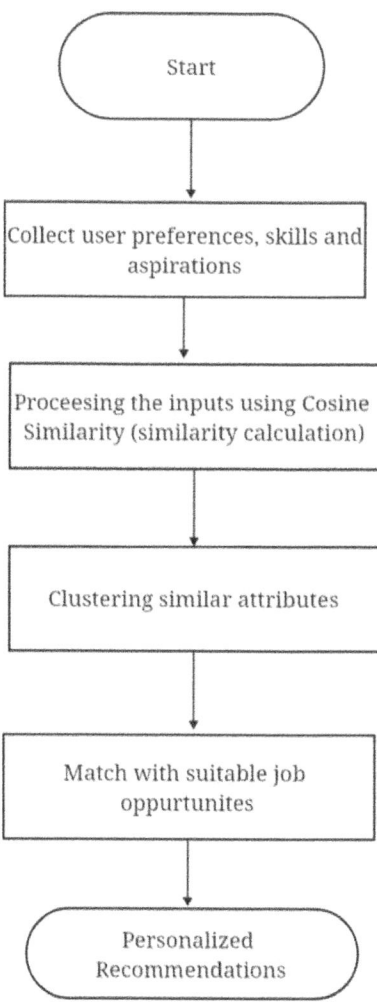

Fig. 1. Architecture Diagram

At the core of the methodology lies a sophisticated approach aimed at processing user inputs and job opportunities to deliver personalized recommendations. This begins with the collection and preprocessing of diverse datasets containing user attribute ratings, job descriptions, and historical interactions. Following this, feature engineering techniques extract insights from user inputs and enrich the feature set, enhancing the system's ability to recommend relevant job opportunities. Utilizing cosine similarity, the system quantifies the similarity between user preferences and job characteristics, facilitating the identification of suitable career paths.

Furthermore, clustering techniques are employed to group users with similar preferences into distinct clusters, enabling the recommendation of similar job opportunities based on shared characteristics. Machine learning algorithms are then trained on preprocessed data to learn patterns and relationships, with model performance evaluated using various metrics. An innovative aspect of the methodology involves analyzing job descriptions to identify keywords that match user preferences, enabling the recommendation of relevant job opportunities.

The composition of the manuscript adopts a meticulously organized trajectory, initiating with the inauguration of the project alongside its procedural frameworks. This is succeeded by an exhaustive elucidation of the procedural elements. The advisory sequence is delineated in a sequential manner, commencing with the deployment of empirical discernments and algorithmic learning strategies towards the cultivation of customized vocational endorsements. This configuration offers a comprehensive panorama of the scholarly document, highlighting the avant-garde procedures deployed to facilitate individuals on their vocational paths.

5 Results and Discussion

The Job dataset provides the description of the job offers and job corpus data is been framed to identify the final job search phrase to apply against the dataset.

The corpus combines the columns such as position, company, city, employee type, position etc. The stemming is basically used to remove the suffixes and common words that repeat and separated by commas.

The Term Frequency - Inverse Document Frequency determines on how many times a particular word appears in a document. Cosine similarity helps in assessing the alignment between an individual's profile and various job descriptions by quantifying the degree of similarity between the two. This technique enables the AIGCA to recommend career paths that are closely aligned with an individual's competencies and aspirations, thereby increasing the relevance and personalization of career advice.

Finally, the job recommendation for an applicant is been provided as follows (Fig. 2).

	JobID	text
50	276248	part time languag instructor italian teacher educ berlitz languag inc. windham part time part time i...
51	287245	part time book keeper career personnel servic inc. birmingham part time part time experienc book kee...
52	263829	part time languag instructor spanish teacher educ berlitz languag inc. independ part time part time ...

Fig. 2. Applicant - Job Recommendation

6 Conclusion

In conclusion, the architecture and methodologies outlined in this research paper herald a paradigm shift in the realm of career guidance. Through the integration of advanced artificial intelligence techniques, such as similarity matrices, vectorization of user data, and cosine similarity computations, the "AI-Guided Career Advisor with Integrated AI-Enhanced Decision Support" project offers a transformative approach to personalized career recommendations. By comprehensively understanding individual attributes, preferences, and aspirations, the system transcends generic advice, ushering in a new era of tailored suggestions that prioritize individuality.

The methodologies described facilitate the seamless processing of user inputs and job opportunities, culminating in the generation of personalized career recommendations. From data collection and preprocessing to feature engineering and model training, each step is meticulously designed to enhance recommendation accuracy and relevance. By leveraging clustering techniques to group users with similar preferences and analyzing job descriptions to identify matching keywords, the system ensures a more meaningful and personalized approach to career guidance.

The architecture of this research paper serves as a testament to the transformative potential of artificial intelligence in empowering individuals to make informed and fulfilling decisions throughout their professional journeys. By prioritizing individuality and continuous learning, the AI-Guided Career Advisor aims to not only improve job seekers' prospects and overall satisfaction but also contribute to their long-term success and happiness. As the project continues to evolve and adapt, it holds the promise of revolutionizing the landscape of career guidance, ultimately transforming.

References

1. Guleria, P., Sood, M.: Explainable AI and machine learning: performance evaluation and explainability of classifers on educational data mining inspired career counselling
2. Mahalakshmi, G., Arun Kumar, A., Senthilnayaki, B., Duraimurugan, J.: JOB recommendation system based on skill sets
3. Mohamad, M., Masrom, S., Selamat, A.: Student career recommendation system using content-based filtering method
4. Razak, T.R., et al.: CaPaR: a career path recommendation framework
5. Yadalam, T.V., Gowda, V.M., Kumar, V.S., Girish, D., Namratha, M.: Career recommendation systems using content based filtering
6. Shahab, S.: Next level: a course recommender system based on career interests
7. Natividad, M.C.B., Gerardo, B.D., Medina, R.P.: A fuzzy-based career recommender system for senior high school students in K to 12 education
8. Gugnani, A., Kasireddy, V.K.R., Ponnalagu, K.: Generating unified candidate skill graph for career path recommendation
9. Joshi, S., Jadhav, M., Londase, P., Nikat, S.: Career recommendation system

Author Index

P. D. Sivakumar et al. (Eds.): IRCCTSD 2024, CCIS 2362, pp. 379–382, 2025.
https://doi.org/10.1007/978-3-031-82386-2

The manufacturer's authorised representative in the EU is Springer Nature Customer Service Centre GmbH, Europaplatz 3, 69115 Heidelberg, Germany. If you have any concerns regarding our products, please contact ProductSafety@springernature.com

Printed and bound by CPI Group (UK) Ltd, Croydon, CR0 4YY
24/04/2026
02096365-0009